Lecture Notes in Computer Science 1133

Edited by G. Goos, J. Hartmanis and J. van Leeuwen

Advisory Board: W. Brauer   D. Gries   J. Stoer

W0049883

Springer

Berlin
Heidelberg
New York
Barcelona
Budapest
Hong Kong
London
Milan
Paris
Santa Clara
Singapore
Tokyo

Jean-Yves Chouinard   Paul Fortier
T. Aaron Gulliver   (Eds.)

Information Theory and Applications II

4th Canadian Workshop
Lac Delage, Québec, Canada, May 28-30, 1995
Selected Papers

 Springer

Series Editors

Gerhard Goos, Karlsruhe University, Germany

Juris Hartmanis, Cornell University, NY, USA

Jan van Leeuwen, Utrecht University, The Netherlands

Volume Editors

Jean-Yves Chouinard
Department of Electrical Engineering, University of Ottawa
161 Louis-Pasteur, Ottawa, Canada, K1N 6N5
E-mail: chouinar@elg.uottawa.ca

Paul Fortier
Département de génie électrique et de génie informatique, Université Laval
Cité Universitaire, Québec, Canada, G1K 7P4
E-mail: fortier@gel.ulaval.ca

T. Aaron Gulliver
Department of Electrical and Electronic Engineering
University of Canterbury
Private Bag 4800, Christchurch, New Zealand
E-mail: gulliver@elec.canterbury.ac.nz

Cataloging-in-Publication data applied for

Die Deutsche Bibliothek - CIP-Einheitsaufnahme

Information theory and applications : ... Canadian workshop ;
proceedings. - Berlin ; Heidelberg ; New York ; London ; Paris
; Tokyo ; Hong Kong ; Barcelona ; Budapest : Springer

4. Lac Delage, Québec, Canada, May 1995 : selected papers.
(Lecture notes in computer science ; 1133)
ISBN 3-540-61748-5
NE: GT

CR Subject Classification (1991): E.4, E.3, F.2, I.4, I.5, I.6, G.2, G.3, B.4

ISSN 0302-9743
ISBN 3-540-61748-5 Springer-Verlag Berlin Heidelberg New York

Typesetting: Camera-ready by author
SPIN 10513584   06/3142 – 5 4 3 2 1 0   Printed on acid-free paper

Preface

The 1995 Canadian Workshop on Information Theory was held in Lac Delage, Québec from May 28 to May 31. This was the fourth workshop to be held under the auspices of the Canadian Society of Information Theory. The previous three workshops took place in Saint-Jovite, Québec in 1987, Sidney, British Columbia in 1989, and Rockland, Ontario in 1993. The purpose of the workshops has been to provide an informal setting for Canadian researchers to meet and exchange ideas on information theory. We were pleased to welcome a number of participants from the United States and France among the 44 attendees, as was the case in 1993. There were 30 regular presentations given at the workshop as well as presentations by three invited speakers.

This volume is the second proceedings to be published from the Canadian Workshop on Information Theory. Following the workshop, participants were asked to submit a manuscript and each paper has been subject to peer review. An extensive list of reviewers appears at the end of the preface. This volume contains 21 papers written by the workshop speakers and co-authors, including the papers from the invited speakers. The papers are grouped into five sections: algebraic coding, cryptography and secure communications, decoding methods and techniques, coding and modulation for fading channels, and signal processing and pattern recognition. A brief summary of the papers in each section is given below.

Algebraic Coding

This section begins with the paper by the invited lecturer, Professor Gérald E. Séguin. This paper presents a study of the algebraic structure of linear codes invariant under a given permutation. Séguin derives expressions for determining the number of such codes as well as methods to decompose them into component codes. In the next paper, Pedersen and Polemi propose a novel method to construct longer codes by combining algebraic geometric Goppa codes. They discuss the minimum distance properties of these new codes. Gulliver and Bhargava present the construction of multiple circulant quasi-cyclic codes over $GF(5)$. In terms of distance properties, they show that several codes are optimal whereas the others provide a lower bound on the maximum possible minimum distance.

Cryptography and Secure Communications

Youssef and Tavares introduce static and dynamic information leakage measures between the inputs and outputs of randomly selected Boolean functions as cryptographic criteria. In particular, they demonstrate that the expected values of information leakages decrease significantly with the number of input variables. Aakvaag, Lacaze and Duverdier investigate a scrambling method to ensure message privacy. By introducing controlled periodic clock changes into transmitted analog signals, the interception of messages can be prevented while the exact knowledge of the corresponding reconstruction filter allows their reconstitution.

Assuming wide sense stationary sources of information, they derive the conditions under which perfect reconstruction can be achieved. In a companion paper, Duverdier, Lacaze and Aakvaag study the application of linear time-varying periodic filtering of stationary processes for message scrambling and introduce a reconstruction method based on their cyclostationary properties. In particular, they show that this method results in an unconventional form of spread spectrum which facilitates the reconstruction of an analog signal and, in the case of a binary signal, can correct errors caused by frequency selective fading.

Decoding Methods and Techniques

The first paper of this section is written by the invited lecturer, Professor Gérard Battail. Professor Battail presents a comprehensive survey of random codes with an emphasis on pseudo-random systematic convolutional codes. He discusses the properties of strongly random-like and weakly random-like codes and shows that by combining several random-like codes (some of which are random with small minimum distance), it is possible to control the tail distribution of the codes and hence increase the channel reliability. Guinand, Lodge and Papke propose two new interleaving methods, one based on number theory and the other on finite projective planes, for iterative maximum a posteriori (MAP) decoding of product block codes. They show that, by using large interleavers, it is possible to eliminate certain problematic error patterns and hence obtain an improved error rate performance. Gravel, Drolet and Rozon present a reduced complexity VLSI decoder design for concatenated codes consisting of an irreducible cyclic inner code and a Reed-Solomon outer code. Using the Berlekamp-Massey algorithms in the time and frequency domains, this concatenation leads to a significant reduction in the hardware complexity of the decoder. Lee and Kschischang present a method, based on atomic span modification, to construct a general class of non-minimal trellises for linear block codes having a regular structure. In this case, the code trellis can be split into a number of structurally identical parallel subtrellises. The authors propose a multiprocessor implementation that uses coset decoding for soft-decision decoding with a parallel trellis search. Esmaeili, Gulliver and Secord explore the trellis complexity and state connectivity of linear codes using trellis oriented generator matrices and atomic codewords. A simple yet comprehensive complexity analysis is given for the minimal L-section trellis diagram of Reed-Muller codes. Jürgensen and Konstantinidis give a mathematical description of discrete channels in which the transmitted symbols are affected by substitution, insertion and deletion errors. The decidability of unique decoding for error correction codes is investigated for non-probabilistic channels.

Coding and Modulation for Fading Channels

This section begins with the paper of Boudreau and Viens who investigate the performance of a reduced complexity linear predictive receiver for 4-ary continuous phase modulation (CPM) signals in Rayleigh flat-fading channels. Using a maximum likelihood sequence estimator (MLSE) with the q-ary soft-output Viterbi algorithm (QSOVA), promising results, in terms of bit error rate and receiver complexity, are obtained for power-limited channels such as mobile satellite channels. In the following paper, Nassar and Soleymani introduce a novel receiver design with a parallel structure for the detection of differentially encoded M-ary phase shift keying (MPSK) symbols in rapidly changing phase environments. The authors demonstrate that this receiver, which approximates maximum likelihood symbol estimation, outperforms DPSK as well as multiple symbol differential detection (MSDD) with an increase in receiver complexity of less than an order of magnitude. D'Amours and Yongaçoğlu study the spectral efficiency of a hybrid code division multiple access system using both direct sequence and frequency hopping schemes (DS/FH-CDMA) that employ M-ary frequency shift keying (MFSK) modulation. Comparison between non-coherent MFSK, combined MFSK and DPSK and wideband multitone FSK (MT-FSK) in Rayleigh fading channels indicates that MFSK with rate 1/2 dual-k coding provides the best spectral efficiency performance.

Signal Processing and Pattern Recognition

This section begins also with the paper of an invited lecturer, Professor H. Vincent Poor. Professor Poor presents an interesting overview of the recent developments of wavelet signal decomposition and reconstruction from a multiresolution signal analysis perspective. He describes cyclic wavelet transforms defined over finite fields and gives an application of multiresolution analysis to the design of multilevel error control coding. In the following paper, Mandal, Panchanathan and Aboulnasr study wavelet searching methods with reduced computational complexity for image source coding applications. The authors also suggest a reduced complexity adaptive algorithm for wavelet packet decomposition which provides good coding performance. Volden, Giraudon and Berthod use information theory to introduce new measures of image redundancy based on the entropy of Markov random fields. They present comparative results between these redundancy measures and the classical correlation coefficient measure for the case of satellite images. Gauvin, Doucet, Gingras and Chevrette present an algorithm for automatic target recognition which makes use of a template matching technique based on distance classifier correlation filters (DCCF). They investigate the object classes discrimination performance as a function of the training set. To improve the functionality of prosthetic devices, Gallant, Morin and Peppard present a new method for the extraction of myoelectric signals and their classification into distinct muscle contraction classes. The approach adopted by the authors, which involves a neural network-based classifier, results in a high correct classification rate for these signals. Liao and Pawlak propose a variant of the method of moments for Chinese character recognition. Using the Legendre moment feature space instead of the Central moment feature space, the authors show that significant improvement in character recognizing ability can be achieved.

Acknowledgements

We would like to express our warm thanks to the following reviewers for generously donating their valuable time. Without their assistance, this volume would not have been possible.

Behnaam Aazang
Tyseer Aboulnasr
Yaser S. Abu-Mostafa
Jakob D. Anderson
Jean Belzile
Sergei V. Bezzateev
Gerald Bolding
Daniel Boudreau
Joseph Boutros
Gilles Brassard
Richard Buz
Lorne Campbell
Jean Conan
Vladimir Cuperman
Claude D'Amours
F. Daneshgaran
Christophe Deutsch
Germain Drolet
Morteza Esmaeili
E. Barry Felstead
Hendrik C. Ferreira
Terrence L. Fine
François Gagnon
Peter J. Gallant
Christian Gehrmann
Allen Gersho
Denis Gingras
Dominic Grenier
Roshdy H.M. Hafez
Joachim Hagenauer
Anwar Hasan
A. S. J. Helberg
Paul Ho
Hideki Imai
Helmut Jürgensen
Tsutomu Kawabata
Hiroshi Kondo
Evangelos Kranakis
Adam Krzyżak

Frank R. Kschischang
Michel Lecours
Shu Lin
Alan Lindsey
John H. Lodge
Lloyd J. Mason
E. Masry
Robert J. McEliece
Laurence B. Milstein
Michael Moher
Lim Nguyen
Haluk Öğmen
Paul C. Van Oorschot
Erik Paaske
Jens Peter Pedersen
S. Eli Posner
Gregory J. Pottie
Sven Riedel
Jean-François Rivest
Ron M. Roth
Leslie A. Rusch
Michael Sablatash
Gérald Séguin
Asrar U. Sheikh
Blanca R. M. Sosa
Elvino Sousa
R. Michael Tanner
Stafford E. Tavares
Desmond P. Taylor
Chokri Trabelsi
Ari Trachtenberg
Marc Tremblay
Victor Keh-Wei Wei
Stephen B. Wicker
Mladen Victor Wickerhauser
Tet Yeap
Abbas Yongaçoğlu
André Zaccarin
Denis Zaccarin

The editors are also grateful to the following organizations for their support of the workshop:

- Canadian Society of Information Theory
- Carleton University
- IEEE Information Theory Society
- IEEE Region 7
- Laval University
- University of Ottawa

June 1996

Jean-Yves Chouinard
Paul Fortier
T. Aaron Gulliver

Table of Contents

Coding and Modulation for Fading Channels

Signal Processing and Pattern Recognition

The Algebraic Structure of Codes Invariant Under a Permutation

Gérald E. Séguin

Department of Electrical and Computer Engineering
Royal Military College of Canada
Kingston, Ontario, Canada K7K 5L0

Abstract. In this paper, we develop the algebraic theory of q-ary linear codes invariant under the action of a permutation σ. We show that the structure of these σ-codes is very similar to that of cyclic codes. In particular, we determine how many such codes there are of length N. We further show that these codes can be very conveniently described using the concatenation function of Jensen which is a simple generalization of the trace description of irreducible cyclic codes.

1   Introduction

The algebraic theory of cyclic codes is well known and may be found in any of the many texts on coding theory. On the other hand, the algebraic theory of quasi-cyclic codes, a natural generalization of cyclic codes, is not at all well known among coding theorists. This is quite apparent when we look at the literature on quasi-cyclic codes.

In this paper, we generalize the results of Séguin and Huynh [3] to σ-codes, i.e. linear codes defined over $\mathbf{F}_q$ which are invariant under a given permutation σ. If we impose the condition that the order of σ be relatively prime to q, then the algebraic theory of σ-codes parallels that of cyclic codes. In particular, we show that a σ-code can be expressed, in a unique way, as the sum of primary σ-codes. This is equivalent to the decomposition of a cyclic code as the sum of minimal cyclic codes (when the length is relatively prime to q). Moreover, we show that σ-codes can be obtained via the trace function. The use of the trace function allows us to obtain, in a very easy fashion, many results about the algebraic theory of σ-codes.

In Section 4, we introduce the notion of a 1-generator σ-code and we show that the structure of these is that of an algebra of polynomials of the form $\mathbf{F}_q[x]/(f(x))$. In the literature on quasi-cyclic codes, most of the codes studied are 1-generator quasi-cyclic codes or their duals. These, however, represent a small sub-class of the class of quasi-cyclic codes.

In Section 5, we show how to obtain the trace description of the dual of a quasi-cyclic code given the trace description of the code itself. This approach is different from that used in references [5,6] and is in fact much simpler.

2   Primary Components of a σ-Code

Let $\mathbf{F}_q$ be the field of size q and $\mathbf{F}_q^N$ the space of N-tuples over $\mathbf{F}_q$. Let σ be a permutation on the N objects $\{0, 1, \ldots, N-1\}$ then σ can be made to act on $\mathbf{F}_q^N$ by setting,

$$\sigma(\underline{a}) = \left(a_{\sigma(0)}, a_{\sigma(1)}, \ldots, a_{\sigma(N-1)}\right) \tag{1}$$

where,

$$\underline{a} = (a_0, a_1, \ldots, a_{N-1}) \in \mathbf{F}_q^N. \tag{2}$$

A q-ary σ-code is then simply a linear subspace $\mathbf{M}$ of $\mathbf{F}_q^N$ which is invariant under σ, i.e.

$$\mathbf{M} = \sigma(\mathbf{M}) = \{\sigma(\underline{a}) | \underline{a} \in \mathbf{M}\}. \tag{3}$$

Let σ be the product of r disjoint cycles of lengths $n_0, n_1, \ldots, n_{r-1}$ and let

$$\mathbf{A}_i = \mathbf{F}_q[x] / (x^{n_i} - 1), \qquad 0 \le i \le r-1. \tag{4}$$

We may then think of $\mathbf{F}_q^N$ as the external direct product of these $\mathbf{A}_i$'s, i.e.

$$\mathbf{F}_q^N = \prod_0^{r-1} \mathbf{A}_i = \{(a_0(x), a_1(x), \ldots, a_{r-1}(x)) | a_i(x) \in \mathbf{A}_i \ \ \forall_i\} \tag{5}$$

We make $\mathbf{F}_q^N$ into an $\mathbf{F}_q[x]$ module by setting

$$a(x)\underline{b} = (a(x)b_0(x), a(x)b_1(x), \ldots, a(x)b_{r-1}(x)) \tag{6}$$

where $\underline{b} = (b_0(x), b_1(x), \ldots, b_{r-1}(x)) \in \mathbf{F}_q^N$ and $a(x) \in \mathbf{F}_q[x]$. It now becomes apparent that; after a suitable rearrangement of the code word coordinates, a σ-code is nothing other than a sub-module of $\mathbf{F}_q^N$. Of course, $\mathbf{A}_i$ itself is an $\mathbf{F}_q[x]$-module with annihilator:

$$\text{Ann } \mathbf{A}_i = \{a(x) \in \mathbf{F}_q[x] | a(x)b(x) = 0 \quad \forall b(x) \in \mathbf{A}_i\} = (X^{n_i} - 1) \tag{7}$$

i.e. the ideal generated by $x^{n_i} - 1$. It is rather obvious that

$$\text{Ann } \mathbf{F}_q^N = \text{Ann} \prod_0^{r-1} \mathbf{A}_i = \bigcap_0^{r-1} \text{Ann } \mathbf{A}_i = \bigcap_0^{r-1} (x^{n_i} - 1) = (h(x)) \tag{8}$$

where

$$h(x) = \operatorname*{lcm}_{0 \le i \le r-1} \ x^{n_i} - 1. \tag{9}$$

We now restrict ourselves to permutations σ whose order $|\sigma|$ is relatively prime to q:

$$\gcd(|\sigma|, q) = 1, \tag{10}$$

which is equivalent to saying that $(n_i, q) = 1$ for $0 \le i \le r-1$. In this case, $x^{n_i} - 1$ has no repeated factors so that $\mathbf{A}_i$ is the direct sum of its minimal ideals and these in turn are finite fields. It follows that $h(x)$ has no repeated factors and so we have that,

$$h(x) = \prod_1^s h_\mu(x) \tag{11}$$

the $h_\mu(x)$'s being distinct irreducible polynomials over $\mathbf{F}_q$. We then write:

$$x^{n_i} - 1 = \prod_{\mu=1}^s h_\mu^{\epsilon_{i\mu}}(x) \qquad 0 \le i \le r - 1 \tag{12}$$

where $\epsilon_{i\mu} = 0$ or 1, $\forall i$, $\forall \mu$. We set,

$$\mathbf{I}_{i\mu} = \langle x^{n_i} - 1/h_\mu^{\epsilon_{i\mu}}(x) \rangle \subset \mathbf{A}_i \tag{13}$$

so that,

$$\mathbf{A}_i = \bigoplus_{\mu=1}^s \mathbf{I}_{i\mu} \tag{14}$$

is the decomposition of $\mathbf{A}_i$ as the direct sum of its minimal ideals (some of the $\mathbf{I}_{i\mu}$'s may be (0)). It now follows that,

$$\mathbf{F}_q^N = \prod_{i=0}^{r-1} \mathbf{A}_i = \prod_{i=0}^{r-1} \bigoplus_{\mu=1}^s \mathbf{I}_{i\mu} = \bigoplus_{\mu=1}^s \left(\prod_{i=0}^{r-1} \mathbf{I}_{i\mu} \right) = \bigoplus_{\mu=1}^s \mathbf{P}_\mu \tag{15}$$

and,

$$\mathbf{P}_\mu = \prod_{i=0}^{r-1} \mathbf{I}_{i\mu} = \{(b_0(x), b_1(x), \ldots, b_{r-1}(x)) \, | b_i(x) \in \mathbf{I}_{i\mu}\} \tag{16}$$

Let $\underline{b} \in \mathbf{F}_q^N$, then the <u>order</u> of $\underline{b}$ is the monic non-zero polynomial $f(x)$ of least degree, such that,

$$f(x)\underline{b} = 0 \tag{17}$$

If $\underline{b} = (b_0(x), b_1(x), \ldots, b_{r-1}(x))$ then it is clear that,

$$\text{ord } (\underline{b}) = \operatorname*{lcm}_{0 \le i \le r-1} \text{ ord } (b_i(x)) \tag{18}$$

and from the theory of cyclic codes,

$$\text{ord } b_i(x) = \frac{X^{n_i} - 1}{(x^{n_i} - 1, b_i(x))}. \tag{19}$$

From the above definition of $\mathbf{P}_\mu$, it is seen that,

$$\mathbf{P}_\mu = \{\underline{b} \in \mathbf{F}_q^N | \text{ord } \underline{b} | h_\mu(x)\} \tag{20}$$

and so,

$$\mathbf{P}_\mu = \frac{h(x)}{h_\mu(x)} \mathbf{F}_q^N \tag{21}$$

The sub-modules $\mathbf{P}_\mu$ are called the primary components of $\mathbf{F}_q^N$ [1, 2, 3]. We may easily determine the cardinality of $\mathbf{P}_\mu$. Let,

$$k_\mu = \deg h_\mu(x), \qquad 1 \leq \mu \leq s \tag{22}$$

then

$$|\mathbf{I}_{i\mu}| = q^{\epsilon_{i\mu} k_\mu} \tag{23}$$

and so,

$$|\mathbf{P}_\mu| = \left| \prod_{i=0}^{r-1} \mathbf{I}_{i\mu} \right| = \prod_{i=0}^{r-1} |\mathbf{I}_{i\mu}| = \prod_i q^{\epsilon_{i\mu} k_\mu} = q^{k_\mu \sum_i \epsilon_{i\mu}} \tag{24}$$

Now as is well-known from elementary module theory [1,2], $\mathbf{F}_q^N$ is a module over $\mathbf{F}_q[x]/(h(x))$ and by the same token $\mathbf{P}_\mu$ is a module over $\mathbf{F}_q[x]/(h_\mu(x))$. But the latter is the finite field $\mathbf{F}_{q^{k_\mu}}$ so that $\mathbf{P}_\mu$ is actually a vector space over $\mathbf{F}_{q^{k_\mu}}$ of dimension $\sum_{i=0}^{r-1} \epsilon_{i\mu}$. Now if $\mathbf{M}$ is a sub-module of $\mathbf{F}_q^N$, then

$$\mathbf{M} = \bigoplus_{\mu=1}^{s} \frac{h(x)}{h_\mu(x)} \mathbf{M} = \bigoplus_{\mu=1}^{s} \mathbf{M}_\mu \tag{25}$$

where,

$$\mathbf{M}_\mu = \frac{h(x)}{h_\mu(x)} \mathbf{M} = \mathbf{M} \bigcap \mathbf{M}_\mu \tag{26}$$

is the collection of elements in $\mathbf{M}$ whose order divides $h_\mu(x)$. The $\mathbf{M}_\mu$'s are the primary components of $\mathbf{M}$ and these are uniquely determined by $\mathbf{M}$. Hence, every sub-module of $\mathbf{F}_q^N$ is a sum of s sub-modules one from each of the primary components $\mathbf{P}_\mu$, $1 \leq \mu \leq s$. Hence, the number of sub-modules of $\mathbf{F}_q^N$, i.e. the number of σ-invariant q-ary linear codes, is

$$\prod_{\mu=1}^{s} \gamma(\mathbf{P}_\mu) \tag{27}$$

where $\gamma(\mathbf{P}_\mu)$ is the number of sub-modules of $\mathbf{P}_\mu$. But this latter number is simply the number of q^{k_μ}-subspaces of $\mathbf{P}_\mu$, a well known number.

Example 1. Let $N = 7$, $\sigma = (0)(123)(456)$ so that $r = 3$, $n_0 = 1$, $n_1 = 3 = n_2$

$$h(x) = \operatorname*{lcm}_{0 \leq i \leq 2} x^{n_i} - 1 = x^3 - 1 = (1+x)(1+x+x^2)$$

therefore, $s = 2$, $h_1(x) = 1 + x$, $h_2(x) = 1 + x + x^2$. We then have that

$$
\begin{aligned}
x - 1 &= h_1(x) & \text{i.e.} \quad & \epsilon_{01} = 1, \epsilon_{02} = 0 \\
x^3 - 1 &= h_1(x) h_2(x) & \text{i.e.} \quad & \epsilon_{11} = 1, \epsilon_{12} = 1 \\
x^3 - 1 &= h_1(x) h_2(x) & \text{i.e.} \quad & \epsilon_{21} = 1, \epsilon_{22} = 1
\end{aligned}
$$

The primary components of $\mathbf{F}_2^7$ are then:

$$\mathbf{P}_1 = \mathbf{I}_{01} \times \mathbf{I}_{11} \times \mathbf{I}_{11} = \{(0,000,000),(0,000,111),(0,111,000),$$
$$(0,111,111),(1,000,000),(1,000,111),(1,111,000),(1,111,111)\}$$

a vector space over $\mathbf{F}_2$ of dimension $\sum \epsilon_{i1} = 3$. The other primary component is:

$$\mathbf{P}_2 = \mathbf{I}_{02} \times \mathbf{I}_{12} \times \mathbf{I}_{22} = \{(0,000,000),(0,000,110),(0,000,011),$$
$$(0,000,101),(0,110,000),(0,110,110),(0,110,011),(0,110,101),$$
$$(0,011,000),(0,011,110),(0,011,011),(0,011,101),(0,101,000),$$
$$(0,101,110),(0,101,011),(0,101,101)\}$$

a vector space over $\mathbf{F}_4$ of dimension $\sum \epsilon_{i2} = 2$. Now,

$$\gamma(\mathbf{P}_1) = \sum_{k=0}^{3} \prod_{j=0}^{k-1} \left(2^3 - 2^j / 2^k - 2^j\right) = 15$$

$$\gamma(\mathbf{P}_2) = \sum_{k=0}^{2} \prod_{j=0}^{k-1} \left(4^2 - 4^j / 4^k - 4^j\right) = 7$$

and so the number of binary codes of block length 7 invariant under σ is:

$$7 \times 15 = 105.$$

$$\square$$

We can obtain an explicit description of the sub-modules of $\mathbf{F}_q^N$ by following the method introduced in 1985 by Jensen [4]. Let $\beta_\mu \in \mathbf{F}_{q^{k_\mu}}$ be chosen so that

$$h_\mu\left(\beta_\mu^{-1}\right) = 0 \tag{28}$$

and let $Tr_1^{k_\mu}$ be the trace function from $\mathbf{F}_{q^{k_\mu}}$ onto $\mathbf{F}_q$. Then

$$\Phi_{i\mu}(a) = \frac{1}{n_i} \sum_{j=0}^{n_i-1} Tr_1^{k_\mu}\left(a\beta_\mu^j\right) X^j \tag{29}$$

is a field isomorphism from $\mathbf{F}_{q^{k_\mu}}^{\epsilon_{i\mu}}$ onto $\mathbf{I}_{i\mu}$. Here we have to introduce the convention that

$$\mathbf{F}_{q^{k_\mu}}^0 = \{0\} \tag{30}$$

which technically speaking is not a field. We may extend this function, in a natural way, to a function from $\prod_{i=0}^{r-1} \mathbf{F}_{q^{k_\mu}}^{\epsilon_{i\mu}}$ onto $\mathbf{P}_\mu$, i.e.

$$\Phi_\mu = \prod_{i=0}^{r-1} \Phi_{i\mu} : \prod_{i=0}^{r-1} \mathbf{F}_{q^{k_\mu}}^{\epsilon_{i\mu}} \to \mathbf{P}_\mu = \prod_{i=0}^{r-1} \mathbf{I}_{i\mu} \tag{31}$$

$$\Phi_\mu\left(\underline{a}\right) = \sum_{i=0}^{r-1} \frac{1}{n_i} \sum_{j=0}^{n_i-1} Tr_1^{k_\mu}\left(a_i\beta_\mu^j\right) X^j Y^i \tag{32}$$

where now we denote the elements of $\mathbf{F}_q^N$ as two dimensional polynomials, i.e. in the form:

$$\sum_{i=0}^{r-1} a_i(x)Y^i, \quad a_i(x) \in \mathbf{A}_i = \mathbf{F}_q[x]/\left(x^{n_i}-1\right). \tag{33}$$

The product $\prod_{i=0}^{r-1} \mathbf{F}_{q^{k_\mu}}^{\epsilon_{i\mu}}$ is a vector space over $\mathbf{F}_{q^{k_\mu}}$ of dimension $\sum_i \epsilon_{i\mu}$. Of course, some of the coordinates of the vectors in $\prod_i \mathbf{F}_{q^{k_\mu}}^{\epsilon_{i\mu}}$ are identically 0. The function Φ_μ is a q^{k_μ}-linear bijective map from $\prod_i \mathbf{F}_{q^{k_\mu}}^{\epsilon_{i\mu}}$ onto $\mathbf{P}_\mu$ which sets up a 1-1 correspondence between the q^{k_μ}-subspaces of $\prod_i \mathbf{F}_{q^{k_\mu}}^{\epsilon_{i\mu}}$ and the submodules of $\mathbf{P}_\mu$. To simplify the notation set,

$$\mathbf{E}_\mu = \prod_{i=0}^{r-1} \mathbf{F}_{q^{k_\mu}}^{\epsilon_{i\mu}}, \quad 1 \le \mu \le s \tag{34}$$

then as indicated previously,

$$\dim_{q^{k_\mu}} \mathbf{E}_\mu = \sum_{i=0}^{r-1} \epsilon_{i\mu} \tag{35}$$

$$\Phi_\mu : \mathbf{E}_\mu \to \mathbf{P}_\mu. \tag{36}$$

So, if $\mathbf{M}$ is an arbitrary submodule of $\mathbf{F}_q^N$, then there exists unique vector spaces $\mathbf{U}_\mu$,

$$\mathbf{U}_\mu \subset \mathbf{E}_\mu \tag{37}$$

such that

$$\mathbf{M} = \bigoplus_{\mu=1}^s \Phi_\mu\left(\mathbf{U}_\mu\right) \tag{38}$$

and, in fact, the $\Phi_\mu(\mathbf{U}_\mu)$ are the primary components of $\mathbf{M}$. We can explicitly describe these $\mathbf{U}_\mu$ since the inverse of the function Φ_μ is:

$$\Psi_\mu : \sum_{i=0}^{r-1} a_i(x)Y^i \to \left(a_0\left(\beta_\mu^{-1}\right), a_1\left(\beta_\mu^{-1}\right), \ldots, a_{r-1}\left(\beta_\mu^{-1}\right)\right) \tag{39}$$

where $a_i(x) \in \mathbf{I}_{i\mu}, 0 \le i \le r-1$. Hence, if $\mathbf{M}$ is a submodule of $\mathbf{F}_q^N$ with primary components $\mathbf{M}_\mu, 1 \le \mu \le s$, then

$$\mathbf{M} = \bigoplus_{\mu=1}^s \Phi_\mu\left(\mathbf{U}_\mu\right), \tag{40}$$

where,

$$\mathbf{U}_\mu = \Psi_\mu\left(\mathbf{M}_\mu\right). \tag{41}$$

Moreover,

$$\dim_q \mathbf{M} = \sum_{\mu=1}^s k_\mu \dim_{q^{k_\mu}} \mathbf{U}_\mu. \tag{42}$$

Example 2. We continue with the last example where $k_1 = 1$, $k_2 = 2$, so that $\mathbf{E}_1 = \mathbf{F}_2^3$ and Φ_1 is simply the identity function. $\mathbf{E}_2 = \mathbf{F}_2^0 \times \mathbf{F}_4 \times \mathbf{F}_4 = \{0\} \times \mathbf{F}_4^2$. Let α be a zero of $1 + x + x^2$ in $\mathbf{F}_4$, then $\mathbf{F}_4 = \{0, 1, \alpha, \alpha^2\}$ and $\beta_2 = \alpha^2$. Let $\mathbf{U}_2 = \{(0,0,0), (0,1,\alpha), (0,\alpha,\alpha^2), (0,\alpha^2,1)\}$ then, $\mathbf{M}_2 = \Phi_2(\mathbf{U}_2) = (0, 000, 000)$, $(0, 011, 101)$, $(0, 101, 110)$, $(0, 110, 011)$ is a submodule of $\mathbf{F}_2^7$ which is primary. On the other hand, every subspace of $\mathbf{P}_1$ is clearly a submodule since it is clearly invariant under σ. If we take for $\mathbf{M}_1$ the space: $(0, 000, 000)$, $(0, 111, 000)$, $(1, 000, 111)$, $(1, 111, 111)$, then $\mathbf{M}_1 \oplus \mathbf{M}_2$ is the Hamming $(7,4)$ code. $\qquad\square$

Example 3. Let $q = 2$, $N = 23$ and $\sigma = (0\ 1\ 2)\ (3\ 4\ 5\ 6\ 7)\ (8\ 9\ 10\ 11\ 12\ 13 \ldots 22)$; hence $r = 3$, $n_0 = 3$, $n_1 = 5$, $n_2 = 15$.

$$x^{n_0} - 1 = x^3 - 1 = (1 + x)(1 + x + x^2)$$
$$x^{n_1} - 1 = x^5 - 1 = (1 + x)(1 + x + x^2 + x^3 + x^4)$$
$$x^{n_2} - 1 = x^{15} - 1 = (1 + x)(1 + x + x^2)(1 + x + x^2 + x^3 + x^4)$$
$$(1 + x + x^4)(1 + x^3 + x^4)$$

and so $h(x) = \underset{0 \le i \le r-1}{\mathrm{lcm}}\ x^{n_i} - 1 = x^{15} - 1$. We then have that $s = 5$ and so the $r \times s = 3 \times 5$ matrix of exponents $\epsilon_{i\mu}$ is

$$\begin{bmatrix} 1\ 1\ 0\ 0\ 0 \\ 1\ 0\ 1\ 0\ 0 \\ 1\ 1\ 1\ 1\ 1 \end{bmatrix}$$

The degrees k_μ are $k_1 = 1$, $k_2 = 2$, $k_3 = 4 = k_4 = k_5$. The primary components of $\mathbf{F}_2^{23}$ are:

$$\mathbf{P}_1 = <x^3 + 1/x + 1><x^5 + 1/x + 1><x^{15} + 1/x + 1>$$
$$\mathbf{P}_2 = <x^3 + 1/x^2 + x + 1><0><x^{15} + 1/x^2 + x + 1>$$
$$\mathbf{P}_3 = <0><x^5 + 1/x^4 + x^3 + x^2 + x + 1>$$
$$<x^{15} + 1/x^4 + x^3 + x^2 + x + 1>$$
$$\mathbf{P}_4 = <0><0><x^{15} + 1/x^4 + x + 1>$$
$$\mathbf{P}_5 = <0><0><x^{15} + 1/x^4 + x^3 + 1>$$

$$\dim_2 \mathbf{P}_1 = \sum \epsilon_{i1} = 3$$
$$\dim_4 \mathbf{P}_2 = \sum \epsilon_{i2} = 2$$
$$\dim_{16} \mathbf{P}_3 = \sum \epsilon_{i3} = 2$$
$$\dim_{16} \mathbf{P}_4 = \sum \epsilon_{i4} = 1$$
$$\dim_{16} \mathbf{P}_5 = \sum \epsilon_{i5} = 1$$

With these numbers, we may easily compute $\gamma(\mathbf{P}_\mu)$ as:

$$\gamma(\mathbf{P}_1) = \sum_{k=0}^{3} \prod_{j=0}^{k-1} \left(2^3 - 2^j/2^k - 2^j\right) = 16$$

$$\gamma(\mathbf{P}_2) = \sum_{k=0}^{2} \prod_{j=0}^{k-1} \left(4^2 - 4^j/4^k - 4^j\right) = 7$$

$$\gamma(\mathbf{P}_3) = \sum_{k=0}^{2} \prod_{j=0}^{k-1} \left(16^2 - 16^j/16^k - 16^j\right) = 19$$

$$\gamma(\mathbf{P}_4) = \sum_{k=0}^{1} \prod_{j=0}^{k-1} \left(16 - 16^j/16^k - 16^j\right) = 2 = \gamma(\mathbf{P}_5)$$

So the number of binary linear codes of length 23 invariant under σ is:

$$\prod_{\mu=1}^{5} \gamma(\mathbf{P}_\mu) = 8,512$$

The number of binary cyclic codes of length 23 is only 8.
Let α be a zero of $1 + x + x^4$ in $\mathbf{F}_{16}$, then the elements of $\mathbf{F}_{16}$, their binary representation with respect to the basis $\{1, \alpha, \alpha^2, \alpha^3\}$ and their trace from $\mathbf{F}_{16}$ into $\mathbf{F}_2$ are:

0	0000	0	α^7	1101	1
1	1000	0	α^8	1010	0
α	0100	0	α^9	0101	1
α^2	0010	0	α^{10}	1110	0
α^3	0001	1	α^{11}	0111	1
α^4	1100	0	α^{12}	1111	1
α^5	0110	0	α^{13}	1011	1
α^6	0011	1	α^{14}	1001	1

The sub-field $\mathbf{F}_4$ is $\{0, 1, \alpha^5, \alpha^{10}\}$ and the trace from $\mathbf{F}_4$ into $\mathbf{F}_2$ of these elements are: $Tr_1^2(0) = Tr_1^2(1) = 0$, $Tr_1^2(\alpha^5) = Tr_1^2(\alpha^{10}) = 1$.
We may take $\beta_2 = \alpha^5$, $\beta_3 = \alpha^3$, $\beta_4 = \alpha^{-1}$, $\beta_5 = \alpha$. We now construct a $(23, 11)$ code invariant under σ which turns out to be the Golay code. For $\mathbf{M}_1$, we take the $\mathbf{F}_2$-space generated by

$$(111, 11111, 00\ldots0).$$

$\mathbf{E}_2 = \mathbf{F}_4 \times \{0\} \times \mathbf{F}_4$ and we pick $\mathbf{U}_2 \subset \mathbf{E}_2$ the $\mathbf{F}_4$-subspace generated by (101). Since 1, α^5 is an $\mathbf{F}_2$-basis for $\mathbf{F}_4$ then $\Psi_2(101)$, $\Psi_2(\alpha^5\, 0\, \alpha^5)$ will be an $\mathbf{F}_2$-basis for $\mathbf{M}_2 = \Psi_2(\mathbf{U}_2)$. These vectors are:

011	00000	011	011	$\ldots$	011
101	00000	101	101	$\ldots$	101

$\mathbf{E}_3 = \{0\} \times \mathbf{F}_{16} \times \mathbf{F}_{16}$ and for $\mathbf{U}_3 \subset \mathbf{E}_3$ we choose the $\mathbf{F}_{16}$-subspace generated by (011). Since $1, \alpha^3, \alpha^6, \alpha^9$ form a basis for $\mathbf{F}_{16}$ over $\mathbf{F}_2$, then $\Psi_3(\alpha^{3i}(011))$, $i = 0, 1, 2, 3$ will form an $\mathbf{F}_2$-basis for $\mathbf{M}_3 = \Psi_3(\mathbf{U}_3)$. These vectors are:

000	01111	01111	01111	01111
000	10111	10111	10111	10111
000	11011	11011	11011	11011
000	11101	11101	11101	11101

$\mathbf{E}_4 = \{0\} \times \{0\} \times \mathbf{F}_{16}$ and for $\mathbf{U}_4$ we choose the $\mathbf{F}_{16}$-subspace generated by $(0\ 0\ 1)$. Since $\{1, \alpha^{-1}, \alpha^{-2}, \alpha^{-3}\}$ form an $\mathbf{F}_2$-basis for $\mathbf{F}_{16}$, then the vectors $\Psi_4(\alpha^{-i}(001))$, $0 \le i \le 3$ form an $\mathbf{F}_2$-basis for $\mathbf{M}_4 = \Psi_4(\mathbf{U}_4)$. These vectors are:

000	00000	111101011001000
000	00000	011110101100100
000	00000	001111010110010
000	00000	000111101011001

For $\mathbf{M}_5$ we choose $\{\underline{0}\}$.

$$
\begin{bmatrix}
111 & 11111 & 000000000000000 \\
011 & 00000 & 011011011011011 \\
101 & 00000 & 101101101101101 \\
000 & 01111 & 011110111101111 \\
000 & 10111 & 101111011110111 \\
000 & 11011 & 110111101111011 \\
000 & 11101 & 111011110111101 \\
000 & 00000 & 111101011001000 \\
000 & 00000 & 001111010110010 \\
000 & 00000 & 000111101011001
\end{bmatrix}
$$

Putting all these vectors together we get the above 11×23 generator matrix. It may be verified that the row-space of this matrix is the binary (23,11) Golay code. $\square$

3   Irreducible σ-Codes

We define an irreducible σ-code in the same way as we define an irreducible cyclic code; namely a σ-code $\mathbf{M}$ is <u>irreducible</u> if $\mathbf{M} \ne (0)$ and if the only σ-subcodes of $\mathbf{M}$ are $\mathbf{M}$ and (0). Since an arbitrary submodule $\mathbf{M}$ of $\mathbf{F}_q^N$ can be expressed as:

$$\mathbf{M} = \bigoplus_{\mu=1}^{s} \Phi_\mu (\mathbf{U}_\mu) \tag{43}$$

where $\mathbf{U}_\mu \subset \mathbf{E}_\mu$ is a subspace, it follows that $\mathbf{M}$ is irreducible if:

$$\mathbf{M} = \Phi_\mu (\mathbf{U}_\mu) \tag{44}$$

for some μ, $1 \le \mu \le s$, and $\mathbf{U}_\mu \subset \mathbf{E}_\mu$ has dimension 1 over $\mathbf{F}_{q^{k_\mu}}$. A submodule $\mathbf{N}$ of $\mathbf{F}_q^N$ is 1-generator (the adjective is cyclic in module theory) if there exists an $\underline{a} \in \mathbf{F}_q^N$ such that,

$$\mathbf{N} = \mathbf{F}_q[x]\underline{a} \tag{45}$$

in which case $\underline{a}$ is called a generator of $\mathbf{N}$. We then have the following theorem:

Theorem 1. *A submodule $0 \ne \mathbf{N}$ of $\mathbf{F}_q^N$ is irreducible if and only if it is primary and 1-generator. The irreducible submodules are all obtained in the form*

$$\mathbf{N}_\mu(\underline{\alpha}) = \left\{ \sum_{i=0}^{r-1} \frac{1}{n_i} \sum_{j=0}^{n_i-1} Tr_1^{k_\mu} \left(\gamma \alpha_i \beta_\mu^j \right) X^j Y^i \, \middle| \, \gamma \in \mathbf{F}_{q^{k_\mu}} \right\} \tag{46}$$

where,

$$\underline{\alpha} = (\alpha_0, \alpha_1, \ldots, \alpha_{r-1}) \in \prod_{i=0}^{r-1} \mathbf{F}_{q^{k_\mu}}^{\epsilon_{i\mu}} \tag{47}$$

and $1 \le \mu \le s$. The number of such modules is

$$\sum_{\mu=1}^{s} \left(q^{k_\mu \sum \epsilon_{i\mu}} - 1 \middle/ q^{k_\mu} - 1 \right). \tag{48}$$

Proof. If $0 \ne \mathbf{N}$ is irreducible then as we have already argued, it is primary i.e. it belongs to $\mathbf{P}_\mu$ for some μ or equivalently its order (the monic generator of its annihilator) is an irreducible polynomial. It is also clear that it has to be 1-generator. Suppose conversely that $\mathbf{N} = \mathbf{F}_q[x]\underline{a} \ne 0$ is primary and let its order be $h_\mu(x)$ i.e. ord $\mathbf{N} =$ ord $\underline{a} = h_\mu(x)$. Let $0 \ne \mathbf{L} \subset \mathbf{N}$ be a sub-module and let $0 \ne f(x)\underline{a} \in \mathbf{L}$. Then it follows that $(f(x), h_\mu(x)) = 1$ and so there exists $s(x)$ such that $s(x)f(x) \equiv 1 \bmod h_\mu(x)$. Since the polynomials multiplying $\underline{a}$ may be reduced modulo $h_\mu(x)$, then $\underline{a} = s(x)f(x)\underline{a} \in \mathbf{L}$ which implies that $\mathbf{L} = \mathbf{M}$ and so $\mathbf{N}$ is irreducible.

We have already shown that the irreducible submodules of $\mathbf{F}_q^N$ are of the form given by (46). For a fixed μ, the number of $\mathbf{N}_\mu(\underline{\alpha})$ is simply the number of subspaces of $\mathbf{E}_\mu$ of dimension 1. This number is:

$$q^{k_\mu \sum \epsilon_{i\mu}} - 1 \middle/ q^{k_\mu} - 1$$

and so the number of irreducible submodules is obtained by summing this last quantity over μ. □

Example 4. For the case $q = 2$, $N = 23$, and σ as in example 3, the number of irreducible σ-invariant binary linear codes of length 23 is:

$$\frac{2^3 - 1}{2 - 1} + \frac{2^{2 \cdot 2} - 1}{2^2 - 1} + \frac{2^{4 \cdot 2} - 1}{2^4 - 1} + \frac{2^{4 \cdot 1} - 1}{2^4 - 1} + \frac{2^{4 \cdot 1} - 1}{2^4 - 1} = 31.$$

By contrast, the number of irreducible binary cyclic codes of length 23 is 3. □

We see that any submodule of $\mathbf{F}_\mu^N$ can be expressed as a direct sum of irreducible submodules in a non-unique way. Such a decomposition can be obtained as follows: let $\mathbf{M} = \oplus \mathbf{M}_\mu$, then $\mathbf{M}_\mu = \Phi_\mu(\mathbf{U}_\mu)$ where $\mathbf{U}_\mu$ is a subspace of $\mathbf{E}_\mu$. If $\{\underline{\alpha}_1, \dots, \}$ is a basis for $\mathbf{E}_\mu$, then using the definition of $\mathbf{N}_\mu(\underline{\alpha})$ given by equation (46) we have that:

$$\mathbf{M}_\mu = \bigoplus_1^{t_\mu} \mathbf{N}_\mu(\underline{\alpha}_\nu) \tag{49}$$

and so

$$\mathbf{M} = \bigoplus_{\mu=1}^s \bigoplus_{\nu=1}^{t_\mu} \mathbf{N}_\mu(\underline{\alpha}_\nu). \tag{50}$$

By using different bases for the $\mathbf{U}_\mu$'s, we will in general get a different decomposition. On the other hand, a cyclic code of length n, $(n, q) = 1$, can be expressed as a direct sum of irreducible cyclic codes in a unique way. This happens in this case because the primary components of a cyclic code are automatically irreducible and are unique.

Example 5. Let $q = 2$, $N = 23$ and σ as in example 3. For this example, the primary components $\mathbf{P}_4$ and $\mathbf{P}_5$ are irreducible. Any subspace of $\mathbf{P}_1$ of dimension 1 is irreducible. If we pick $(011) \in \mathbf{E}_3$ (as we did in example 3), then $\mathbf{N}_3(011)$ is irreducible. It is the linear span of the following matrix:

$$\begin{bmatrix} 000 & 01111 & 01111 & 01111 & 01111 \\ 000 & 10111 & 10111 & 10111 & 10111 \\ 000 & 11011 & 11011 & 11011 & 11011 \\ 000 & 11101 & 11101 & 11101 & 11101 \end{bmatrix}$$

and is a σ-invariant subcode of the Golay code. □

Remark. Since any q-ary linear code is a σ-code for every σ in its automorphism group, then any linear code can be decomposed as a direct sum of irreducible σ-codes. As we have seen, these irreducible σ-codes are closely connected with irreducible cyclic codes. Hence, we see that linear codes which possess a nontrivial, auto-morphism group are intimately connected with cyclic codes.

4 One Generator σ-Codes

Cyclic codes are generated by a single element. Most of the quasi-cyclic codes studied in the literature are generated by a single element or they are the duals of such codes. These codes have rather simple canonical generator matrices defined in terms of circulant (or partial circulant) matrices.

We say that the submodule $\mathbf{M} \subset \mathbf{F}_q^N$ is a 1-generator submodule (called a cyclic submodule in the module literature) if there exists an $\underline{a} \in \mathbf{F}_q^N$ such that:

$$\mathbf{M} = \mathbf{F}_q[x]\underline{a} = \langle \underline{a} \rangle = \{g(x)\underline{a} \mid g(x) \in \mathbf{F}_q[x]\}. \tag{51}$$

Naturally $\underline{a}$ is called a generator of $\mathbf{M}$. Let $f(x)$ be the order of $\underline{a}$, i.e. the monic polynomial of least degree such that $f(x)\underline{a} = 0$. It is easy to see that $f(x)$ divides $h(x)$ where Ann $\mathbf{F}_q^N = \mathbf{F}_q[x]h(x)$. Naturally $\langle \underline{a} \rangle$ is a module over the factor ring $\mathbf{F}_q[x]/(f(x))$ obtained by setting:

$$\overline{s(x)}b(x)\underline{a} = s(x)b(x)\underline{a}$$

where $\overline{s(x)} = s(x) + (f(x)) \in \mathbf{F}_q[x]/(f(x))$ and $b(x) \in \mathbf{F}_q[x]$. The structure of 1-generator submodules of $\mathbf{F}_q^N$ is summarized in the next theorem.

Theorem 2. *Let $\underline{a} \in \mathbf{F}_q^N$ have order $f(x)$ with prime factorization $f(x) = h_1(x)h_2(x)\ldots h_t(x)$. Then the function:*

$$\eta : \mathbf{F}_q[x]/\left(f(x)\right) \to \langle \underline{a} \rangle \tag{52}$$
$$\eta : \overline{s(x)} \to s(x)\underline{a}$$

is a bijective module homo-morphism which sets up a 1-1 correspondence between the ideals of $\mathbf{F}_q[x]/(f(x))$ and the submodules of $\langle \underline{a} \rangle$; in particular:

$$|\langle \underline{a} \rangle| = q^{\deg \operatorname{ord} \underline{a}}. \tag{53}$$

The primary components of $\langle \underline{a} \rangle$ are irreducible and so the decomposition of $\langle \underline{a} \rangle$ as a sum of irreducible modules is unique. More specifically, this decomposition is obtained as:

$$\langle \underline{a} \rangle = \bigoplus_1^t \eta\left(\left\langle \overline{f(x)/h_\nu(x)} \right\rangle\right) = \bigoplus_1^t \left\langle \frac{f(x)}{h_\nu(x)}\underline{a} \right\rangle \tag{54}$$

All the 1-generator submodules of $\mathbf{F}_q^N$ are obtained in the form:

$$\mathbf{M} = \bigoplus_1^s \langle \underline{a}_\mu \rangle \tag{55}$$

where $\underline{a}_\mu \in \mathbf{P}_\mu$ and $\underline{a} = \sum \underline{a}_\mu$ is a generator of $\mathbf{M}$. Finally, the number of 1-generator submodules of $\mathbf{F}_q^N$ is given by:

$$\prod_{\mu=1}^s \left[\frac{q^{k_\mu \sum_i \epsilon_{i\mu}} - 1}{q^{k_\mu} - 1} + 1 \right] \tag{56}$$

and are obtained as:

$$\bigoplus_{\mu=1}^s \mathbf{N}_\mu\left(\underline{\alpha}_\mu\right) \tag{57}$$

where,

$$\underline{\alpha}_\mu \in \prod_i \mathbf{F}_{q^{k_\mu}}^{\epsilon_{i\mu}}, \qquad 1 \le \mu \le s. \tag{58}$$

Proof. It is a routine matter to verify that η is a module homomorphism which is clearly surjective. If $\eta\left(\overline{s_1(x)}\right) = \eta\left(\overline{s_2(x)}\right)$, then $s_1(x)\underline{a} = s_2(x)\underline{a}$ which implies that $\underline{(s_1(x) - s_2(x))\,\underline{a} = \underline{0}}$ and so $f(x) = \text{ord } \underline{a}\,|s_1(x) - s_2(x)$ which means that $s_1(x) = s_2(x)$; hence η is bijective. So η sets up a 1-1 correspondence between the submodules of $\mathbf{F}_q[x]/(f(x))$ and those of $\langle \underline{a}\rangle$; but the submodules of $\mathbf{F}_q[x]/(f(x))$ are its ideals. Hence, the submodules of $\langle \underline{a}\rangle$ are $\langle s(x)\underline{a}\rangle$ where $s(x)\,|f(x)$ and the primary components of $\langle \underline{a}\rangle$ correspond to the minimal ideals of $\mathbf{F}_q[x]/(f(x))$ i.e. they are $\left\langle \frac{f(x)}{h_\nu(x)}\underline{a}\right\rangle$, $\nu = 1, 2, \ldots, t$. If $\underline{a}_\nu = \frac{f(x)}{h_\nu(x)}\underline{a}$, then $\langle \underline{a}\rangle = \bigoplus_{\nu=1}^{t} \langle \underline{a}_\nu\rangle$.

Conversely, suppose $\underline{a}_\mu \in \mathbf{P}_\mu$, $1 \leq \mu \leq s$, and let $\mathbf{M} = \bigoplus_{1}^{s} \langle \underline{a}_\mu\rangle$. We claim that $\mathbf{M} = \langle \underline{a}\rangle$ where $\underline{a} = \sum_\mu \underline{a}_\mu$. Firstly, we see that $\underline{a} \in \mathbf{M}$, therefore $\langle \underline{a}\rangle \subset \mathbf{M}$. Multiplying $\underline{a}$ by $h(x)/h_\nu(x) = g_\nu(x)$, we get $g_\nu(x)\underline{a} = g_\nu(x)\underline{a}_\nu$. Now, there exists $s_\nu(x)$ such that $s_\nu(x)g_\nu(x) \equiv 1 \bmod h_\nu(x)$ and so $\underline{a}_\nu = s_\nu(x)g_\nu(x)\underline{a}_\nu = s_\nu(x)g_\nu(x)\underline{a} \in \langle \underline{a}\rangle$ and hence $\mathbf{M} = \langle \underline{a}\rangle$. It now follows that all the 1-generator submodules of $\mathbf{F}_q^N$ are given by (57) and their number by (56). $\qquad\square$

Example 6. Let $q = 2$, $N = 12$ and $\sigma = (0\ 1\ 2)\ (3\ 4\ 5\ 6\ 7\ 8\ 9\ 10\ 11)$, so that $n_0 = 3$, $n_1 = 9$, $r = 2$.

$$h(x) = \text{lcm}(x^3 - 1, x^9 - 1) = x^9 - 1 = (1 + x)(1 + x + x^2)(1 + x^3 + x^6)$$
$$x^3 - 1 = (1 + x)(1 + x + x^2)$$
$$x^9 - 1 = (1 + x)(1 + x + x^2)(1 + x^3 + x^6)$$

and so $s = 3$ and the 2×3 matrix of exponents $\epsilon_{i\mu}$ is

$$\begin{bmatrix} 1 & 1 & 0 \\ 1 & 1 & 1 \end{bmatrix}.$$

We also have that

$$k_1 = 1, k_2 = 2, k_3 = 6,$$

$$\dim_2 \mathbf{P}_1 = 3$$
$$\dim_4 \mathbf{P}_2 = 2$$
$$\dim_6 \mathbf{P}_3 = 1$$

and so the function γ, as defined in section 2, is computed to be:

$$\gamma(\mathbf{P}_1) = \sum_{k=0}^{3} \prod_{j=0}^{k-1} (2^3 - 2^j/2^k - 2^j) = 15$$

$$\gamma(\mathbf{P}_2) = \sum_{k=0}^{2} \prod_{j=0}^{k-1} (4^3 - 4^j/4^k - 4^j) = 7$$

$$\gamma(\mathbf{P}_3) = \sum_{k=0}^{1} \prod_{j=0}^{k-1} (64 - 64^j/64^k - 64^j) = 2.$$

Consequently, there are: 210 binary linear codes of length 9 invariant under σ. The number of these which are irreducible, computed as per (48), is

$$\frac{2^3 - 1}{2 - 1} + \frac{4^2 - 1}{4 - 1} + \frac{64^1 - 1}{64^1 - 1} = 7 + 5 + 1 = 13.$$

The number which are 1-generator is:

$$8 \cdot 6 \cdot 2 = 96.$$

Now, $\mathbf{E}_1 = \mathbf{F}_2 \times \mathbf{F}_2, \mathbf{E}_2 = \mathbf{F}_4 \times \mathbf{F}_4, \mathbf{E}_3 = \{0\} \times \mathbf{F}_{64}$ and,

$$\mathbf{P}_1 = \langle x^3 + 1/x + 1 \rangle \langle x^9 + 1/x + 1 \rangle$$
$$\mathbf{P}_2 = \langle x^3 + 1/x^2 + x + 1 \rangle \langle x^9 + 1/x^2 + x + 1 \rangle$$
$$\mathbf{P}_3 = \langle 0 \rangle \langle x^9 + 1/x^6 + x^3 + 1 \rangle.$$

We choose,

$$\underline{a}_1 = (000, 111111111)$$
$$\underline{a}_2 = (101, 101101101)$$
$$\underline{a}_3 = (000, 100100000)$$

and, $\underline{a} = \underline{a}_1 + \underline{a}_2 + \underline{a}_3 = (101, 110110010)$.

Then, ord $\underline{a}$ = ord $\underline{a}_1$ ord $\underline{a}_2$ ord $\underline{a}_3 = (1 + x)(1 + x + x^2)(1 + x^3 + x^6) = 1 + x^9$ and so $\langle \underline{a} \rangle$ has dimension 9. A generator matrix for this code is obtained as:

$$\begin{bmatrix} \begin{bmatrix} 1 & 0 & 1 \\ 1 & 1 & 0 \\ 0 & 1 & 1 \end{bmatrix} & \begin{bmatrix} 1 & 1 & 0 & 1 & 1 & 0 & 0 & 1 & 0 \\ 0 & 1 & 1 & 0 & 1 & 1 & 0 & 0 & 1 \\ 1 & 0 & 1 & 1 & 0 & 1 & 1 & 0 & 0 \end{bmatrix} \\ \begin{bmatrix} 1 & 0 & 1 \\ 1 & 1 & 0 \\ 0 & 1 & 1 \end{bmatrix} & \begin{bmatrix} 0 & 1 & 0 & 1 & 1 & 0 & 1 & 1 & 0 \\ 0 & 0 & 1 & 0 & 1 & 1 & 0 & 1 & 1 \\ 1 & 0 & 0 & 1 & 0 & 1 & 1 & 0 & 1 \end{bmatrix} \\ \begin{bmatrix} 1 & 0 & 1 \\ 1 & 1 & 0 \\ 0 & 1 & 1 \end{bmatrix} & \begin{bmatrix} 1 & 1 & 0 & 0 & 1 & 0 & 1 & 1 & 0 \\ 0 & 1 & 1 & 0 & 0 & 1 & 0 & 1 & 1 \\ 1 & 0 & 1 & 1 & 0 & 0 & 1 & 0 & 1 \end{bmatrix} \end{bmatrix}$$

where we have identified the naturally occurring submatrices which are circulants. The primary components of $\langle \underline{a} \rangle$ are $\langle \underline{a}_1 \rangle$, $\langle \underline{a}_2 \rangle$ and $\langle \underline{a}_3 \rangle$, of respective dimensions 1, 2 and 6. So another generator matrix for $\langle \underline{a} \rangle$ is obtained as:

$$\begin{bmatrix} 000 & 111111111 \\ 101 & 101101101 \\ 110 & 110110110 \\ 000 & 100100000 \\ 000 & 010010000 \\ 000 & 001001000 \\ 000 & 000100100 \\ 000 & 000010010 \\ 000 & 000001001 \end{bmatrix}$$

$\square$

The maximal dimension of 1-generator σ-codes, according to Theorem 2, is

$$\deg h(x)$$

and these are obtained as,

$$\bigoplus_{\mu=1}^{s} \mathbf{N}_\mu\left(\underline{\alpha}_\mu\right)$$

where $0 \neq \underline{\alpha}_\mu \in \mathbf{E}_\mu$, $1 \leq \mu \leq s$, and there are,

$$\prod_{\mu=1}^{s}\left(q^{k_\mu \sum \epsilon_{i\mu}} - 1 \Big/ q^{k_\mu} - 1\right) \tag{59}$$

of these. For the last example, this number is $7 \cdot 5 \cdot 1 = 35$.

5 The Dual of a Quasi-Cyclic Code

A σ-code is a quasi-cyclic code when the cycles of σ all have the same length; i.e. when,

$$n_i = n, \qquad 0 \leq i < r \tag{60}$$

and we continue to impose the condition $(n, q) = 1$. In this case,

$$\mathbf{A}_i = \mathbf{A} = \mathbf{F}_q[x]/(x^n - 1) \tag{61}$$

$$\mathbf{F}_q^N = \mathbf{A}^r = \{\,(a_0(x), a_1(x), \ldots, a_{r-1}(x))|\, a_i(x) \in \mathbf{A}\} \tag{62}$$

$$h(x) = \operatorname*{lcm}_{0 \leq i \leq r} x^{n_i} - 1 = x^n - 1 = \prod_{\mu=1}^{s} h_\mu(x) \tag{63}$$

$$\mathbf{P}_\mu = \langle x^n - 1 / h_\mu(x)\rangle^r$$

$$\mathbf{E}_\mu = \mathbf{F}_{q^{k_\mu}}^r, \qquad 1 \leq \mu \leq s. \tag{64}$$

Let $\mathbf{M} \subset \mathbf{F}_q^N$ be a submodule; we define its σ-dual $\mathbf{M}^*$ by setting:

$$\mathbf{M}^* = \left\{ \sum_0^{r-1} a_i(x)Y^i \,\middle|\, \sum_0^{r-1} a_i(x)b_i(x) = 0, \right. \tag{65}$$

$$\left. \text{for every } \sum_0^{r-1} b_i(x)Y^i \in \mathbf{M} \right\}$$

It is easy to verify that $\mathbf{M}^*$ is a submodule of $\mathbf{F}_q^N$. Next we define the σ-reciprocal operator R on $\mathbf{F}_q^N$ by setting:

$$R\left(\sum_0^{r-1} a_i(x)Y^i\right) = \sum_0^{r-1} x^n a_i\left(x^{-1}\right) Y^i. \tag{66}$$

We then have the following result:

Theorem 3. $\mathbf{M}^\perp = R(\mathbf{M}^*)$.

Proof. The proof is identical to that for cyclic codes since $\mathbf{M}$ is clearly a sub-module of $\mathbf{F}_q^N$. $\qquad\qquad\square$

The σ-dual was introduced by Drolet [5] who used it to describe the dual of 1-generator quasi-cyclic codes. The results of Drolet were then generalized by Conan and Séguin [6]. Our approach here is different from that of [6]. In the product $\sum_0^{r-1} a_i(x)b_i(x)$ we may write $a_i(x) = \sum_{\mu=1}^s a_{i\mu}(x)$, $b_i(x) = \sum_{\mu=1}^s b_{i\mu}(x)$ where $a_{i\mu}(x), b_{i\mu}(x) \in \langle x^n - 1/h_\mu(x) \rangle$ and so,

$$\sum_0^{r-1} a_i(x)b_i(x) = \sum_{\mu=1}^s \sum_{i=0}^{r-1} a_{i\mu}(x)b_{i\mu}(x)$$

since $a_{i\mu}(x)b_{i\nu}(x) = 0$ whenever $\mu \neq \nu$. Hence, it follows that $\sum_0^{r-1} a_i(x)b_i(x) = 0$ if and only if,

$$\sum_{i=0}^{r-1} a_{i\mu}(x)b_{i\mu}(x) = 0, \qquad 1 \leq \mu \leq s \tag{67}$$

This shows that if, $\mathbf{M}^* = \bigoplus_1^s \mathbf{M}_\mu^*$ then, $\mathbf{M} = \bigoplus_1^s \mathbf{M}_\mu$ where,

$$\mathbf{M}_\mu^* = \left\{ \sum_0^{r-1} b_{i\mu}(x)Y^i \in \mathbf{P}_\mu \middle| \sum_0^{r-1} a_{i\mu}(x)b_{i\mu}(x) = 0, \right. \tag{68}$$
$$\left. \text{for every } \sum_0^{r-1} b_{i\mu}(x)Y^i \in \mathbf{M}_\mu \right\}$$

This allows us to establish the following characterization of $\mathbf{M}^*$:

Theorem 4. *If* $\mathbf{M} = \bigoplus_1^s \Phi_\mu(\mathbf{U}_\mu)$, $\mathbf{U}_\mu \subset \mathbf{F}_{q^{k_\mu}}^r$ *a subspace, then*

$$\mathbf{M}^* = \bigoplus_1^s \Phi_\mu\left(\mathbf{U}_\mu^\perp\right) \tag{69}$$

where $\mathbf{U}_\mu^\perp \subset \mathbf{F}_{q^{k_\mu}}^r$ *is the dual of* $\mathbf{U}_\mu$ *with respect to the usual dot product.*

Proof. Let $\mathbf{M}_\mu = \Phi_\mu(\mathbf{U}_\mu)$, then $\sum \Phi_\mu(\gamma_i) Y^i \in \mathbf{M}_\mu^*$, $\underline{\gamma} \in \mathbf{F}_{q^{k_\mu}}^r$, if and only if for every $\underline{\alpha} \in \mathbf{U}_\mu$, we have that

$$0 = \sum_0^{r-1} \Phi_\mu(\gamma_i) \Phi_\mu(\alpha_i) = \Phi_\mu\left(\sum_0^{r-1} \gamma_i \alpha_i\right)$$

which is so if and only if $\sum_0^{r-1} \gamma_i \alpha_i = 0$ i.e. if and only if $\underline{\gamma} \in \mathbf{U}_\mu^\perp$. $\qquad\square$

Example 7. Let $q = 2$, $N = 15$, $n = 3$, then

$$x^3 + 1 = (1 + x)(1 + x + x^2)$$

so that $r = N/n = 5$, $s = 2$ and so $\mathbf{E}_1 = \mathbf{F}_2^5$, $\mathbf{E}_2 = \mathbf{F}_4^5$. Let $\mathbf{F}_4 = \{0, 1, \alpha, \alpha^2 = 1 + \alpha\}$ and choose $\mathbf{U}_2 = LS\{(1\ 1\ \alpha\ \alpha^2\ 1), (\alpha\ 1\ \alpha\ 1\ \alpha)\}$. Then $\mathbf{M}_2$ is the binary linear span of the set $\{\Phi_2(1\ 1\ \alpha\ \alpha^2\ 1), \Phi_2(\alpha\ \alpha\ \alpha^2\ 1\ \alpha), \Phi_2(\alpha\ 1\ \alpha\ 1\ \alpha), \Phi_2(\alpha^2\ \alpha\ \alpha^2\ \alpha\ \alpha^2)\}$ which has generator matrix

$$\begin{bmatrix} 0\ 1\ 1\ 0\ 1\ 1\ 1\ 1\ 0\ 1\ 0\ 1\ 0\ 1\ 1 \\ 1\ 1\ 0\ 1\ 1\ 0\ 1\ 0\ 1\ 0\ 1\ 1\ 1\ 1\ 0 \\ 1\ 1\ 0\ 0\ 1\ 1\ 1\ 1\ 0\ 0\ 1\ 1\ 1\ 1\ 0 \\ 1\ 0\ 1\ 1\ 1\ 0\ 1\ 0\ 1\ 1\ 1\ 0\ 1\ 0\ 1 \end{bmatrix} \tag{70}$$

$\mathbf{U}_2^{\perp} = LS\ \{(0\ \alpha\ 1\ 0\ 0), (\alpha^2\ 0\ 0\ 1\ 0), (1\ 0\ 0\ 0\ 1)\}$ and so by the last theorem $\mathbf{M}_2^*$ is the binary linear span of the set $\{\Phi_2(0\ \alpha\ 1\ 0\ 0), \Phi_2(0\ \alpha^2\ \alpha\ 0\ 0), \Phi_2(\alpha^2\ 0\ 0\ 1\ 0), \Phi_2(1\ 0\ 0\ \alpha\ 0), \Phi(1\ 0\ 0\ 0\ 1), \Phi_2(\alpha\ 0\ 0\ 0\ \alpha)\}$ which is the row space of the following matrix:

$$\begin{bmatrix} 0\ 0\ 0\ 1\ 1\ 0\ 0\ 1\ 1\ 0\ 0\ 0\ 0\ 0\ 0 \\ 0\ 0\ 0\ 1\ 0\ 1\ 1\ 1\ 0\ 0\ 0\ 0\ 0\ 0\ 0 \\ 1\ 0\ 1\ 0\ 0\ 0\ 0\ 0\ 0\ 0\ 1\ 1\ 0\ 0\ 0 \\ 0\ 1\ 1\ 0\ 0\ 0\ 0\ 0\ 0\ 1\ 1\ 0\ 0\ 0\ 0 \\ 0\ 1\ 1\ 0\ 0\ 0\ 0\ 0\ 0\ 0\ 0\ 0\ 0\ 1\ 1 \\ 1\ 1\ 0\ 0\ 0\ 0\ 0\ 0\ 0\ 0\ 0\ 0\ 1\ 1\ 0 \end{bmatrix}$$

Applying the σ-reciprocal operator to the rows of this latter matrix, we get a generator matrix for $\mathbf{M}_2^{\perp} \in \mathbf{P}_2$. This is:

$$\begin{bmatrix} 0\ 0\ 0\ 0\ 1\ 1\ 1\ 1\ 0\ 0\ 0\ 0\ 0\ 0\ 0 \\ 0\ 0\ 0\ 1\ 0\ 1\ 0\ 1\ 1\ 0\ 0\ 0\ 0\ 0\ 0 \\ 1\ 0\ 1\ 0\ 0\ 0\ 0\ 0\ 0\ 1\ 1\ 0\ 0\ 0\ 0 \\ 1\ 1\ 0\ 0\ 0\ 0\ 0\ 0\ 0\ 0\ 1\ 1\ 0\ 0\ 0 \\ 1\ 1\ 0\ 0\ 0\ 0\ 0\ 0\ 0\ 0\ 0\ 0\ 1\ 1\ 0 \\ 0\ 1\ 1\ 0\ 0\ 0\ 0\ 0\ 0\ 0\ 0\ 0\ 0\ 1\ 1 \end{bmatrix} \tag{71}$$

If $\mathbf{M} = \mathbf{M}_2$, then $\mathbf{M}^{\perp} = \mathbf{P}_1 \oplus \mathbf{M}_2^{\perp}$ so we must augment the latter matrix by a basis for $\mathbf{P}_1$ to get a generator matrix of $\mathbf{M}^{\perp}$. Suppose that $\mathbf{U}_1 \subset \mathbf{F}_2^5$ is the linear span of $\{(10000), (10010)\}$ then $\mathbf{M}_1$ is the row space of the following matrix:

$$\begin{bmatrix} 1\ 1\ 1\ 0\ 0\ 0\ 0\ 0\ 0\ 0\ 0\ 0\ 0\ 0\ 0 \\ 1\ 1\ 1\ 0\ 0\ 0\ 0\ 0\ 0\ 1\ 1\ 1\ 0\ 0\ 0 \end{bmatrix} \tag{72}$$

$\mathbf{U}_1^{\perp}$ is the linear span of $\{(0\ 1\ 0\ 0\ 0), (0\ 0\ 1\ 0\ 0), (0\ 0\ 0\ 0\ 1)\}$ and so $\mathbf{M}_1^{\perp}$ (in $\mathbf{P}_1$) is the row space of:

$$\begin{bmatrix} 0\ 0\ 0\ 1\ 1\ 1\ 0\ 0\ 0\ 0\ 0\ 0\ 0\ 0\ 0 \\ 0\ 0\ 0\ 0\ 0\ 0\ 1\ 1\ 1\ 0\ 0\ 0\ 0\ 0\ 0 \\ 0\ 0\ 0\ 0\ 0\ 0\ 0\ 0\ 0\ 0\ 0\ 0\ 1\ 1\ 1 \end{bmatrix} \tag{73}$$

A generator matrix for $\mathbf{M} = \mathbf{M}_1 \oplus \mathbf{M}_2$ is obtained by augmenting (70) by (72) and a generator matrix for $\mathbf{M}^{\perp}$ is obtained by augmenting (71) by (73). □

6   Conclusions

In this paper, we have shown that the algebraic theory of σ-codes is very similar to that of cyclic codes. In particular, we were able to obtain expressions allowing us to determine how many such codes there are. We showed that a σ-code can be decomposed in several ways: as a direct sum of primary σ-codes or as a direct sum of irreducible codes or finally as a direct sum of 1-generator σ-codes.

It is hoped that a better understanding of the algebraic structure of σ-codes, and in particular quasi-cyclic codes will lead to constructive classes of σ-codes i.e. BCH type σ-codes.

References

1. N. Jacobson, *Lectures in Abstract Algebra*, Vol. II. D. Van Nostrand Co., 1953.
2. C.W. Curtis and I. Reiner, *Representation Theory of Finite Groups and Associative Algebra*, John Wiley and Sons, 1962.
3. G.E. Séguin and H.T. Huynh, "Quasi-Cyclic Codes: A Study", report published by the *Laboratoire de Radiocommunications et de Traitement du Signal*, Université Laval, Québec, Canada, 1985.
4. J.M. Jensen, "On the Concatenated Structure of Cyclic and Abelian Codes", *IEEE Trans. Inform. Theory*, Vol. 31, No. 6, pp. 788–793, Nov. 1985.
5. G. Drolet, "Sur les codes quasi-cycliques à base cyclique", Thèse de Ph.D., Université Laval, Québec, Canada, 1990.
6. J. Conan and G.E. Séguin, "Structural Properties and Enumeration of Quasi-Cyclic Codes", *Applicable Algebra in Engineering, Communication and Computing*, Vol. 4, pp. 25–39, 1993.
7. R.L. Townsend and E.J. Weldon, "Self-Orthogonal Quasi-cyclic Codes", *IEEE Trans. Inform. Theory*, Vol. IT-13, No. 2, pp. 183–195, April 1967.
8. M. Karlin, "New Binary Coding Results by Circulants", *IEEE Trans. Inform. Theory*, Vol. IT-15, No. 1, pp. 81–92, Jan. 1969.
9. C.L. Chen and W.W. Peterson, "Some Results on Quasi-Cyclic Codes", *Information and Control*, Academic Press, Vol. 15, No. 5, pp. 407–423, Nov. 1969.
10. M. Karlin, "Decoding of Circulant Codes", *IEEE Trans. Inform. Theory*, Vol. IT-16, No. 6, pp. 797–802, Nov. 1970.
11. T. Kasami, "A Gilbert-Varshamov Bound for Quasi-Cyclic Codes of Rate 1/2", *IEEE Trans. Inform. Theory*, Vol. IT-20, No. 5, p. 679, Sept. 1974.
12. J.M. Stein, V.K. Bhargava and S.E. Tavares, "Weight Distribution of Some Best $(3m, 2m)$ Binary Quasi-Cyclic Codes", *IEEE Trans. Inform. Theory*, Vol. IT-21, No. 6, pp. 708–711, Nov. 1975.
13. T.A. Gulliver and V.K. Bhargava, "Two New Rate 2/p Binary Quasi-Cyclic Codes", *IEEE Trans. Inform. Theory*, Vol. IT-40, No. 5, pp. 1667–1668, Sept. 1994.

A Method for Combining Algebraic Geometric Goppa Codes

Jens Peter Pedersen[1] and Despina Polemi [*2]

[1] Department of Mathematics and Computer Science,
Aalborg University, Denmark, jpp@iesd.auc.dk
[2] Department of Mathematics, State University of New York,
Farmingdale, NY 11735, polemid@snyfarva.bitnet

Abstract. Various methods for combining codes to obtain new ones have been described by MacWilliams and Sloane [10]. In this paper we present another method for combining algebraic geometric (a.g.) Goppa codes to construct new longer codes using function field extensions. We also prove that if one starts with an a.g. code with minimum distance better than expected, then the new code, obtained by this method, will also have minimum distance better than expected. Furthermore, we give an estimate on the improvement of the minimum distance of the new code.

KEY WORDS AND PHRASES. algebraic geometric Goppa codes, algebraic function fields of one variable, elementary abelian extensions

1 Introduction.

The area of coding theory which deals with algebraic geometric (a.g.) Goppa codes up until now has been lacking in constructive examples. One might inquire as to how known codes (a.g.) can be used to generate new codes. MacWilliams and Sloane (see [10], chapter 18) have presented various methods for combining codes to obtain new ones.

Our first goal in this paper is to present a new method for combining a.g. Goppa codes to construct new longer a.g. Goppa codes. Our method will be based on a different approach from the ones used by MacWilliams and Sloane, i.e. the use of elementary abelian extensions. We start with an a.g. Goppa code and then "lift" it to a longer code of higher dimension. In particular we start with an a.g. Goppa code C_1 from a function field F_1. Then we consider an elementary abelian extension F_2 of F_1 and lift the code C_1 to a code C_2 from F_2.

A crucial and difficult problem that arises in the theory of error-correcting codes is to determine the minimum distance of a given code. Goppa's construction of a.g. codes (see [5], [6], [7], [8]), yielded a lower bound for the minimum

---

[*] Supported by NSA and by Dr. Nuala McGann Drescher Foundation

distance, the so-called *designed minimum distance*. Since then many authors (see e.g. [2], [1], [4], [3], [9]) have considered the problem of determining the minimum distance of a.g. codes. In their work concepts like Weistrass gaps and gonality are used to obtain lower bounds for the minimum distance.

The second goal of our paper is to prove that if one starts with an a.g. code with minimum distance better than expected, then the new code, obtained by the new method, will also have minimum distance better than expected. In particular if we assume that the code C_1 has minimum distance at least a better than the designed minimum distance. Then we prove that under certain conditions the new code C_2 has minimum distance at least c better than the designed minimum distance.

We will also estimate the minimum distance of the new a.g. code by exploiting the conclusions we already have of the minimum distance of the old code. Specifically we will prove that the improvement c is lower bounded by a and the genera of F_1 and F_2.

2 Construction.

In this section we will describe a method for combining algebraic geometric (a.g.) Goppa codes in order to obtain new a.g. codes. Our construction will be based on function field extensions. For the main properties of algebraic geometric codes and function fields we refer the reader to Stichtenoth [11].

Let F be a function field over $\mathbf{F}_q$ of genus g and choose divisors $D, G \in \mathrm{Div}(F/\mathbf{F}_q)$ such that $D = Q_1 + \ldots + Q_n$, where the Q_i's are pairwise distinct $\mathbf{F}_q$-rational places, and $\mathrm{Supp}G \cap \mathrm{Supp}D = \emptyset$. Consider the vector space $\mathcal{L}(G) = \{z \in F/(z) \geq -G\} \cup \{0\}$ which is finite dimensional meaning that $\mathcal{L}(G) = \mathrm{Span}_{\mathbf{F}_q}\{z_1, ..., z_k\}$. The first definition $C_{\mathcal{L}}(D, G)$ of an *algebraic geometric (a.g.) Goppa code* is based on the injection

$$\Phi : \mathcal{L}(G) \longrightarrow \mathbf{F}_q^n$$
$$z \longrightarrow (z(Q_1), ..., z(Q_n)).$$

In particular the a.g. Goppa code $C_{\mathcal{L}}(D, G)$ is the image $\Phi(\mathcal{L}(G)) \subset \mathbf{F}_q^n$ of the map Φ.

The second definition of an a.g. Goppa code is using differentials, for that we consider the vector space $\Omega_F(G) = \{\eta \in \Omega_F/(\eta) \geq G\} \cup \{0\}$. The image of the injection

$$\Psi : \Omega_F(G - D) \longrightarrow \mathbf{F}_q^n$$
$$\eta \longrightarrow (\mathrm{res}_{Q_1}(\eta), ..., \mathrm{res}_{Q_n}(\eta))$$

yields the a.g. Goppa code $C_\Omega(D, G)$, i.e. $\Psi(\Omega_F(G - D)) = C_\Omega(D, G)$.

We can summarize the two definitions of the a.g. Goppa codes associated with the divisors D and G as follows:

$$C_{\mathcal{L}}(D, G) = \{(z(Q_1), ..., z(Q_n)) \mid z \in \mathcal{L}(G)\}$$
$$C_{\Omega}(D, G) = \{(\text{res}_{Q_1}(\eta), ..., \text{res}_{Q_n}(\eta)) \mid \omega \in \Omega(G - D)\}.$$

The two codes are dual to each other, meaning that the matrix

$$\begin{pmatrix} z_1(Q_1) & \cdots & z_1(Q_n) \\ \vdots & & \vdots \\ z_r(Q_1) & \cdots & z_r(Q_n) \end{pmatrix},$$

is the parity check matrix for $C_{\Omega}(D, G)$, and the generator matrix of $C_{\mathcal{L}}(D, G)$.

In this paper we consider the case where $\mathbf{F}_q$ is of characteristic 2. For our construction we first define two codes C_1 and C_1' as follows:

Definition 1. Let F_1 be a function field over $\mathbf{F}_q$ of genus g_1 and let $D_1 = Q_1 + \ldots + Q_n$, where the Q_i's are distinct $\mathbf{F}_q$-rational places of F_1. Let Q be a $\mathbf{F}_q$-rational place of F_1 distinct from $Q_1, \ldots, Q_n$. For any nonnegative integer l we define the code

$$C_1 := C_{\Omega}(D_1, G_1), \text{ where } G_1 = lQ.$$

Let $y_1 \in F_1 \backslash \mathbf{F}_q$ with $(y_1)_\infty = mQ$, m is odd, and define the code

$$C_1' := C_{\Omega}(D_1, G_1'), \text{ where } G_1' = (l - \frac{m+1}{2})Q.$$

Remark.
Obviously, $C_1 \subset C_1'$. Furthermore, if we let $d_1^* = l - 2g_1 + 2$ (resp. $d_1'^* = l - \frac{m+1}{2} - 2g_1 + 2$) denote the designed minimum distance of C_1 (resp. C_1'), then C_1 is an $[n, k, d_1]$ code over $\mathbf{F}_q$ with

$$n - k \leq r := \dim G_1$$
$$d_1 \geq d_1^*$$

and C_1' is a $[n, k', d_1']$ code over $\mathbf{F}_q$ with

$$n - k' \leq r' := \dim G_1'$$
$$d_1' \geq d_1'^*.$$

We will now *lift* the code C_1 to obtain a new a.g. code C_2 by means of the *conorm*, Con_{F_2/F_1} ([11], p.63).

Definition 2. Let $F_2 := F_1(y_2)$ where $y_2^2 + y_2 = y_1$ and assume that any $Q_i \in \text{Supp}D_1$ splits into two places $P_{i,1}$ and $P_{i,2}$ in F_2. By the *lift* of C_1 we understand the code

$$C_2 := C_{\Omega}(D_2, G_2),$$

where $D_2 := \text{Con}_{F_2/F_1} D_1 = P_{1,1} + P_{1,2} + \ldots + P_{n,1} + P_{n,2}$ and $G_2 := \text{Con}_{F_2/F_1} G_1$.

Remarks.

1. Note that Q is the only place of F_1 which ramifies in F_2 and that the ramification index is 2. Therefore, there is only one place P in F_2 lying above Q and $G_2 = \mathrm{Con}_{F_2/F_1} G_1 = \mathrm{Con}_{F_2/F_1} lQ = 2lP$.

2. F_2 is an elementary abelian extension of F_1, and by [[11], p.115] we have that the genus g_2 of F_2 is

$$g_2 = 2g_1 + \frac{m-1}{2}.$$

3. The code C_1' will be used in the next section to estimate the parameters of the new code C_2.

Example 1A.

Consider the elliptic function field (i.e. $g_1 = 1$) $F_1 = \mathbf{F}_4(x, y_1)$ defined by the equation $y_1^2 + y_1 = x^3 + 1$. Choose $l = m = 3$. Let $D_1 = Q_{\alpha,0} + Q_{\beta,0} + Q_{0,1} + Q_{1,1}$, where $\mathbf{F}_4 = \{0, 1, \alpha, \beta\}$ and $Q_{\delta,\lambda}$ is the place corresponding to the solution $x = \delta$ and $y_1 = \lambda$ to the defining equation for F_1. Let $G_1 = 3Q_\infty$ and $G_1' = Q_\infty$. In this case $C_1 = C_\Omega(D_1, G_1)$ is a $[4, 1, 4]$ and $C_1' = C_\Omega(D_1, G_1')$ is a $[4, 3, 2]$ code.

We consider the extension F_2 of F_1 where F_2 is a function field defined by the equation $y_2^2 + y_2 = y_1$; it is of genus 3, i.e. $g_2 = 3$. Under this extension the point Q_∞ will be mapped to P_∞, thus the divisor $G_2 = 6P_\infty$. The divisor $D_2 = P_{a,0,1} + P_{a,0,0} + P_{b,0,1} + P_{b,0,0} + P_{0,1,a} + P_{0,1,b} + P_{1,1,a} + P_{1,1,b}$, where $P_{\delta,\lambda,\gamma}$ is the place corresponding to the solution $x = \delta$ and $y_1 = \lambda$, $y_2 = \gamma$ to the defining equation for F_2. These are the points lying above $Q_{\delta,\lambda}$. We can now construct the code $C_2 = C_\Omega(D_2, G_2)$, over $\mathbf{F}_4$. In the next section we will compute the parameters of the new code C_2.

3 Parameters of New Codes.

In the previous section we constructed a new a.g. code C_2 by lifting the code C_1. In the present section we will prove that C_2 is a longer code and its minimum distance is better than the expected minimum distance. We will use the notation and definitions of the previous section.

We first want to prove that the functions in the basis of the new vector space $\mathcal{L}(G_2)$, where $G_2 = 2lP$, can be constructed using the functions in the basis of the vector spaces $\mathcal{L}(G_1)$ and $\mathcal{L}(G_1')$, where $G_1 = lQ$ and $G_1' = (l - \frac{m+1}{2})Q$. This is described in the next lemma.

Lemma 3. *Let* $\mathcal{L}(lQ) = \mathrm{Span}\{z_1, ..., z_r\}$ *and* $\mathcal{L}((l - \frac{m+1}{2})Q) = \mathrm{Span}\{z_1, ..., z_{r'}\}$. *Then*

$$\mathcal{L}(2lP) = \mathrm{Span}\{z_1, ..., z_r, y_2 z_1, ..., y_2 z_{r'}\}.$$

Proof. Obviously, we have $r' \leq r$ and by the construction, described in the previous section, we have that the elements $z_1, ..., z_r, y_2 \in F_2$ have poles only at

P. Let ν_P (resp. ν_Q) denote the discrete valuation at P (resp. Q). Then

$$\nu_P(z_i) = 2\nu_Q(z_i) \geq -2l \quad \text{for} \quad i = 1, ..., r \quad \text{and}$$

$$\nu_P(y_2 z_j) = \nu_P(y_2) + 2\nu_Q(z_j)$$

$$\geq -m - 2(l - \frac{m+1}{2})$$

$$= -2l + 1 > -2l \quad \text{for} \quad j = 1, ..., r'.$$

Therefore, $z_i, y_2 z_j \in \mathcal{L}(2lP)$ for $i = 1, ..., r$ and $j = 1, ..., r'$. Now, since $\mathcal{L}(lQ) =$ Span$\{z_1, ..., z_r\}$ we have $\nu_Q(z_i) \neq \nu_Q(z_j)$ for $i \neq j$. This, together with the assumption that m is odd, implies that the functions $z_1, ..., z_r, y_2 z_1, ..., y_2 z_{r'}$ have different pole orders at P, and therefore they are linearly independent.

It remains to prove that these functions span all of $\mathcal{L}(2lP)$. By Weierstrass's Gap Theorem (see [11], p.32) it is sufficient to consider the case when $l = g_2$. Here we have to show that

$$\dim(2lP) = \dim(lQ) + \dim[(l - \frac{m+1}{2})Q].$$

We notice that $2l = 2g_2 > 2g_2 - 2$, and Riemann-Roch theorem implies that $\dim(2lP) = g_2 + 1$. Furthermore,

$$l = g_2 > g_2 - \frac{m+1}{2} = 2g_1 + \frac{m-1}{2} - \frac{m+1}{2} > 2g_1 - 2,$$

which implies that

$$\dim(lQ) + \dim[(l - \frac{m+1}{2})Q] = (g_2 + 1 - g_1) + (g_2 - \frac{m+1}{2} + 1 - g_1) = g_2 + 1.$$

$\square$

Remark.
As a consequence of Lemma 3.1 we have that if $z \in \mathcal{L}(2lP)$ then z can be written uniquely as $z = v_1 + y_2 v_2$, where $v_1 \in \mathcal{L}(lQ)$ and $v_2 \in \mathcal{L}((l - \frac{m+1}{2})Q)$. Furthermore, if $(c_1, c_2, ..., c_n) \in C_1'$ then

$$(c_1, c_1, c_2, c_2, ..., c_n, c_n) \in C_2.$$

In this way C_1 and C_1' can be seen as subcodes of C_2.

We will now compute the parameters $[n_2, k_2, d_2]$ of the new code C_2. We will also prove that if the code C_1 (res. C_1') have minimum distance d_1 (res. d_1') better than the designed minimum distance d_1^* (res. $d_1'^* = 1$), then the minimum distance d_2 of C_2 is at least c better then its designed minimum distance d_2^*. The improvement c will be proven to depend only on the two subcodes C_1 and C_1', and the genera of F_1 and F_2. These results will be stated in our next main theorem.

Theorem 4. *Let a and b be nonnegative integers such that*

$$d_1 \geq d_1^* + a,$$
$$d_1' \geq {d_1'}^* + b.$$

Then C_2 is a $[2n, k_2, d_2]$ code over $\mathbf{F}_q$ with

$$k_2 = k + k',$$
$$d_2 \geq d_2^* := 2l - 2g_2 + 2.$$

Furthermore, if $2(g_2 - g_1) + a - l \geq 0$ then

$$d_2 \geq d_2^* + c,$$

where $c = \min\{2b, \; 2(g_2 - g_1) + a - l\}$.

Proof. In the first statement of the theorem only the expression for k_2 is non trivial. First we want to prove $k_2 = k + k'$. From [11] (pp. 43,47) we have that

$$n - k = \dim(lQ) - \dim(lQ - D_1)$$
$$n - k' = \dim[(l - \frac{m+1}{2})Q] - \dim[(l - \frac{m+1}{2})Q - D_1].$$

If we add the above equations, we obtain

$$2n - k_2 = \dim(2lP) - \dim(2lP - D_2).$$

Lemma 3.1 implies that $\dim(lQ) + \dim[(l - \frac{m+1}{2})Q] = \dim(2lP)$. Therefore the last equation becomes:

$$k_2 = k + k' + \dim(2lP - D_2) - \dim(G_1 - D_1) - \dim((l - \frac{m+1}{2})Q - D_1).$$

Let $z = v_1 + y_2 v_2 \in \mathcal{L}(2lP)$, where $v_1 \in \mathcal{L}(lQ)$ and $v_2 \in \mathcal{L}((l - \frac{m+1}{2})Q)$. We have $v_1(P_{i,1}) = v_1(P_{i,2}) = v_1(Q_i) := \alpha_i$ and $v_2(P_{i,1}) = v_2(P_{i,2}) = v_2(Q_i) := \beta_i$, and since F_2 is an elementary abelian extension of F_1 we get $y_2(P_{i,1}) := \gamma_i$ which implies that $y_2(P_{i,2}) = \gamma_i + 1$ (see [11], p.115). Therefore, $z \in \mathcal{L}(2lP - D_2)$ implies that $0 = z(P_{i,1})$ and $0 = z(P_{i,2})$ for $i = 1, ..., n$. These yield that $0 = \alpha_i + \gamma_i \beta_i$ and $0 = \alpha_i + (\gamma_i + 1)\beta_i$ for $i = 1, ..., n$. Thus $0 = \alpha_i = \beta_i$ which yield that $v_1(P_{i1}) = v_2(P_{i2}) = 0$ for $i = 1, ..., n$. Therefore $v_1 \in \mathcal{L}(lQ - D_1)$ and $v_2 \in \mathcal{L}((l - \frac{m+1}{2})Q - D_1)$ implying that $\dim(2lP - D_2) = \dim(lQ - D_1) + \dim((l - \frac{m+1}{2})Q - D_1)$, which enables us to conclude that $k_2 = k + k'$.

From now on we assume that $2(g_2 - g_1) + a - l \geq 0$. If we put $\alpha_{i,j} := z_i(P_{j,1}) = z_i(P_{j,2})$ we get that the parity check matrix of C_2 is:

$$\mathbf{H}_2 = \begin{pmatrix} \alpha_{1,1} & \alpha_{1,1} & \cdots & \alpha_{1,n} & \alpha_{1,n} \\ \vdots & \vdots & & \vdots & \vdots \\ \alpha_{r,1} & \alpha_{r,1} & \cdots & \alpha_{r,n} & \alpha_{r,n} \\ \gamma_1\alpha_{1,1} & (\gamma_1+1)\alpha_{1,1} & \cdots & \gamma_n\alpha_{1,n} & (\gamma_n+1)\alpha_{1,n} \\ \vdots & \vdots & & \vdots & \vdots \\ \gamma_1\alpha_{r',1} & (\gamma_1+1)\alpha_{r',1} & \cdots & \gamma_n\alpha_{r',n} & (\gamma_n+1)\alpha_{r',n} \end{pmatrix}.$$

In order to prove $d_2 \geq d_2^* + c$ we have to show that any $w := d_2^* + c - 1$ columns in $\mathbf{H}_2$ are linearly independent. We choose the columns corresponding to the places $P_{1,1}, P_{1,2}, ..., P_{s,1}, P_{s,2}, P_{s+1,1}, ..., P_{s+t,1}$, where $w = 2s + t$, and arrange them in a matrix $\mathbf{H}$ of the form:

$$\mathbf{H} = \begin{pmatrix} \alpha_{1,1} & \alpha_{1,1} & \cdots & \alpha_{1,s} & \alpha_{1,s} & \alpha_{1,s+1} & \cdots & \alpha_{1,s+t} \\ \vdots & \vdots & & \vdots & \vdots & \vdots & & \vdots \\ \alpha_{r,1} & \alpha_{r,1} & \cdots & \alpha_{r,s} & \alpha_{r,s} & \alpha_{r,s+1} & \cdots & \alpha_{r,s+t} \\ \gamma_1\alpha_{1,1} & (\gamma_1+1)\alpha_{1,1} & \cdots & \gamma_s\alpha_{1,s} & (\gamma_s+1)\alpha_{1,s} & \gamma_{s+1}\alpha_{1,s+1} & \cdots & \gamma_{s+t}\alpha_{1,s+t} \\ \vdots & \vdots & & \vdots & \vdots & \vdots & & \vdots \\ \gamma_1\alpha_{r',1} & (\gamma_1+1)\alpha_{r',1} & \cdots & \gamma_s\alpha_{r',s} & (\gamma_s+1)\alpha_{r',s} & \gamma_{s+1}\alpha_{r',s+1} & \cdots & \gamma_{s+t}\alpha_{r',s+t} \end{pmatrix}.$$

If we add the column pairs corresponding to $P_{i,1}$ and $P_{i,2}$ for $i = 1, ..., s$ and rearrange the columns we get that $\mathbf{H}$ is equivalent to the matrix

$$\begin{pmatrix} \alpha_{1,1} & \cdots & \alpha_{1,s+t} & 0 & \cdots & 0 \\ \vdots & & \vdots & \vdots & & \vdots \\ \alpha_{r,1} & \cdots & \alpha_{r,s+t} & 0 & \cdots & 0 \\ \gamma_1\alpha_{1,1} & \cdots & \gamma_{s+t}\alpha_{1,s+t} & \alpha_{1,1} & \cdots & \alpha_{1,s} \\ \vdots & & \vdots & \vdots & & \vdots \\ \gamma_1\alpha_{r',1} & \cdots & \gamma_{s+t}\alpha_{r',s+t} & \alpha_{r',1} & \cdots & \alpha_{r',s} \end{pmatrix}.$$

Hence, we have

$$\text{rank}\mathbf{H} \geq \text{rank}\mathbf{M} + \text{rank}\mathbf{M}',$$

where

$$\mathbf{M} = \begin{pmatrix} \alpha_{1,1} & \cdots & \alpha_{1,s+t} \\ \vdots & & \vdots \\ \alpha_{r,1} & \cdots & \alpha_{r,s+t} \end{pmatrix}$$

and

$$\mathbf{M}' = \begin{pmatrix} \alpha_{1,1} & \cdots & \alpha_{1,s} \\ \vdots & & \vdots \\ \alpha_{r',1} & \cdots & \alpha_{r',s} \end{pmatrix}.$$

M is a collection of $s + t$ columns from the parity check matrix of C_1. We note that

$$s + t \leq w = d_2^* + c - 1 \leq d_2^* + 2(g_2 - g_1) + a - l - 1,$$

and since $d_2^* = 2l - 2g_2 + 2$, the last inequality implies that $s + t \leq l - 2g_1 + 2 + a - 1$. But $l - 2g_1 + 2 = d_1^*$; therefore

$$s + t \leq d_1^* + a - 1 < d_1^* + a \leq d_1.$$

We conclude that $s + t < d_1$. Since the minimum distance of C_1 is d_1 then any $d_1 - 1$ columns are linearly independent, thus the strict inequality $s + t < d_1$ implies that $s + t$ columns of M are linearly independent, i.e. rank$\mathbf{M} = s + t$.

Similarly, $\mathbf{M}'$ is a collection of s columns from the parity check matrix of the code C_1'. We note that

$$2s \leq w = d_2^* + c - 1 \leq d_2^* + 2b - 1.$$

But since $d_2^* = 2l - 2g_2 + 2$ and $g_2 = 2g_1 + \frac{m-1}{2}$, the last inequality implies that $2s \leq 2l - (m + 1) - 4g_1 + 4 + 2b - 1$. Since $2l - (m + 1) - 4g_1 + 4 = 2d_1'^*$, the last inequality becomes $2s \leq 2d_1'^* + 2b - 1 < 2d_1'^* + 2b \leq 2d_1'$. Thus $s < d_1'$. Since the minimum distance of C_1' is d_1' then any $d_1' - 1$ columns are linearly independent, since $s < d_1'$ we conclude that the s columns are linearly independent, i.e. rank$\mathbf{M}' = s$. $\qquad\square$

Example 1B.

In Example 1A we constructed the codes C_1 with parameters $[4, 1, 4]$ and the code C_1' with parameters $[4, 3, 2]$. The designed minimum distance of C_1 (res. C_1') is $d_1^* = \deg G_1 - 2g_1 + 2 = 3$ (res. $d_1'^* = 1$). Since $4 = d_1 = d_1^* + a$, we conclude that $a = 1$, and since $2 = d_1' = d_1'^* + b$, we have that $b = 1$. The codes C_1 and C_1' yield the code C_2. Let us estimate the parameters of the new code using Theorem 3.1, i.e. the length of the code $n_2 = 8$, the designed minimum distance $d_2^* = 2$, the dimension $k_2 = 4$. The integer $c = \min\{2b, 2(g_2 - g_1) + a - l\} = 2$. Thus $d_2 \geq d_2^* + c = 4$. This is an MDS code and since no MDS code exist with minimum distance ≥ 5, we conclude that $C_2 = C_\Omega(D_2, 6P_\infty)$ is a $[8, 4, 4]$ code over $\mathbf{F}_4$.

References

1. G.L. Feng and R.R.N. Rao. A novel approach for contruction of algebraic geometric codes from affine plane curves. *University of Southwestern Louisiana*, preprint, 1992.
2. G.L. Feng and R.R.N. Rao. Decoding of algebraic geometric codes up to designed minimum distance. *IEEE Trans. Inf. Theory*, 39:37–46, 1993.
3. A. Garcia, S.J. Kim, and R.F. Lax. Goppa codes and Weistrass gaps. *Journal of Pure and Applied Algebra*, 84:199–207, 1992.
4. A. Garcia and R.F. Lax. Goppa codes and Weistrass gaps. *Lecture Notes in Mathematics, Springer Verlag*, 1518:33–42, 1992.

5. V.D. Goppa. Codes on algebraic curves. *Soviet Math. Doklady*, 24:170–172, 1981.
6. V.D. Goppa. Algebraic geometric codes. *Mathematics of the USSR Izvestiya*, 21, no.1:75–91, 1983.
7. V.D. Goppa. Codes and information. *Russian Mathematical Surveys*, 39, no.1:87–141, 1984.
8. V.D. Goppa. *Geometry and Codes*. Kluwer Academic Publishers, The Netherlands, 1988.
9. C. Kirfel and R. Pellikaan. The minimum distance of codes in an array comming from telescopic semigroups. *Eindhoven University of Technology*, 1993. preprint.
10. F.J. MacWilliams and N.J.A Sloane. *The Theory of Error Correcting Codes*. North Holland, New York, 1977.
11. H Stichtenoth. *Algebraic Function Fields and Codes*. Springer-Verlag, New York, 1993.

Some Best Rate 1/p Quasi-Cyclic Codes over GF(5)[1]

T. Aaron Gulliver[1] and Vijay K. Bhargava[2]

[1] Department of Systems & Computer Engineering, Carleton University
1125 Colonel By Dr., Ottawa, ON, Canada K1S 5B6, gulliver@sce.carleton.ca
[2] Department of Electrical & Computer Engineering, University of Victoria
P.O. Box 3055, MS 8610, Victoria, BC, Canada V8W 3P6, bhargava@sirius.uvic.ca

Abstract. The class of quasi-cyclic (QC) codes has been proven to contain many good codes. In this paper, new rate $1/p$ QC codes over GF(5) are constructed using integer linear programming and heuristic combinatorial optimization. Many of these attain the maximum possible minimum distance for a linear code, and so are optimal. The others provide a lower bound on the maximum minimum distance. Power residue and self-dual QC codes are also presented.

1 Introduction

A fundamental and challenging problem in coding theory is to find a linear (n, k) code over GF(q) achieving the maximum possible minimum Hamming distance, d. This value is denoted as $d_q(n, k)$, and linear codes which have a minimum distance equal to $d_q(n, k)$ are called *optimal*. A related problem is to find the smallest value of n for which there exists an (n, k) code with minimum distance d. This is denoted as $n_q(d, k)$. For a given value of q, solving one of these problems is equivalent to solving the other.

For $q = 2$, $n_q(d, k)$ has been determined for $k \leq 7$. For $q = 3$, $n_q(d, k)$ has been determined for $k \leq 5$, and for $q = 4$, $n_q(d, k)$ has been determined for $k \leq 4$. Many other values of $n_q(d, k)$ are also known, and Brouwer [1] maintains an up to date table of upper and lower bounds on d for $q = 2$, $k \leq n \leq 264$, and $q = 3$ and 4, $k \leq n \leq 132$. Conversely, little is known about the bounds for $q = 5$. This problem is addressed in this paper.

The class of quasi-cyclic (QC) codes has been shown to contain many optimal codes over GF(2), GF(3) and GF(4) [2, 3]. In fact, QC codes meet a modified version of the Gilbert-Varshamov bound [4], unlike many other classes of codes. In this paper, new rate $1/p$ QC codes over GF(5) are constructed using integer

---

[1] This research was supported in part by the Natural Sciences and Engineering Research Council of Canada.

linear programming, as in [5] and heuristic combinatorial optimization [6]. Many of these establish the maximum possible minimum distance for a linear code with the same parameters, and so are optimal. The others provide a lower bound on the maximum minimum distance. Since bounds on linear codes over GF(5) have not been tabulated, these results provide a benchmark and starting point for future research.

A *best* code is defined as one which achieves the maximum possible minimum distance for a given class of linear codes. A *good* code is defined as one which has the maximum known minimum distance for a given n and k, i.e., it attains (or improves) the known lower bound on the minimum distance. In general, to find a best (n, k) linear code requires an almost exhaustive search, which is intractable for all but the smallest code dimensions. In fact, this problem falls into the class of NP-hard combinatorial optimization problems [7]. While the restriction to QC codes considerably reduces the search, an exhaustive search for the best QC codes also becomes intractable. To further restrict the search, only the subclass of QC codes formed from circulant matrices is considered in this paper. These are called multiple circulant codes in [8].

2 Rate 1/p Quasi-Cyclic Codes

QC codes are a generalization of cyclic codes whereby a cyclic shift of a codeword by p positions results in another codeword. Thus if $p = 1$ we have the class of cyclic codes. The blocklength, n, of a QC code, is a multiple of p, $n = mp$. Many of the results on QC codes presented in the literature concern those for which a generator matrix can be constructed from $m \times m$ circulant matrices (with a suitable permutation of coordinates). In this case, the generator matrix of a rate $1/p$ QC code is given by

$$G = [C_0, C_1, C_2, \ldots, C_{p-1}]. \tag{1}$$

The C_i are $m \times m$ circulant matrices of the form

$$C = \begin{bmatrix} c_0 & c_1 & c_2 & \cdots & c_{m-1} \\ c_{m-1} & c_0 & c_1 & \cdots & c_{m-2} \\ c_{m-2} & c_{m-1} & c_0 & \cdots & c_{m-3} \\ \vdots & \vdots & \vdots & & \vdots \\ c_1 & c_2 & c_3 & \cdots & c_0 \end{bmatrix}, \tag{2}$$

where each successive row is a right cyclic shift of the previous one. These codes are a subclass of the more general 1-generator QC codes [9], which is in turn a subclass of all QC codes. The algebra of $m \times m$ circulant matrices over GF(q) is isomorphic to the algebra of polynomials in the ring $f[x]/x^m - 1$ if C_i is mapped onto the polynomial, $c_i(x) = c_{0,i} + c_{1,i}x + c_{2,i}x^2 + \cdots + c_{m-1,i}x^{m-1}$, formed from the entries in the first row of C_i [8]. Thus G can also be represented by

$$G = [c_0(x),\ c_1(x),\ c_2(x),\ \ldots,\ c_{p-1}(x)]. \tag{3}$$

To conduct a search for good QC codes, a representative set of generator polynomials is required. Consider the set, A, of polynomials of degree $m - 1$ or less, with $|A| = q^m$ elements. Two polynomials, $c_j(x)$ and $c_i(x)$ can be said to belong to the same equivalence class if

$$c_j(x) = ax^l c_i(x) \bmod (x^m - 1),$$

for some integer, $l > 0$ and scalar $a \in GF(q)\backslash\{0\}$. This means that two polynomials are in the same class if one can be obtained from the other by a cyclic shift, multiplying by a nonzero scalar, or both. Only one polynomial from each class need be considered when constructing QC codes since polynomials from the same class produce equivalent codes [10]. This equivalence relation is induced by the action of a finite group on the set of m-tuples. Distinct equivalence classes correspond to distinct orbits under the action of this group and so can be enumerated using Burnside's Lemma [10]. The number of classes, $N_1(m)$, for $q = 5$ are given in Table 1.

Table 1. The Number of Defining Polynomials for Rate $1/p$ QC Codes over GF(5)

m	$N_1(m)$
1	2
2	5
3	12
4	45
5	158
6	665
7	2792

Once a set of defining polynomials has been constructed, the weight distributions of the codes generated by the corresponding circulant matrices, C_i, must be computed. This task can be simplified since the Hamming weight of $i_j(x)c_i(x) \bmod (x^m - 1)$ is equal to the weight of $ai_j(x)x^l c_i(x) \bmod (x^m - 1)$ for all $a \in GF(q)/\{0\}$ and $0 \le l < m$, so these redundant weights can be eliminated.

Arranging the remaining weights in a matrix [2] gives

$$
D = \begin{array}{c|ccccccc}
 & i_1(x) & i_2(x) & \cdots & i_j(x) & \cdots & i_y(x) \\
\hline
c_1(x) & w_{11} & w_{12} & \cdots & w_{1j} & \cdots & w_{1y} \\
c_2(x) & w_{21} & w_{22} & \cdots & w_{2j} & \cdots & w_{2y} \\
\vdots & \vdots & \vdots & & \vdots & & \vdots \\
c_k(x) & w_{k1} & w_{k2} & \cdots & w_{kj} & \cdots & w_{ky} \\
\vdots & \vdots & \vdots & & \vdots & & \vdots \\
c_z(x) & w_{z1} & w_{z2} & \cdots & w_{zj} & \cdots & w_{zy}
\end{array} \tag{4}
$$

where $i_j(x)$ is the jth information polynomial, $c_k(x)$ is the kth generator polynomial, and w_{kj} is the Hamming weight of $i_j(x)c_k(x) \bmod (x^m - 1)$. Since the $i_j(x)$ and $c_k(x)$ correspond to the $N_1(m) - 1$ non-zero defining polynomials, D is a square matrix with dimension $y = z = N_1(m) - 1$. It should be noted that D is also a symmetric matrix.

The complete weight distribution for any QC code can be constructed from D. For example, with $q = 5$ and $m = 3$, $N_1(m) - 1$ is 11 and D is given in Table 2. The coefficients of the eleven defining polynomials are given on the left, with

Table 2. D Matrix of Weights for $(3p, 3)$ QC Codes

	$i_j(x)$										
$c_k(x)$	001	011	012	013	014	111	112	113	114	123	132
001	1	2	2	2	2	3	3	3	3	3	3
011	2	3	3	3	2	3	3	3	1	2	2
012	2	3	3	2	3	3	2	2	3	3	1
013	2	3	2	3	3	3	2	2	3	1	3
014	2	2	3	3	3	0	2	2	2	3	3
111	3	3	3	3	0	3	3	0	3	3	3
112	3	3	2	2	2	3	1	3	2	3	3
113	3	3	2	2	2	0	3	3	3	2	2
114	3	1	3	3	2	3	2	3	3	2	2
123	3	2	3	1	3	3	3	2	2	2	3
132	3	2	1	3	3	3	3	2	2	3	2

the least significant coefficient on the right. With linear programming, the best (3p,3) QC codes can easily be found. For instance, with $p = 8$, a best (24,3) QC code was constructed using the following defining polynomials: 1, 11, 12, 13, 112, 114, 123 and 132. It is easily determined from Table 2 that $d = 19$ for this code, since the codeword weights are given by the column sums of the 8 rows corresponding to the 8 defining polynomials. The weight distribution of this code is

Weight	Count
0	1
19	96
20	24
24	4

Using linear programming to find the maximum possible minimum distance for a rate $1/p$ QC code becomes infeasible for $m > 3$ because of the rapid growth of $N_1(m)$. In this case, heuristic combinatorial optimization techniques [6] can be used. Although the resulting code is not guaranteed to be the best possible, comparison with a bound can provide an indication of the quality of the code. The Griesmer bound [11] provides a lower bound on the length, n, of a linear code for a given k, d and q,

$$n(k, d) \geq G(k, d) = \sum_{i=0}^{k-1} \lceil \frac{d}{q^i} \rceil, \tag{5}$$

where $\lceil x \rceil$ denotes the smallest integer greater than or equal to x. For the problem considered in this paper, if d is the maximum minimum distance found for an (n, k) QC code, it is optimal if $G(k, d+1) > n$. For example, the (24,3) code given above is an optimal linear code based on the fact that $G(3, 20) = 25 > 24$. Many of the QC codes presented in this paper achieve this bound, and are denoted by ^{o} in the tables.

The search for a rate $1/p$ QC code is initialized with an arbitrary code by choosing p generator polynomials. Subsets of the polynomials from the power residue (PR) codes found in the next section were used in some cases to initialize the search. Clearly the minimum distance of this code is the minimum of the column sums of the corresponding p rows of D, since the weight of a minimum distance codeword must be contained in these sums. To improve the code, a new $c_k(x)$ is found to replace one presently in the code so that the minimum distance, or the column sum of the p rows, is increased. If one is not found, a selection algorithm is used which provides some degree of randomness, as in [2].

This process is repeated until the required minimum distance is achieved, or a limit on the number of iterations is reached. With this algorithm, many rate $1/p$ QC codes with large minimum distances up to $m = 7$ have been constructed. The maximum known minimum distances for these codes are compiled in Table 3. Due to space limitations, the generator polynomials for these codes are not given, but can be obtained from the authors.

3 Power Residue Codes

Power Residue (PR) codes are cyclic codes which can be transformed into QC codes using the Normal Basis Theorem [12, 13]. Let m be the order of $q \bmod n$, $(q^m \equiv 1 \bmod n)$, n a prime. Then if m divides $(n-1)$, i.e., $n = ems + 1$, a cyclic (n, em), es-th PR code exists, as does a rate $1/s$, $(n-1, em)$ QC code formed of $m \times m$ circulant matrices. A normal basis can be formed from the roots of a primitive polynomial of degree m with linearly independent roots, as found in [13].

To illustrate the construction of PR codes, consider the following example. Let $n = 13$ and $q = 5$, then $4^5 = 1024 \equiv 1 \bmod 13$. Thus we have an $(13,4)$ PR code over GF(5) composed of three 4×4 circulant matrices and an all 1's column, (in this case $e = 1, s = 3$ and $m = 4$). By definition [12], this is a cyclic code over GF(5) with generator matrix

$$G = \left[1, \alpha, \alpha^1, \alpha^2, \ldots, \alpha^{12} \right], \tag{6}$$

where α is a primitive 13-th root of unity over GF(5). To form the circulants, the columns of G must be rearranged according to the cyclic classes mod 13 over GF(5), i.e.,

$$\{x_1, qx_1, q^2x_1, q^3x_1\}; \{x_2, qx_2, q^2x_2, q^3x_2\}; \{x_3, qx_3, q^2x_3, q^3x_3\}.$$

Substituting $x_1 = 1, x_2 = 2, x_3 = 4$ and $q = 5$, results in

$$\{1, 5, 12, 8\}; \{2, 10, 11, 3\}; \{4, 7, 9, 6\}.$$

Thus G becomes

$$G = \left[1, \alpha, \alpha^5, \alpha^{12}, \alpha^8, \alpha^2, \alpha^{10}, \alpha^{11}, \alpha^3, \alpha^4, \alpha^7, \alpha^9, \alpha^6 \right]. \tag{7}$$

Now, if these columns are represented in terms of a normal basis, α^5 becomes a cyclic shift of α, α^{12} becomes a cyclic shift of α^5, etc. Then G has the form

$$G = [1, C_1, C_2, C_3],$$

$$= \begin{bmatrix} 1 & 3040 & 2100 & 1032 \\ 1 & 0304 & 0210 & 2103 \\ 1 & 4030 & 0021 & 3210 \\ 1 & 0403 & 1002 & 0321 \end{bmatrix}. \tag{8}$$

The weight distribution of G is

Weight	Count
0	1
8	52
9	104
10	208
11	104
12	104
13	52

The related (12,4) QC code, found by deleting the all 1's column from (8), has weight distribution

Weight	Count
0	1
7	32
8	92
9	192
10	136
11	112
12	60

Note that the QC code has minimum distance one less than the corresponding PR code.

Table 4 gives the parameters of the PR codes and related QC codes over $GF(5)$. From an (n, k) PR code with minimum distance d, an $(n - 1, k)$ QC code can be constructed which appears to always have minimum distance $d - 1$, as was shown in the example above. The duals of the QC codes have the same minimum distances as the duals of the PR codes. The circulant matrices from these PR codes can be used to initialize the search for good QC codes. This use of PR codes is justified by the fact that the class of PR codes has been shown to contain many excellent (often optimal) codes [14].

4   Self-Dual Codes

Two codes C and C' over $GF(p)$ are *equivalent* if there exists an n by n monomial matrix P over $GF(p)$ with $C' = C \cdot P = \{xP \mid x \in C\}$. The dual code $C^\perp$ of C is defined as $C^\perp = \{x \in GF(p)^n \mid x \cdot y = 0$ for all $y \in C\}$. C is *self-dual* if $C = C^\perp$. There has been much in the literature on the subject of self-dual codes [15]. Much less work has been done considering self-dual codes over $GF(5)$. In this section, self-dual QC codes are presented for lengths $n \leq 16$. Because the dimensions of C and $C^\perp$ must be equal, these codes are double circulant codes of the form

$$[\,I\,,\,C\,].$$

By exhaustive search, all distinct double circulant self-dual codes over $GF(5)$ have been found with a minimum distance which is highest among all double circulant self-dual codes for each length. These codes are given below.

The [6,3,4] self-dual QC code has the following weight distribution

Weight	Count
0	1
4	60
5	24
6	40

The coefficients of the polynomial $c(x)$ is 331, with the highest degree coefficient on the right. The [8,4,4] self-dual QC code has the following weight distribution

Weight	Count
0	1
4	48
5	32
6	288
7	128
8	128

The polynomial is 4111. The [10,5,4] self-dual QC code has the following weight distribution

Weight	Count
0	1
4	20
5	40
6	440
7	320
8	1280
9	640
10	384

The polynomials are 24321 and 42231. The following four weight distributions exist for [12,6,4] self-dual QC codes

Weight	W_1 Count	W_2 Count	W_3 Count	W_4 Count
0	1	1	1	1
4	120	24	12	60
5	48	48	0	0
6	80	272	416	320
7	0	384	480	288
8	3600	3120	2700	2940
9	2880	2496	2688	2880
10	5376	5376	5544	5544
11	1920	2688	2544	2160
12	1600	1216	1240	1432

For distribution 1, the polynomial is 303010 For distribution 2, the polynomials are 410110 and 243120. For distribution 3, the polynomials are 212210 and 242310. For distribution 4, the polynomials are 311111 and 414121. The [14,7,6] self-dual QC code has the following weight distribution

Weight	Count
0	1
6	252
7	392
8	3472
9	4872
10	16324
11	15848
12	22708
13	10528
14	3728

The polynomials are 1424110 and 4344110. 3414410 and 3221211. The [16,8,6] self-dual QC code has the following weight distribution

Weight	Count
0	1
6	160
7	192
8	2880
9	5568
10	26848
11	37824
12	89568
13	84480
14	91392
15	39936
16	11776

The polynomials are 1424110 and 4344110.

All but the [12,6,4] codes are optimal, since a [12,6] self-dual code with minimum distance 6 is known [15]. Note that these weight distributions are approximately binomial. This phenomena will be the subject of a future paper.

5 Summary

The construction of multiple circulant quasi-cyclic codes over GF(5) has been presented. Many of these codes attain the Griesmer bound, and so are optimal. The other codes provide a benchmark for the construction of linear codes over GF(5). It is expected that many of these will also be shown to be optimal, once the upper bounds on minimum distance are improved. To illustrate how good the minimum distances of the best QC codes are, a comparison was made with the best cyclic codes over GF(5) of length less than 39. Of the 31 cyclic codes with parameters matching those in Table 3, 10 have minimum distances which equal the best QC codes, and none exceed the best QC codes. In addition, 2 best cyclic codes have a minimum distance which is 12 less than the corresponding best QC codes. Thus it is clear that the class of QC provides a good starting point for determining optimal linear code parameters.

Power residue codes over GF(5) were also presented which include several optimal QC codes. All double circulant self-dual QC codes were given up to $n = 16$. All except the length 12 codes are optimal self-dual codes.

The techniques used in this paper to find good codes can be used for other classes of codes for which a group of code components (in this case circulant matrices) can be defined.

References

1. A.E. Brouwer, Tables of minimum-distance bounds for linear codes over GF(2), GF(3) and GF(4), lincodbd server, aeb@cwi.nl, Eindhoven University of Technology, Eindhoven, the Netherlands

2. T.A. Gulliver and V.K. Bhargava, Some best rate $1/p$ and rate $(p-1)/p$ systematic quasi-cyclic codes, *IEEE Trans. Inf. Theory*, **37** (1991) 552-555.

3. T.A. Gulliver and V.K. Bhargava, Some best rate $1/p$ and $(p-1)/p$ quasi-cyclic codes over GF(3) and GF(4), *IEEE Trans. Inform. Theory*, **38** (1992) 1369-1374.

4. T. Kasami, A Gilbert-Varshamov bound for quasi-cyclic codes of rate $1/2$, *IEEE Trans. Inf. Theory* **20** (1974) 679.

5. H.C.A. van Tilborg, On quasi-cyclic codes with rate $1/m$, *IEEE Trans. Inf. Theory*, **24** (1978) 628-629.

6. G.L. Nemhauser and L.A. Wolsey, *Integer and Combinatorial Optimization*, New York: Wiley, 1988.

7. E.H.L. Aarts and P.J.M. van Laarhoven, Local search in coding theory, *Discrete Math.* **106/107** (1992) 11-18.

8. F.J. MacWilliams and N.J.A. Sloane, *The Theory of Error-Correcting Codes*, New York: North-Holland Publishing Co., 1977.

9. G.E. Séguin and G. Drolet, The theory of 1-generator quasi-cyclic codes, preprint, Royal Military College of Canada, Kingston, ON, 1991.

10. T.A. Gulliver, New optimal ternary linear codes, *IEEE Trans. Inf. Theory*, **41** (1995) 1182-1185.

11. G. Solomon and J.J. Stiffler, Algebraically punctured cyclic codes, *Inf. and Control* 8 (1965) 170-179.

12. C.L. Chen, W.W. Peterson, and E.J. Weldon, Jr., Some results on quasi-cyclic codes, *Inf. and Control* 15 (1969) 407-423.

13. T.A. Gulliver, M. Serra and V.K. Bhargava, The generation of primitive polynomials in GF(q) with independent roots and their applications for power residue codes, VLSI testing and finite field multipliers using normal basis, *Int. J. Elect.* **71** (1991) 559-576.

14. E.R. Berlekamp, *Algebraic Coding Theory*, New York: McGraw Hill, 1969.

15. J.S. Leon, V. Pless and N.J.A. Sloane, Self-dual codes over GF(5), *J. Comb. Theory A* **32** (1982) 178-194.

Table 3. Maximum Minimum Distances for Rate (pm, m) Systematic QC Codes over GF(5)

m	2	3	4	5	6	7	8	9	10	11	12	13	14	15	16	17	18	19	20	21	22
2	3°	4	6°	8°	10°	11°	13°	14	16°	18°	20°	21°	23°	24	26°	28°	30°	31°	33°	34	36°
3	4°	6°	8	11°	13	16°	19°	21°	24°	25°	28°	30°	33°	35°	38°	40°	43°	45°	48°	50°	52°
4	4	7	11°	14	18°	20	23°	26	30°	33	36	40°	42	46	50°	53°	56°	59	62	65	68
5	5	9	13	17	20	24	28	32	36	40	44	47	51	56	59	64	68	71	75	79	83
6	6	10	15	19	24	28	32	37	42	46	50	55	60	65	69	74	78	83	87	92	97
7	6	11	16	21	26	31	37	42	47	52	58	63	68	74	79	84	89	95	100	106	111

m	23	24	25	26	27	28	29	30	31	32	33	34	35	36	37	38	39	40	41	42	43
2	38°	40°	41°	43°	44	46°	48°	50°	51°	53°	54	56°	58°	60°	61°	63°	64	66°	68°	70°	71°
3	55°	57°	60°	62°	65°	67°	70°	72°	75°	76°	79°	81°	84°	86°	89°	91°	94°	96°	99°	100°	103°
4	72	75°	79°	81	84	88	91	95°	98	101°	104	107	111°	115	117	120	122	125	130°	133°	136°
5	87	91	95	99	103	107	111	115	119	122	126	130	134	138	142	146	150	154	158	162	166
6	102	106	111	115	120	125	129	134	139	144	148	152	158	162	167	172	176	181	186	191	196
7	116	121	127	132	137	143															

m	44	45	46	47	48	49	50	51	52	53	54	55	56	57	58	59	60	61	62	63	64
2	73°	74	76°	78°	80°	81°	83°	84	86°	88°	90°	91°	93°	94	96°	98°	100°	101°	103°	104	106°
3	105°	108°	110°	113°	115°	118°	120°	123°	125°	127°	130°	132°	135°	137°	140°	142°	145°	147°	150°	151°	154°
4	139	142	145	148	152°	155°	158	161	164	166	170	173	178°	180	183	186	190	193	196	200°	202
5	170	173	177	182	186	190	193	198	201	205	209	213	217	221	225	229	233	237	241	245	249

The superscript $^\circ$ denotes an optimal code.

Table 4. Power Residue Codes, Duals and Related Quasi-Cyclic Codes Over GF(5)

PR code	d_{min}	dual code	d_{min}	m	rate	QC code	d_{min}	dual code	d_{min}
$(11,5)^{Qo}$	6	$(11,6)°$	5	5	1/2	$(10,5)°$	5		
$(13,4)$	8	$(13,9)$	4	4	1/3	$(12,4)$	7	$(12,8)$	4
$(19,9)^{Q}$	8	$(19,10)$	7	9	1/2	$(18,9)$	7		
$(31,3)°$	25	$(31,28)$	3	3	1/10	$(30,3)°$	24	$(30,27)$	3
$(31,6)$	19	$(31,25)$	4	3	1/5	$(30,6)$	18	$(30,24)$	4
$(71,5)$	50^{d5}	$(71,66)$	3	5	1/14	$(70,5)$	49	$(70,65)$	3
$(71,10)$	44	$(71,61)$	6	10	1/7	$(70,10)$	43	$(60,50)$	6
$(313,8)$	235	$(313,305)$	4	8	1/39	$(312,8)$	234	$(312,304)$	4
$(521,10)$	370	$(521,511)$	4	10	1/52	$(520,10)$	369	$(520,510)$	4
$(829,9)$	635^{d5}	$(829,820)$	3	9	1/92	$(828,9)$	634	$(828,819)$	3
$(19531,7)°$	15625	$(19531,19524)$	3	7	1/2790	$(19530,7)°$	15624	$(19530,7)$	3

Notes: n^Q denotes a Quadratic Residue code
n^{dz} denoted weights divisible by z
n^o denotes an optimal code

Gordon and Retkin [9] conjectured that good substitution boxes (S-boxes) may be built by choosing a random reversible mapping of sufficient size. Their argument is based on the observation that the probability of accidental linearity occurring in such S-boxes decreases dramatically as the size of the S-box increases. Here in this paper, we provide further evidence that bigger S-boxes (by bigger we mean S-boxes with a larger number of inputs) are better by showing that the expected value of information leakage of a randomly selected boolean function decreases rapidly with the number of input variables.

It is worth noting that Brynielsson[3] gives an approximate expression for the expected value of the mutual information between the output and input subvectors for multi-output boolean functions. Here we follow the definition of information leakage given in [24] and give an exact expression for the expected value of different forms of these information leakages for a randomly selected multi-output boolean function and for some other combinatorial structures of interest such as regular mappings, and injective mappings.

2 Definitions

Throughout this paper, let Y be the output of a boolean function $f : Z_2^n \rightarrow Z_2^m$, then we have:

Static Information Leakage: the static information leakage of Y, given input subvector $X_k \in Z_2^k$ (i.e., given that we know k bits of the n-bit input vector) , is defined by

$$SL(Y|X_k) = m - H(Y|X_k), \tag{1}$$

where $H(Y \mid X_k)$ is the conditional entropy of Y given X_k.

Remark: It is easy to show that

$$SL(Y|X_k) = m - H(Y) + I(Y; X_k), \tag{2}$$

where $I(Y; X_k)$ is the mutual information between Y and X_k. Note that if the mutual information $I(Y; X_k)$ is used to define the static information leakage, then the minimum of $I(Y; X_k)$ can be achieved while $H(Y) = 0$ which contradicts our objective. For a good general reference for information theory, see [5].

Dynamic Information Leakage: the dynamic information leakage of ΔY, given the input change vector ΔX is defined by:

$$DL(\Delta Y|\Delta X) = m - H(\Delta Y|\Delta X), \tag{3}$$

where $\Delta Y = Y(X) \oplus Y(X \oplus \Delta X)$.

The self static information leakage of Y is defined as:

$$SSL(Y) = m - H(Y). \tag{4}$$

It is clear that $SSL(Y) = SL(Y \mid X_0)$. We note that the static information leakage $SL(Y \mid X_k) = 0$ is achieved by k^{th} order resilient functions (see [4], [18] for the definition and properties of resilient functions), while zero dynamic information leakage

Information Leakage of a Randomly Selected Boolean Function

A. M. Youssef and S. E. Tavares

Department Of Electrical and Computer Engineering
Queen's University
Kingston, Ontario, Canada, K7L 3N6
email : tavares@ee.queensu.ca

Abstract. It is argued that a boolean function $f : Z_2^n \rightarrow Z_2^m$ is resistant to statistical analysis if there is no significant static and dynamic leakage between its inputs and outputs. In this paper, we derive expressions for the expected value of the information leakage of randomly selected boolean functions and for the interesting cases of randomly selected balanced, and randomly selected injective boolean functions. It is shown that the expected value of different forms of information leakage decreases dramatically with the number of input variables n. For example, for a single output boolean function, we show that the expected value of different forms of leakage goes down exponentially with n.

1 Introduction

Several cryptographic criteria have been previously proposed as a measure of the strength of cryptographic functions. Among these criteria are balance, correlation immunity[18], resiliency[4], nonlinearity[12], Strict Avalanche Criterion (SAC)[21], higher order SAC[7], Propagation Criterion (PC), higher order PC[14], Bit Independence Criterion [20], and Completeness[10].

The above set of cryptographic criteria are not independent of each other and a cryptographic function that satisfies all these criteria would be a golden one. Unfortunately, it can be proven that no function can satisfy all the above set of criteria simultaneously. This can be considered as the main motive for proposing a new set of criteria based on information theory.

Several design criteria, based on information theory, have been proposed in [6],[8], [19], and [24].

In [24] Information leakage was proposed as a measure of the performance of cryptographic functions. Information Leakage can be classified into two classes: Static information leakage and dynamic information leakage. It is argued in [24] that a boolean function is resistant to statistical analysis (e.g., differential cryptanalysis [2], linear cryptanalysis [11], and Siegenthaler's correlation attack [17]) if there is no significant static and dynamic information leakage between its inputs and outputs. In [22], the authors studied the relation between the spectral properties and information leakage of multi-output boolean functions.

for all values of $\Delta X \neq 0$ is achieved only by perfect nonlinear functions (see [13], [16] for the definition and properties of both bent and perfect nonlinear functions).

Let Y be the output of a boolean function $f(X)$ and define

$$
\begin{aligned}
N_y &= \#\{X \in Z_2^n | f(X) = y\}, \\
N_{\dot{x}y} &= \#\{X \in Z_2^n \mid X_k = \dot{x}, Y = y\}, \\
N_{\Delta x \Delta y} &= \#\{X \in Z_2^n \mid f(X \oplus \Delta x) \oplus f(X) = \Delta y\},
\end{aligned}
\tag{5}
$$

where $\dot{x} \in Z_2^k$, $\Delta x \in Z_2^n$, $y \in Z_2^m$, $\Delta y \in Z_2^m$.

Assuming that all input vectors are equally probable, we have:

$$
\begin{aligned}
SSL(Y) &= m - \sum_{y \in Z_2^m} \frac{N_y}{2^n} log_2 \left(\frac{2^n}{N_y} \right), \\
SL(Y \mid X_k) &= m - 2^{-k} \sum_{\substack{y \in Z_2^m \\ \dot{x} \in Z_2^k}} \frac{N_{\dot{x}y}}{2^{n-k}} log_2 \left(\frac{2^{n-k}}{N_{\dot{x}y}} \right), \\
DL(\Delta Y \mid \Delta X) &= m - 2^{-n} \sum_{\substack{\Delta x \in Z_2^n \\ \Delta y \in Z_2^m}} \left(\frac{N_{\Delta x \Delta y}}{2^n} \right) log_2 \left(\frac{2^n}{N_{\Delta x \Delta y}} \right).
\end{aligned}
\tag{6}
$$

The problem of finding the expected values of the above forms of information leakage is now reduced to finding the marginal probability distribution of the random variables $N_y, N_{\dot{x}y}, N_{\Delta x \Delta y}$.

3 Information Leakage Of a Randomly Selected Boolean Function

Lemma 3.1:

Let Y be the output of a randomly selected boolean function $f : \; Z_2^n \rightarrow Z_2^m$ then we have the following probabilities:

$$
\begin{aligned}
P(N_y = i) &= \binom{2^n}{i} \left(\frac{1}{2^m} \right)^i \left(1 - \frac{1}{2^m} \right)^{2^n - i}, \\
P(N_{\dot{x}y} = i) &= \binom{2^{n-k}}{i} \left(\frac{1}{2^m} \right)^i \left(1 - \frac{1}{2^m} \right)^{2^{n-k} - i}, \\
P(N_{\Delta x \Delta y} = 2i) &= \binom{2^{n-1}}{i} \left(\frac{1}{2^m} \right)^i \left(1 - \frac{1}{2^m} \right)^{2^{n-1} - i}, \quad \Delta x \neq \underline{0}.
\end{aligned}
\tag{7}
$$

Proof: The proof of the above Lemma follows by noting that N_y, $N_{\dot{x}y}$, and $N_{\Delta x \Delta y}/2$ follow the multi-nomial distribution. $\qquad \Box$

Theorem 3.1:

The expected values of the static and dynamic information leakage of a randomly selected boolean function $f : Z_2^n \rightarrow Z_2^m$ are given respectively by

$$\overline{SL(Y \mid X_k)} = m - 2^m \sum_{i=0}^{2^{n-k}} P(N_{\dot{x}y} = i) \left(\frac{i}{2^{n-k}}\right) log_2 \left(\frac{2^{n-k}}{i}\right), \quad 0 \le k \le n,$$

$$\overline{DL(\Delta Y \mid \Delta X)} = m - \frac{2^m (2^n - 1)}{2^n} \sum_{i=0}^{2^{n-1}} P(N_{\Delta x \Delta y} = 2i) \left(\frac{i}{2^{n-1}}\right) log_2 \left(\frac{2^{n-1}}{i}\right).$$

(8)

Proof: Theorem 3.1 follows directly from the definition of the expected value, and (for part 2) by noting that $\Delta Y = 0$ for $\Delta X = 0$, and hence $H(\Delta Y \mid 0) = 0$. $\square$

Figs. 1 and 2 show the expected value of the self static information leakage and the expected value of the static leakage given that half the input bits are known. From these graphs , it is clear that the relative dimensions of the boolean functions (i.e., the ratio between n, m) greatly affect different forms of information leakage.

Based on the results above, one can not conclude that S-boxes with $n > m$ are better than S-boxes with $n < m$ because of the method we used in the normalization step (dividing by the number of output bits to get information leakage per output bit). Moreover, S-boxes with $n < m$ provide better diffusion characteristics, and may be used in SPNs with no permutation layers [1] which leads to faster software implementation. The conclusion that we can make at this time is that all forms of information leakage seem to decrease with the number of input variables.

Using theorem 3.1, one can derive an upper bound for the information leakage of a single output boolean function. Single output functions are of practical interest especially for the combining functions in stream ciphers.

Corollary 3.1

Let Y be the output of a single output boolean function, then the expected values of both the static leakage and dynamic leakage are bounded by

$$\overline{SL(Y \mid X_k)} \quad \le \quad \frac{1}{2^{n-k}}, \qquad 0 \le k \le n,$$

(9)

$$\overline{DL(\Delta Y \mid \Delta X)} \quad \le \quad \frac{3}{2^n}.$$

Proof: The above corollary follows by direct substitution into theorem 3.1 and by noting that for the binary entropy function

$$h(t) = -t \, log_2(t) - (1 - t) log_2(1 - t),$$

(10)

we have

$$h(t) \ge 4t - 4t^2, \qquad 1 \ge t \ge 0 .$$

(11)

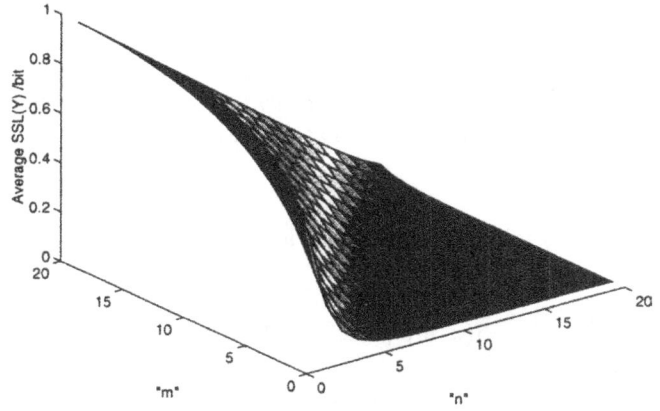

Figure 1 Expected value of $SSL(Y)$

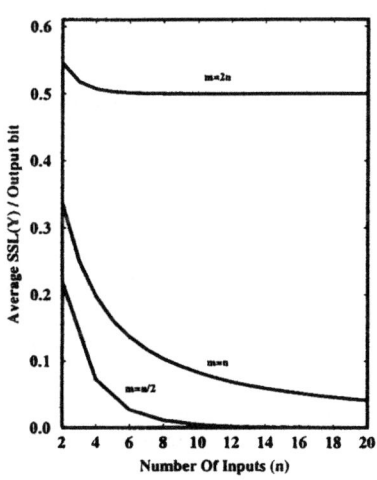

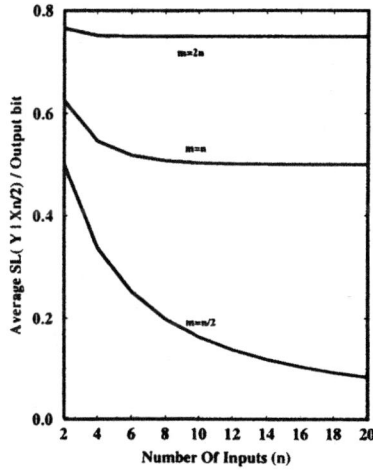

Figure 2(a) Expected Value of $SSL(Y)$ **Figure 2(b)** Expected Value of $SL\left(Y \mid X_{n/2}\right)$

4 Information Leakage Of a Randomly Selected Balanced Boolean Function

In this section, we calculate the expected values of both the dynamic leakage and the static leakage for regular (also called balanced) functions. For balanced functions, every output symbol appears an equal number of times as the input varies through all possible values.

Lemma 4.1

Let Y be a randomly selected balanced function $f : Z_2^n \rightarrow Z_2^m$, $n \geq m$, then we have

$$P(N_{\hat{x}y} = i) = \frac{1}{B(n,m)} \binom{2^{n-k}}{i} \binom{2^n - 2^{n-k}}{2^{n-m} - i} \frac{(2^n - 2^{n-m})!}{(2^{n-m}!)^{(2^m - 1)}} , \qquad (12)$$

where $B(n,m) = \frac{2^n!}{(2^{n-m}!)^{2^m}}$ is the number of $n \times m$ balanced boolean functions.

Proof: For $n \times m$ balanced boolean functions, we have $N_y = 2^{n-m}$. If we fix k input variables, then there are $\binom{2^{n-k}}{i} \binom{2^n - 2^{n-k}}{2^{n-m} - i}$ ways of arranging the output such that $Y = y$ when $X_k = \hat{x}$ for i times. The remaining $(2^n - 2^{n-m})$ outputs , of which there are only $(2^{n-m} - 1)$ distinct ones, can be permuted in $\frac{(2^n - 2^{n-m})!}{(2^{n-m}!)^{(2^m - 1)}}$ ways.   □

Corollary 4.1

Let Y be a randomly selected bijective mapping $\pi : Z_2^n \rightarrow Z_2^n$ then the expected value of the static information leakage of Y given the input subvector X_k, $0 \leq k \leq n$, is given by

$$\overline{SL(Y \mid X_k)} = k. \qquad (13)$$

Proof: The proof follows directly by substituting (12), with $n = m$, into (8). A simpler proof (independent of Lemma 4.1) follows by noting that for any arbitrary bijective function, $\pi : Z_2^n \rightarrow Z_2^n$, if we fix k input bits, we will have 2^{n-k} different output symbols with $H(Y \mid X_k) = n - k$.   □

In Lemmas 4.2, 4.3, and 4.4 we will derive an expression for the marginal probability density function of the random variable $N_{\Delta x \Delta y}$ for a randomly selected balanced boolean function.

Let

$$G(k_1, k_2, ..., k_{2^m - 1}) = C(k; k_1, k_2, ..., k_{2^m - 1}) C(2^n - 2k; l_1, l_1, l_2, l_2, ..., l_{2^m - 1}, l_{2^m - 1}) ,$$

$$(14)$$

where $C(k; k_1, k_2, ..., k_{2^m - 1}) = \frac{k!}{k_1! \, k_2! \, \cdots \, k_{2^m - 1}!}$, and $l_i = 2^{n-m} - k_i$, $l_i \geq 0$, $k_i \geq 0$, then we have:

Lemma 4.2

The number of balanced functions with $N_{\Delta x \Delta y} \geq 2k$, $\Delta x \neq 0, \Delta y \neq 0$, is upper bounded by

$$\Psi_{n,m}(k) = \binom{2^{n-1}}{k} 2^k \sum_{\sum k_i = k} G(k_1, k_2, ..., k_{2^m-1}) , \qquad (15)$$

where $G(k_1, k_2, ..., k_{2^m-1})$ is given by (14).

Proof: By noting that we have 2^m distinct output symbols, and each of them is repeated 2^{n-m} times, it is easy to see that for a given $\Delta y \neq 0, \Delta x \neq 0$ we have 2^{m-1} distinct XOR pairs, each of them is repeated 2^{n-m} times. Group each of these 2^{n-m} pairs into one set.

There is only one way to choose k_i pairs from the set $s, s = 1, 2, ..., 2^{m-1}$ (as the pairs within a given set are indistinguishable). These k pairs can be permuted in $C(k; k_1, k_2, ..., k_{2^m-1})$ ways. The remaining $2^n - 2k$ output symbols can be permuted into $C(2^n - 2k; l_1, l_1, l_2, l_2, ..., l_{2^m-1}, l_{2^m-1})$ ways, where $l_i = 2^{n-m} - k_i$. Note that there are two possible orders for each pair, giving 2^k total possible orders, and $\binom{2^{n-1}}{k}$ possible choices for the X positions of these k pairs.

The construction approach described above does not guarantee that these balanced functions are all distinct, and so $\Psi_{n,m}(k)$ is an upper bound. $\qquad\square$

Let

$$D(k_1, k_2, ..., k_{2^m-1}) = C(k; k_1, k_2, ..., k_{2^m})C(2^n - 2k; l_1, l_2, ..., l_{2^m}) , \qquad (16)$$

and $l_i = 2^{n-m} - 2k_i$, $l_i \geq 0$, $k_i \geq 0$ then we have:

Lemma 4.3

The number of balanced functions with $N_{\Delta x 0} \geq 2k$, $\Delta x \neq 0, \Delta y = 0$, is upper bounded by

$$\Phi_{n,m}(k) = \binom{2^{n-1}}{k} \sum_{\sum k_i = k} D(k_1, k_2, ..., k_{2^m}) , \qquad (17)$$

where $D(k_1, k_2, ..., k_{2^m})$ is given by (16).

Proof: Similar to the proof of Lemma 4.2. $\qquad\square$

Lemma 4.4

The exact number of balanced functions with $N_{\Delta x \Delta y} = 2k$, $\Delta x \neq 0, \Delta y \neq 0$,, and the exact number of balanced functions with $N_{\Delta x \Delta y} = 2k$, $\Delta x \neq 0, \Delta y = 0$, are given respectively by

$$\Lambda_{n,m,\Delta y}(k) = \sum_{i=k}^{2^n-1} (-1)^{i-k} \binom{i}{k} \Psi_{n,m}(i) ,$$

$$\Lambda_{n,m,0}(k) = \sum_{i=k}^{2^n-1} (-1)^{i-k} \binom{i}{k} \Phi_{n,m}(i) .$$

(18)

Proof: Follows by using the inclusion-exclusion principle [15]. □

By direct substitution, the expected value of the dynamic leakage of a randomly selected balanced function is given by

Theorem 4.1

$$\overline{DL(\Delta Y \mid \Delta X)} = m$$

$$- \frac{(2^n - 1)(2^m - 1)}{2^n} \sum_{i=0}^{2^n-1} \frac{\Lambda_{n,m,\Delta y}(i)}{B(n,m)} \left(\frac{i}{2^{n-1}}\right) log_2\left(\frac{2^{n-1}}{i}\right)$$

$$- \frac{(2^n - 1)}{2^n} \sum_{i=0}^{2^n-1} \frac{\Lambda_{n,m,0}(i)}{B(n,m)} \left(\frac{i}{2^{n-1}}\right) log_2\left(\frac{2^{n-1}}{i}\right) .$$

(19)

Corollary 4.2

Let Y be a randomly selected bijective mapping $\pi : Z_2^n \to Z_2^n$ then the expected value of the dynamic information leakage given the input change vector ΔX, is given by

$$\overline{DL(\Delta Y \mid \Delta X)} = n - \frac{(2^n - 1)^2}{2^n} \sum_{i=0}^{2^n-1} \frac{\Lambda_{n,n,\Delta y}(i)}{n!} \left(\frac{i}{2^{n-1}}\right) log_2\left(\frac{2^{n-1}}{i}\right) .$$

(20)

where $\Lambda_{n,n,\Delta y}$ is given by (18).

Proof: Corollary 4.2 is a special case (with $n = m$) of theorem 4.1. □

Remark: Note that $\Lambda_{n,n,0} = 0$ as each output symbol occurs once. Note also that, by substitution into (15) with $n = m$, $\Psi_{n,n}$ can be simplified to

$$\Psi_{n,n}(i) = \binom{2^{n-1}}{i}^2 i! \, 2^i \, (2^n - 2i)!$$

(21)

Fig.3 shows a comparison between the expected value of dynamic information leakage of a randomly chosen $n \times n$ bijective mapping and that of a randomly chosen function of the same dimensions.

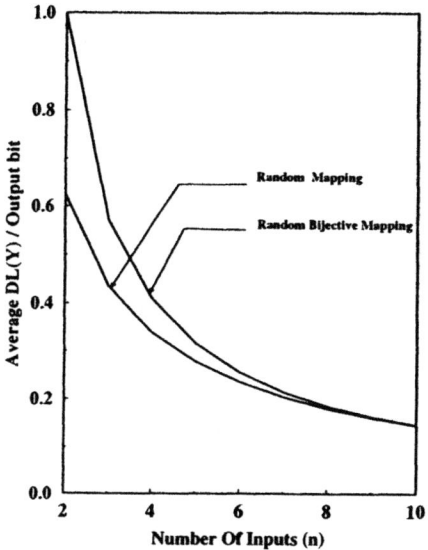

Figure 3 Expected value of $DL(Y)$ for an $n \times n$
random mapping and an $n \times n$ random bijective mapping

5 Information Leakage Of a Randomly Selected Injective Boolean Function

In this section, we calculate the expected values of both the dynamic leakage and the static leakage for injective functions.

Theorem 5.1

Let Y be the output of a randomly selected injective function $f : Z_2^n \rightarrow Z_2^m$, $n \leq m$, then

$$\overline{SL(Y \mid X_k)} = (m - n) + k. \tag{22}$$

Proof: The theorem follows by noting that for any arbitrary injective function, $f : Z_2^n \rightarrow Z_2^m$, if we fix k input bits, we will have 2^{n-k} different output symbols with $H(Y \mid X_k) = n - k$. $\qquad \Box$

Lemma 5.1

The number of injective functions with $N_{\Delta x \Delta y} \geq 2k, \Delta x \neq 0, \Delta y \neq 0$, is upper bounded by

$$\Psi_{n,m}(k) = \binom{2^{n-1}}{k}\binom{2^{m-1}}{k} 2^k \ In(2^m - 2k, 2^n - 2k) \tag{23}$$

where

$$In(u, v) = \prod_{i=0}^{(v-1)} (u - i). \tag{24}$$

Proof: Lemma 5.1 follows by using an argument similar to that used to prove Lemma 4.2. $\qquad \square$

Lemma 5.2

The exact number of injective functions with $N_{\Delta x \Delta y} = 2k, \Delta x \neq 0, \Delta y \neq 0$, is given by

$$\Lambda_{n,m,\Delta y}(k) = \sum_{i=k}^{2^{n-1}} (-1)^{i-k} \binom{i}{k} \Psi_{n,m}(i) . \tag{25}$$

Remark: Note that $N_{\Delta x\, 0} = 0$ for $\Delta x \neq 0$, and hence $\Lambda_{n,m,0} = 0$.

Theorem 5.2

$$\overline{DL(\Delta Y; \Delta X)} = m$$
$$- \frac{(2^n - 1)(2^m - 1)}{2^n} \sum_{i=0}^{2^{n-1}} \frac{\Lambda_{n,m,\Delta y}(i)}{In(2^m, 2^n)} \left(\frac{i}{2^{n-1}}\right) log_2 \left(\frac{2^{n-1}}{i}\right), \tag{26}$$

where $In(2^m, 2^n)$, the number of $n \times m$ injective boolean functions, is given by (24).

Numerical substitution into the theorem 5.2 shows that the dynamic information leakage of a randomly selected injective function decreases with the number of input variables. This rate of decrease is very similar to that of a randomly selected boolean function with the same number of inputs and outputs, especially for $n \lll m$. This can be explained by noting that for $n \lll m$, a randomly selected function is most likely to be injective.

6 Conclusion

Many of the previously known cryptographic criteria are related to information leakage. Most of these criteria require zero information leakage in some domain. However, they often constrain the function to such an extent that large information leakage of other types become likely. These leakages provide useful information for the cryptanalyst to develop attacks on the cipher. This motivates the minimization of information leakage as a general criterion for cryptographic functions.

We have derived expressions for the expected values of the static and dynamic information leakage of randomly selected boolean functions and for randomly selected balanced, and injective boolean functions. Based on this we showed that the expected values of the information leakages decrease dramatically with the number of input variables. In some cases, we showed that this decrease is exponential. With the same approach developed in this paper, one can show that the variance of different forms of information leakage also decreases dramatically with the number of input variables. Using an approach similar to the one developed in [23] one can also show that the expected maximum value of different forms of information leakage decrease with the number of input variables. This indicates that cryptographically strong boolean functions may be obtained by choosing random mappings of sufficiently large dimensions.

References

1. C.M. Adams. *A Formal and Practical Design Procedure for Substitution-Permutation Network Cryptosystems*. PhD thesis, Queen's University, Kingston, Ontario, Canada, September, 1990.

2. E. Biham and A. Shamir. Differential cryptanalysis of DES-like cryptosystems. *Journal of Cryptology, vol. 4, no. 1, pp. 3–72, 1991.*

3. L. Brynielsson. The information leakage through a randomly generated function. *Advances in Cryptology: Proc. of EUROCRYPT '91, Springer-Verlag, Berlin, pp. 552–553, 1991.*

4. B. Chor, O. Goldreich, J. Hastad, J. Friedman, S. Rudich, and R. Smolensky. The bit extraction problem or t-resilient functions. *Proc. 26th IEEE Symposium on Foundation of Computer Science, pp. 396–407, 1985.*

5. T.M. Cover and J.A. Thomas. *Elements of Information Theory.* John Wiely & Sons Inc, 1991.

6. M.H. Dawson and S.E. Tavares. An expanded set of S-box design criteria based on information theory and its relation to differential attacks. *Advances in Cryptology: Proc. of EUROCRYPT '91, Springer-Verlag, pp. 352–365, 1992.*

7. R. Forré. The strict avalanche criterion: Spectral properties of boolean functions and an extended definition. *Advances in Cryptology: Proc. of CRYPTO '88, Springer-Verlag, pp. 450–468, 1989.*

8. R. Forré. Methods and instruments for designing S-boxes. *Journal of Cryptology, Vol .2, No.3 pp. 115–130, 1990.*

9. J. Gordon and H. Retkin. Are big S-boxes best ? *Lecture Notes in Computer Science : Proc. of the Workshop on Cryptography, Springer-Verlag, Berlin, pp.257–262, 1982.*

10. J.B. Kam and G.I. Davida. Structured design of substitution-permutation encryption networks. *IEEE Trans. Comp. C-28, pp.747–753, 1979.*

11. M. Matsui. Linear cryptanalysis method for DES cipher. *Advances in Cryptology: Proc. of EUROCRYPT '93, Springer-Verlag, Berlin. pp. 386–397, 1994.*

12. W. Meier and O. Staffelbach. Nonlinearity criteria for cryptographic functions. *Advances in Cryptology: Proc. of EUROCRYPT' 89, Springer-Verlag, pp. 549–562, 1990.*

13. K. Nyberg. Perfect nonlinear S-boxes. *Advances in Cryptology: Proc. of EURO-CRYPT '91 , Springer-Verlag, pp. 378–386, 1992.*

14. B. Preneel, W.V. Leekwijk, L.V. Linden, R.Govaerts, and J. Vandewalle. Propagation charchteristic of boolean functions. *Advances in Cryptology: Proc. of EUROCRYPT '90, Springer-Verlag, pp. 161–173*, 1991.

15. F.S. Roberts. *Applied Combinatorics*. Englewood Cliffs, N.J.: Prentice-Hall, 1984.

16. O.S Rothaus. On bent functions. *Journal of Combinatorial Theory, Vol. 20(A):300–305*, 1976.

17. T. Siegenthaler. Decrypting a class of stream ciphers using ciphertext only. *IEEE Trans. Comput., Vol.C-34, No. 1, pp. 81:85*, 1985.

18. T. Siegenthaler. Correlation-immunity of nonlinear combining functions forcryptographic applications. *IEEE Trans. on Inform. Theory, Vol.IT-30, No.5, pp. 776:780*, Sept. 1984.

19. M. Sivabalan, S.E. Tavares, and L.E. Peppard. On the design of SP networks from an information theoretic point of view. *Advances in Cryptology: Proc. of CRYPTO '92, Springer-Verlag, Berlin, pp. 260–279*, 1993.

20. A.F. Webster. Plaintext / ciphertext bit dependencies in cryptographic systems. Master's thesis, Queen's University, Kingston, Ontario, Canada, December, 1985.

21. A.F. Webster and S.E. Tavares. On the design of S-boxes. *Advances in Cryptology : Proc. of CRYPTO '85 , Springer-Verlag, pp. 523–534*, 1986.

22. A.M .Youssef and S.E. Tavares. Spectral properties and information leakage of multi-output boolean functions. *In Proceedings of the IEEE International Symposium On Information Theory. Whistler, B.C., Canada, Sep. 17–22*, 1995.

23. A.M. Youssef, S.E. Tavares, S. Mister, and C.M. Adams. Linear approximation of injective s-boxes. *IEE Electronics Letters, Vol. 31, No. 25, pp.2168-2169*, 1995.

24. M. Zhang, S.E. Tavares, and L.L. Campbell. Information leakage of boolean functions and its relationship to other cryptographic criteria. *Proceedings of 2nd ACM Conference on Computer and Communications Security, Fairfax, Virgina, pp. 156-165.*, 1994.

On the Use of Periodic Timebase Companding in the Scrambling of Stationary Processes

N.D.Aakvaag, B.Lacaze and A.Duverdier

National Polytechnics Institute of Toulouse
LEN7/GAPSE, 2 rue Camichel, 31071 Toulouse, France
tel: (33) 61 58 83 67   Fax:(33) 61 58 82 37
email: aakvaag@len7.enseeiht.fr

Abstract. In a number of applications, such as commercial television or military communication, it is necessary to construct mechanisms preventing a tranmitted message from being intercepted. This article presents a novel method for the scrambling of wide sense stationary sources. The method is based on the controlled introduction of a periodic clock change, i.e. a periodic compression/expansion of the signal timebase. We derive the conditions under which perfect reconstruction may be achieved and present an example highlighting the schemes most important properties.

1 Introduction

In numerous communication applications it is desirable to scramble the contents of a transmitted message, either for commercial reasons or in the interest of privacy [1]. Various schemes have been devised for privacy, where some are applied to the digital signal and others are applied directly to the analogue signal. This paper presents a method belonging to the second category, in that its implementation may be entirely analogue. It is based on the controlled introduction of a periodic clock change (PCC) at the transmitter side [4]. At the receiving end, reconstruction is achieved by a linear filtering operation. The element of privacy in this scheme is introduced by the fact that exact knowledge of the reconstruction filter is necessary in order to reproduce the original process. In choosing a scrambling scheme two, often contradictory, parameters are of paramount importance; the degree of privacy obtained and the simplicity of implementing the scheme. These topics are briefly commented in [7].

The paper is organized as follows. In the next section the basic principles of clock changes and their use in scrambling is presented. Next, the autocorrelation function and power spectral density of the scrambled process is deduced. Section 4 provides the details of the reconstruction process, including the theoretical treatment, practical considerations and simulation examples. Finally, section 5 treats the possibility of performing reconstruction from non-baseband components.

2   Periodic clock changes and scrambling

In what follows, the original process, i.e. the process to be scrambled will be denoted $Z(t)$. Let the $Z = \{Z(t), t \in \mathbb{R}\}$ be a (in general complex) continuous wide sense stationary, stochastic process of zero mean and spectral representation $\Theta_Z(\omega)$ having a Cramér-Loève representation:

$$Z(t) = \int_{\mathbb{R}} e^{i\omega t} d\Theta_Z(\omega) \tag{1}$$

In addition, we define the spectrum $S_Z(\omega)$ and the autocorrelation function $K_Z(\tau)$ satisfying:

$$K_Z(\tau) = \int_{\mathbb{R}} e^{i\omega\tau} dS_Z(\omega) \tag{2}$$

Finally, we define two functions, $f(t)$ and $g(t)$, sufficiently regular and periodic with the same period $T = \frac{2\pi}{\omega_0}$, where $f(\cdot)$ is real and $g(\cdot)$ is (in general) complex. Armed with these definitions, we may now define the scrambled process, $U(t)$ [4]:

$$U(t) = g(t + \varphi) Z(t - f(t + \varphi)) \tag{3}$$

where the phase φ is a random variable, independent of $Z(t)$ and uniformly distributed on the interval $(0, T)$. Its inclusion in the definition of $U(t)$ renders the scrambled process stationary, a property we will be needing in subsequent sections. In the case where this phase is known, $U(t)$ becomes cyclostationary [3]. This topic has been treated by [7], where the latter applies cyclostationary techniques to spread spectrum communication.

The basic idea of the scrambler/descrambler is to subject the original process to the transformation given by equation (3). In the following sections, the reconstruction is presented and it is shown in particular that, under certain conditions, perfect reconstruction of the parent process is possible.

3   Signal spectrum and autocorrelation

This section deduces the autocorrelation function and the power spectral density of the scrambled process. From first principles, the autocorrelation function of process $U(t)$ is readily expressed as a function of $Z(t)$ and the two periodic functions, $f(\cdot)$ and $g(\cdot)$:

$$
\begin{aligned}
K_U(\tau) &= E\{U(t)U^*(t - \tau)\} \\
&= E\{g(t + \varphi) Z(t - f(t + \varphi)) g^*(t + \varphi - \tau) Z^*(t - \tau - f(t + \varphi - \tau))\}
\end{aligned}
\tag{4}
$$

Using the independence of φ and $Z(t)$ and the method of conditional mathematical expectation, $K_U(\tau)$ may be written:

$$
\begin{aligned}
K_U(\tau) &= E\{E\{g(\varphi) g^*(\varphi - \tau) Z(t) Z^*(t - \tau + f(t + \varphi) - f(t + \varphi - \tau)) | \varphi\}\} \\
&= E\left\{\int_{\mathbb{R}} e^{i\omega(\tau - f(t+\varphi) + f(t+\varphi-\tau))} g(\varphi) g^*(\varphi - \tau) dS_Z(\omega)\right\}
\end{aligned}
\tag{5}
$$

Interchanging the order of integration and expectation, we find

$$K_U(\tau) = \int_{\mathbf{R}} e^{i\omega\tau} \Phi(\tau,\omega) \, dS_Z(\omega) \tag{6}$$

where $\Phi(\tau,\omega)$ is defined as the expectation with respect to the random variable φ over the interval $(0,T)$, i.e.

$$\Phi(\tau,\omega) = E\left\{ g(\varphi) g^*(\varphi-\tau) e^{-i\omega(f(t+\varphi)-f(t+\varphi-\tau))} \right\}$$
$$= \frac{1}{T} \int_0^T g(\varphi) g^*(\varphi-\tau) e^{-i\omega(f(t+\varphi)-f(t+\varphi-\tau))} d\varphi \tag{7}$$

Because $f(\cdot)$ and $g(\cdot)$ are periodic with the same period and sufficiently regular, the expression $g(\varphi) e^{-i\omega f(t+\varphi)}$ can be written in terms of its Fourier series expansion

$$g(\varphi) e^{-i\omega f(t+\varphi)} = \sum_{k=-\infty}^{\infty} \Psi_k(\omega) e^{ik\omega_0\varphi} \tag{8}$$

where the coefficients $\Psi_k(\omega)$ are given by

$$\Psi_k(\omega) = \frac{1}{T} \int_0^T g(\varphi) e^{-i\omega f(\varphi)-i\varphi k\omega_0} d\varphi \tag{9}$$

Taking the Fourier transform of the autocorrelation function in equation (6) and using the above relations, one obtains the power spectral density of the process $U(t)$

$$dS_U(\omega) = \sum_{k=-\infty}^{\infty} |\Psi_k(\omega-k\omega_0)|^2 \, dS_Z(\omega-k\omega_0) \tag{10}$$

In other words, the effect of the periodic clock change function has been to reproduce weighted versions of the original spectrum at integer multiples of the clock change frequency ω_0. It is worth noting that the weighting functions are the modulus squared of the Fourier series of $g(\varphi) e^{-i\omega f(t+\varphi)}$. Figure 1 below shows a realization of this spectrum for an $N.R.Z$ signal with $\omega_0 = 26$ where the appearance of the new spectral contributions are clearly indicated.

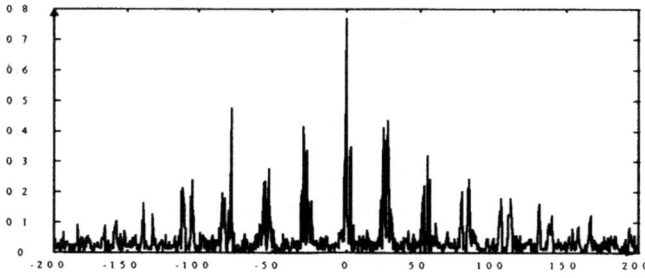

Figure 1, Example of signal spectrum

Using the same kind of development as above, it is straightforward to obtain the cross correlation between the two processes $Z(t)$ and $U(t)$. Again, starting with the definition we have

$$K_{ZU}(\tau) = E\{Z(t)U^*(t-\tau)\}$$

$$= \int_{\mathbf{R}} e^{i\omega\tau}\Psi_0^*(\omega)\,dS_Z(\omega) \tag{11}$$

4 Reconstruction with known $f(\cdot)$ and $g(\cdot)$

Having deduced the spectral behavior of the process subjected to a periodic clock change, our next task is to derive the best linear reconstruction of $Z(t)$ from the observation of the transmitted signal, $U(t)$. Denote this reconstructed signal $\widetilde{Z}(t)$. This section derives the necessary conditions for exact reconstruction, presents simulation results and discusses certain practical issues involved in the implementation.

4.1 Theoretical treatment

From Hilbert geometry, it is well known that the linear estimation which minimizes the reconstruction error from the observation of $U(t)$,

$$\sigma^2 = E\left\{\left|Z(t) - \widetilde{Z}(t)\right|^2\right\} \tag{12}$$

is the projection of $Z(t)$ on the Hilbert space spanned by $U(t)$. Let $H(U)$ denote this space. The projection may be written

$$E\left\{\left(Z(t) - \widetilde{Z}(t)\right)U^*(u)\right\} = 0 \qquad \forall(t,u) \in \mathbb{R}^2 \tag{13}$$

Figure 2 below gives a pictorial illustration of this orthogonal projection.

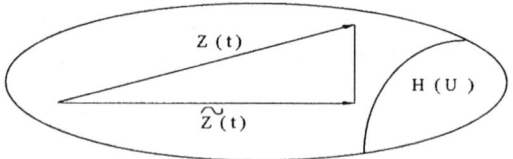

Figure 2, Orthogonal projection of $Z(t)$ on $H(U)$

We now define a new space, $H(S_U)$, spanned by the complex exponential $\exp(i\omega t)$ and define **I**, the linear application which connects $H(U)$ and $H(S_U)$, by associating $U(t)$ with $\exp(i\omega t)$

$$U(t) \quad \xleftarrow{\quad I \quad} \quad \exp(i\omega t) \tag{14}$$

I is an isometry because it leaves the distances between two elements in $H(U)$ and their corresponding elements in $H(S_U)$ unchanged [2]. Using **I** greatly facilitates the evaluation of the projection. Denote $\mu_t(\omega)$ the element in $H(S_U)$ corresponding to $\tilde{Z}(t)$. Using the uniqueness of **I** and the absolute continuity of S_Z with respect to S_U, we may write equation (13) as

$$\int_{\mathbf{R}} e^{i\omega(t-u)}\Psi_0^*(\omega)\, dS_Z(\omega) = \int_{\mathbf{R}} \mu_t(\omega)\, e^{-i\omega u}dS_U(\omega)$$

$$\Downarrow$$

$$\mu_t(\omega) = e^{i\omega t}\Psi_0^*(\omega)\frac{dS_Z}{dS_U}(\omega) \tag{15}$$

Inspection of equation (15) reveals that the projection of $Z(t)$ onto $H(U)$ is simply the result of the linear filtering of $U(t)$ by the filter of (in general complex) frequency response

$$H(\omega) = \Psi_0^*(\omega)\frac{dS_Z}{dS_U}(\omega) \tag{16}$$

The reconstructed process, $\tilde{Z}(t)$, may thus be written, with the aid of the Wiener-Lee relations [5]

$$\tilde{Z}(t) = \int_{\mathbf{R}} e^{i\omega t}\Psi_0^*(\omega)\frac{dS_Z}{dS_U}(\omega)\, d\Theta_U(\omega) \tag{17}$$

where $\Theta_U(\omega)$ is the spectral representation of $U(t)$. We note that this projection is the result of a time invariant filtering operation. This is a direct consequence of the randomizing phase introduced in the definition of $U(t)$.

4.2   Error performance

The reconstruction error is defined by equation (12). It may be evaluated by noting that the reconstruction error is orthogonal to $\tilde{Z}(t)$, i.e.

$$E\left\{\tilde{Z}(t)\left(Z(t) - \tilde{Z}(t)\right)^*\right\} = 0 \tag{18}$$

With the aid of this relation, it is a simple exercise to show that the reconstruction error is given by

$$\sigma^2 = \int_{\mathbf{R}} \left(1 - |\Psi_0(\omega)|^2\frac{dS_Z}{dS_U}(\omega)\right) dS_Z(\omega) \tag{19}$$

From equation (19) it is seen that σ^2 is zero if there is no overlap between the different spectral components in (10). A sufficient condition to avoid overlap is that the following requirement be fulfilled

$$dS_Z(\omega) \equiv 0 \qquad \forall \omega \notin \left(-\frac{\omega_0}{2}, \frac{\omega_0}{2}\right) \tag{20}$$

Using equation (10) we find that, in the case of periodic clock change and a sufficiently bounded spectrum, a perfect reconstruction is obtained by filtering $U(t)$ by the filter of (in general complex) frequency response

$$H(\omega) = \Psi_0^{-1}(\omega) \tag{21}$$

subject to the condition that $\Psi_0(\omega)$ does not cancel on the support of $dS_Z(\omega)$.

4.3 Implementation

In a practical implementation of this scrambler, the mathematical expressions given in the preceding paragraphs need to be transformed into a physical realization. For the encoder, this requires a sampling process followed by a tapped delay line. The delay line obviously needs to span twice the maximum PCC amplitude. Having access to past values of the process, it is possible to construct a demultiplexor sweeping the delay line in accordance with the wanted PCC. It should be noted that this procedure introduces a time delay in the scrambler corresponding to $\max\{f(\cdot)\}$. The encryption operation is shown in the below figure.

The decoding, being a simple time invariant linear filter is simple to implement. The filter frequency response, given by equation (6), needs to be realized. This should not, in general, be problematic. The only problems that may occur is the appearance of zeros in the passband of $Z(t)$, a problem discussed is section (4.5).

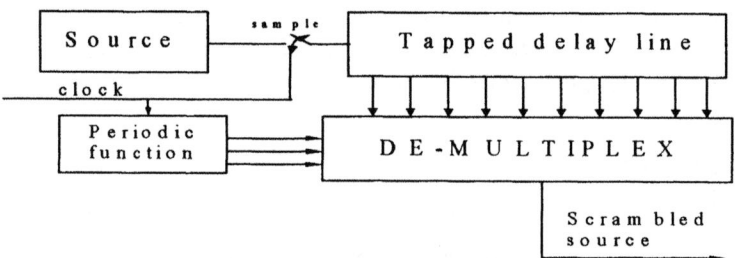

Figure 3, Encryption of the source

4.4 Simulation results

In this section, a particular example of the proposed scrambling scheme is studied. It is based on the following functions

$$f(t) = \alpha \sin(\omega_0 t)$$
$$g(t) = 1 \tag{22}$$

With this definition, the weighting functions, $\Psi_k(\omega)$, are given by

$$\Psi_k(\omega) = \frac{1}{T} \int_0^T e^{-i\alpha\omega \sin(\varphi\omega_0) - i\varphi k\omega_0} d\varphi$$
$$= J_k(\alpha\omega) \tag{23}$$

where $J_k(\omega)$ is the k'th order Bessel function of the first kind [5]. The original process is taken as a sequence of *N.R.Z* symbols [6] in the absence of additive noise. Note that there is no reason to limit the study to discrete valued processes. All the generation and reconstruction formulae above remain valid for all wide sense stationary mean square continuous processes.

The set of figures below show a realization of the scrambler/descrambler operation for 25 bits (each of duration 1), with $\omega_0 = 26$ and $\alpha = 1.5$. With α in this range, substantial intersymbol interference is introduced into the bit stream. Figure 4 shows the original bit stream. Figure 5 depicts the process after the application of the periodic clock change. It is apparent that the original message is completely scrambled by the PCC. In figure 6 the reconstruction of the process is shown overlaid with the original sequence. We note that there is a good correspondence between the two curves in this figure. The discrepancies that do appear are discussed in the section below. Finally, figure 7 gives a result of the filtering of $Z(t)$ with an ideal low-pass filter (ideal in the sense of ideal cut-off frequency). In this figure there is no apparent relation between the original and reconstructed processes. In other words; simple low pass filtering is insufficient. Exact knowledge of the reconstruction filter is thus required for adequate signal reconstruction.

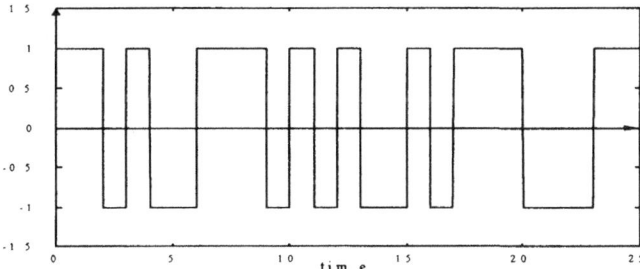

Figure 4, Original bit sequence

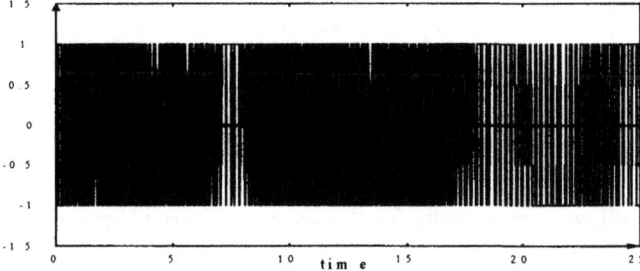

Figure 5, Scrambled sequence

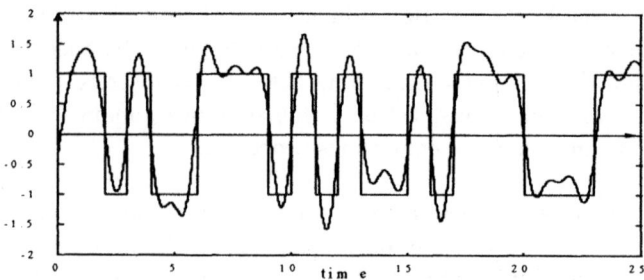

Figure 6, Reconstruction with $\Psi_0^{-1}(\omega)$

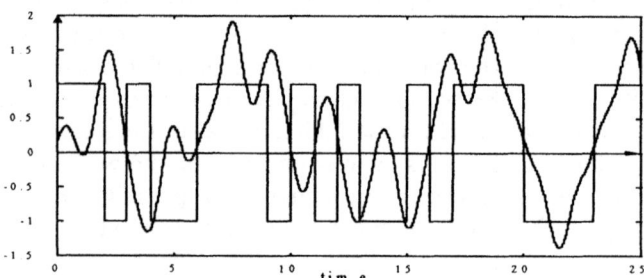

Figure 7, Reconstruction with low pass filter

Seeing as we are dealing with a binary signal in this example, other error measures than the mean squared error may be considered. A simple way of reconstructing the original signal (after filtering) consists of taking the sign. From this signal, a reasonable error measure is the mean time of disagreement between the two signals. Figure 8 below shows the simulation results of this error as a function of the "depth" α. We notice, not surprisingly, that the reconstruction error increases as α increases.

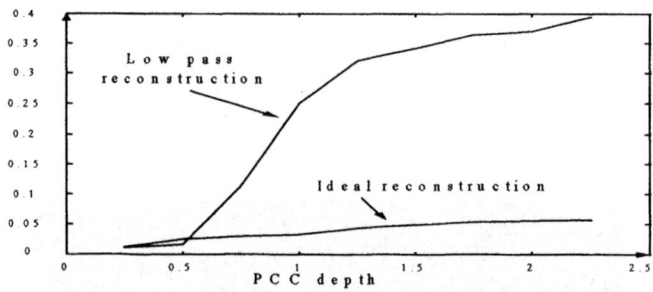

Figure 8, Error versus α

The corresponding error measure for the low pass filtered signal increases rapidly with α, with the most marked rise in the range $\alpha \in (0.5, 1.0)$. This observation is hardly surprising because $\alpha > \frac{1}{2}$ implies that symbols separated by one bit will interfere.

4.5 Comments

The deviation from perfect reconstruction observed in figure 3 has two principal sources stemming from conditions that are not fulfilled. The first problem lies in the choice of the original process. An $N.R.Z$ signal has a power spectral density given by [6]

$$dS_Z(\omega) = \text{sinc}^2\left(\frac{\omega}{2}\right) \tag{24}$$

and is thus of infinite support. Spectral overlap is therefore bound to occur, yielding non-ideal results.

Secondly, the zero order Bessel function needed for the reconstruction has several zeros in the passband of $Z(t)$. In our implementation this problem has been solved by hard-limiting the gain of $\Psi^{-1}(\omega)$. This procedure obviously produces inaccuracies. However, as it can be seen from figure 3, adequate message decoding is still possible.

5 Reconstruction from non-baseband component

In the previous sections we saw how the original signal may be reproduced from the baseband version of the jittered signal. The signal may also be reconstructed on the basis of any one of the spectral reproductions of $U(t)$. Starting from the Cramér-Loève representation of $U(t)$ we have

$$U(t) = g(t+\varphi)\int_{\mathbf{R}} e^{i\omega(t-f(t+\varphi))}d\Theta_Z(\omega)$$

$$= \sum_{k=-\infty}^{\infty} e^{ik\omega_0(t+\varphi)}\int_{\mathbf{R}} e^{i\omega t}\Psi_k(\omega)\,d\Theta_Z(\omega) \tag{25}$$

If one wishes to reconstruct from the part of the spectrum centered at $n\omega_0$, it will be necessary to bring this part of the signal down to baseband by the multiplication with the complex exponential

$$m(t) = e^{-in\omega_0(t+\varphi)} \tag{26}$$

The only practical problem involved in this operation lies in the tracking of the unknown phase φ. Assuming this may be achieved efficiently, the resulting baseband signal, $U_n(t)$, is given by

$$U_n(t) = \int_{\mathbf{R}} e^{i\omega t}\Psi_k(\omega)\,d\Theta_Z(\omega) \tag{27}$$

From equation (27) we see that, having obtained $U_n(t)$, the final stage of the reconstruction process consists simply of filtering $U_n(t)$ by the filter of (in general complex) frequency response $\Psi_k^{-1}(\omega)$. The two figures below show the results of the descrambling with the same parameters as in the previous section except that the decoding was based on the energy centered around ω_0. Figure 9 gives the ideal reconstruction and figure 10 the signal obtained with a low pass filter. Comparing with previous simulation results (figures 5 and 6), we observe that the descrambler performs as well in this band as it did at baseband.

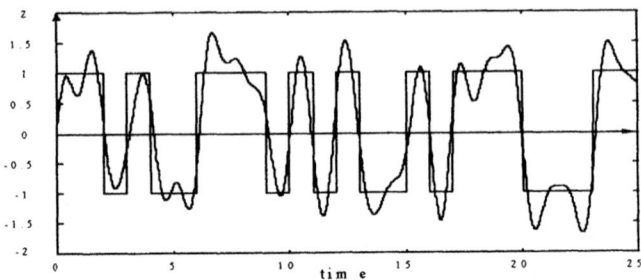

Figure 9, Passband reconstruction, ideal filter

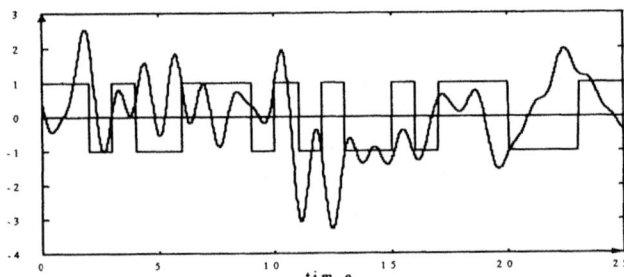

Figure 10, Passband reconstruction, low pass filter

Note that all conditions for perfect reconstruction deduced in the previous section still hold with the modification that the power of the reconstruction error in equation (19) now depends on the k'th order coefficient, $\Psi_k(\omega)$ instead of $\Psi_0(\omega)$.

6 Conclusion

In this paper we have defined a new method for the scrambling of wide sense stationary processes by means of periodic clock changes. It was shown, in particular, that if the parent process is sufficiently band limited compared with the clock change frequency, then perfect reconstruction is possible. The feasibility of the method was demonstrated with the aid of a simple simulation example. Finally, the method was generalized by extending the basis for reconstruction to non-baseband portions of the signal spectrum.

References

1. W.Diffie and M.H.Hellman, "New directions in cryptography", *IEEE Transactions Inform. Theory*, pp. 644-654, November 1976

2. J.L.Doob, *Stochastic processes*, Wiley, 1967
3. W.A.Gardner and L.E.Franks, "Characterization of cyclostationary random signal processes", *IEEE Transactions Inform. Theory*, pp. 4-14, 1975
4. B.Lacaze, "A new binary modulation process", *Proceedings of SPIE Advanced Signal Processing: Algorithms, Architectures and Implementation*, pp 254-259, San-Diego 1994
5. A.Papoulis, *Probability, random variables and stochastic processes*, MacGraw-Hill, 1965
6. H.Taub and D.L.Schilling, *Principles of communication systems*, McGraw-Hill, 1986
7. A.Duverdier, B.Lacaze and N.Aakvaag, *A novel approach to spread spectrum communication using linear time-varying periodic filters*, Applications on information theory, Springer-Verlag, 1996

A Novel Approach to Spread Spectrum Communication Using Linear Periodic Time-Varying Filters

A. Duverdier, B. Lacaze and N. D. Aakvaag

National Polytechnics Institute of Toulouse
LEN7/GAPSE, 2 rue Camichel, 31071 Toulouse, France
tel: (33) 61 58 83 67   Fax:(33) 61 58 82 37
email: duverdie@len7.enseeiht.fr

Abstract. A stationary process subjected to a Linear Periodically Time-Varying (LPTV) transformation becomes cyclostationary. In this article, we use a particular continuous-series representation of this output process to compute its spectrum and verify its cyclostationarity. We then prove that Periodic Clock Changes (PCC) are a particular case of the LPTV transformation. A method of linear reconstruction for any LPTV filtering is then introduced. We apply it to scramble stationary processes. It corresponds to an unconventional case of spread spectrum.

1 Introduction

In modern telecommunications, it is often necessary to encrypt information. Clearly, these days most encryption is done digitally. Yet analogue scramblers [1] [2] have advantages in some cases – in particular, when the bandwidth cannot be significantly increased (as in speech transmissions [3]). Linear periodic time-varying filtering [4] is a typical model in communications [5] and in mechanics [6]. In scrambling applications, discrete linear periodic time-varying techniques have been used successfully [7]. In this paper, we present continuous linear periodic time-varying filtering applied to scrambling and we show that it corresponds to an unconventional case of spread spectrum [8].

In the first section, we recall some definitions. Then we show that a stationary process subjected to a linear periodic filter becomes cyclostationary [9]. It can be represented by a continuous-time series representation. For the applications we demonstrate that periodic clock changes [10] constitute a subset of linear periodic filters. Next we compute a linear reconstruction of the input process when the filter is known. We obtain a scrambling method that spreads the spectrum. Finally, we present some examples using our novel spread spectrum method.

2 Definitions

In this section, we present the main definitions concerning the spectral representation of stationary processes and Linear Periodic Time-Varying (LPTV) filtering [4].

2.1 Stationary process

In what follows, the original signal, i.e. the process to be scrambled will be denoted $Z(t)$. We let $Z = \{Z(t), \ t \in \mathbb{R}\}$ be a random stationary process, zero mean and mean square continuous. $\Theta_Z(\omega)$ is the Cramér-Loève spectral representation [11] of $Z(t)$ and the two are related by:

$$Z(t) = \int_{-\infty}^{+\infty} e^{i\omega t} d\Theta_Z(\omega) \tag{1}$$

We note $R_Z(\tau)$ its autocorrelation function and $S_Z(\omega)$ its spectrum defined by:

$$R_Z(\tau) = E\left[Z(t)Z^*(t-\tau)\right] = \int_{-\infty}^{+\infty} e^{i\omega t} dS_Z(\omega) \tag{2}$$

2.2 Linear periodic time-varying filter

The basic idea of the scrambling system is to subject the original process to a linear periodic time-varying filter. Let $\tilde{h}$ be a continuous-time linear filter. Its response $X(t)$ to an input $Z(t)$ may in general be written as:

$$X(t) = \int_{-\infty}^{+\infty} h(t,s)Z(s)ds \tag{3}$$

where $h(t,s)$ is the filter impulse response. We will study the case where $\tilde{h}$ is an LPTV system [4], i.e. where there exists a period $T = 2\pi/\omega_0$ such that:

$$h(t+T, s+T) = h(t,s) \tag{4}$$

The time-varying frequency response of the LPTV filter $\tilde{h}$ is defined by:

$$H_t(\omega) = \int_{-\infty}^{+\infty} h(t, t-\tau)e^{-i\omega\tau} d\tau \tag{5}$$

It is worth noting that $H_t(\omega)$ is periodic in t. We then define its Fourier development, that we assume to be sufficiently regular, by:

$$H_t(\omega) = \sum_{k=-\infty}^{+\infty} \psi_k(\omega)e^{ik\omega_0 t} \tag{6}$$

where the Fourier coefficients are expressed as:

$$\psi_k(\omega) = \frac{1}{T} \int_0^T H_t(\omega)e^{-ik\omega_0 t} dt \tag{7}$$

3 Response of a stationary process through an LPTV filter

Let $X(t)$ be the response of the stationary process $Z(t)$ through the LPTV filter of impulse response $h(t,s)$. By means of a continuous-time series representation, we will present an easy way to prove that $X(t)$ is a cyclostationary process (it can be stationarized by a random phase) [9].

3.1 Continuous-time series representation of $X(t)$

$X(t)$ is the filtering of $Z(t)$ by the LPTV filter $\tilde{h}$ of impulse response $h(t,s)$ and frequency response $H_t(\omega)$. Using (1) and (5), the expression of X given by equation (3) becomes:

$$
\begin{aligned}
X(t) &= \int\limits_{-\infty}^{+\infty} h(t,s)Z(s)ds \\
&= \int\limits_{-\infty}^{+\infty}\int\limits_{-\infty}^{+\infty} h(t,s)e^{i\omega s}d\Theta_Z(\omega)ds \\
&= \int\limits_{-\infty}^{+\infty} H_t(\omega)e^{i\omega t}d\Theta_Z(\omega)
\end{aligned}
\tag{8}
$$

The Fourier representation of $H_t(\omega)$ (6) allows us to define a continuous-series representation of $X(t)$ such that:

$$
X(t) = \sum_{k=-\infty}^{+\infty} e^{ik\omega_0 t}G_k(t) \quad \text{with} \quad G_k(t) = \int\limits_{-\infty}^{+\infty} e^{i\omega t}\psi_k(\omega)d\Theta_Z(\omega)
\tag{9}
$$

$G_k(t)$ is the response to $Z(t)$ through the time-invariant linear filter whose frequency response is $\psi_k(\omega)$. We have:

$$
E\left[G_m(t)G_n^*(t-\tau)\right] = \int\limits_{-\infty}^{+\infty} e^{i\omega\tau}\psi_m(\omega)\psi_n^*(\omega)dS_Z(\omega)
\tag{10}
$$

The $\{G_k(t)\}_{k\in\mathbf{Z}}$ are also jointly stationary processes.

3.2 Stochastic parameters of $X(t)$

Since $Z(t)$ is zero mean, then all $G_k(t)$ are zero mean. Therefore the above decomposition shows that $X(t)$ is a zero mean process.

From the Wiener-Lee relations, we can easily obtain the autocorrelation function of $X(t)$ and its two-dimensional spectral density:

$$R_X(t, \tau) = \sum_{m=-\infty}^{+\infty} e^{im\omega_0 t} \int_{-\infty}^{+\infty} \sum_{l=-\infty}^{+\infty} e^{i(\omega+l\omega_0)\tau} \psi_{m+l}(\omega)\psi_l^*(\omega) dS_Z(\omega) \qquad (11)$$

$$dS_X(t, \omega) = \sum_{m=-\infty}^{+\infty} e^{im\omega_0 t} \sum_{l=-\infty}^{+\infty} \psi_{m+l}(\omega - l\omega_0)\psi_l^*(\omega - l\omega_0) dS_Z(\omega - l\omega_0) \qquad (12)$$

They are both periodic in t. $X(t)$ is thus cyclostationary in the wide sense [9].

3.3 Stationarization by phase randomization

If $X(t)$ is observed with a random phase φ uniformly distributed on $[0, 2\pi]$, then $X(t + \varphi)$ is stationarized and its autocorrelation function and spectrum correspond to the stationary part of equations (11) and (12) [9], i.e. the infinite sum reduced to $m = 0$.

4 Case of the Periodic Clock Changes

In this section, we prove that Periodic Clock Changes (PCC) [10] correspond to a particular case of LPTV filtering. Furthermore, we show that sums of PCC are also LPTV filters. As an example, we take the case of Pulse-Amplitude-Modulation (PAM).

4.1 Linear filter associated with a PCC

The general form of a stationary process $Z(t)$ subjected to a PCC was defined in [10]. The resulting scrambled signal is given by:

$$U(t) = Z[t - f(t)]g(t) \qquad (13)$$

where $f(t)$ is a real measurable function and $g(t)$ is a real or complex integrable function, $f(t)$ and $g(t)$ being $T = 2\pi/\omega_0$ periodic. Using the Cramér-Loève representation of $Z(t)$, we can write the process subjected to a PCC as:

$$U(t) = \int_{-\infty}^{+\infty} e^{i\omega(t-f(t))} g(t) d\Theta_Z(\omega) \qquad (14)$$

Because of the periodicity of $f(t)$ and $g(t)$, $e^{-i\omega f(t)}g(t)$ is periodic in t. A PCC is then an LPTV filter of frequency response:

$$H_t(\omega) = e^{-i\omega f(t)} g(t) \qquad (15)$$

If we denote by $\delta(t)$ the Dirac function, then, in the sense of distributions, the PCC corresponds to an LPTV filter whose impulse response is:

$$h(t, s) = g(t)\delta(t - f(t) - s) \qquad (16)$$

4.2   Sum of periodic clock changes

For a sum of periodic clock changes, we consider the following extended definition:

$$V(t) = \sum_{n=1}^{M} Z(t - f_n(t)) g_n(t) \tag{17}$$

The functions $f_n(t)$ and $g_n(t)$ are assumed to be periodic with the same period $T_n = 2\pi/\omega_n$. We choose the frequencies such that:

$$\forall (m,n) \in \{1..M\}^2, \quad \frac{\omega_n}{\omega_m} = \frac{p}{q} \quad (p,q) \in N \times N^* \tag{18}$$

There then exists a frequency λ, the smallest multiple of $\{\omega_n\}_{n\in\{1..M\}}$, such that the functions $f_n(t)$ and $g_n(t)$ are periodic in $T_\lambda = 2\pi/\lambda$.

A sum of LPTV filters is an LPTV system, due to the linearity of the operator. Its impulse response is the sum of the individual impulse responses. $V(t)$ therefore corresponds to the LPTV filter such that:

$$H_t(\omega) = \sum_{n=1}^{M} e^{-i\omega f_n(t)} g_n(t) \tag{19}$$

or, in the sense of distributions:

$$h(t,s) = \sum_{n=1}^{M} g_n(t)\delta(t - f_n(t) - s) \tag{20}$$

4.3   Example: Pulse-Amplitude Modulation

Let $Z(t)$ be a stationary process and $q(t)$ be a finite-energy pulse. The PAM pulse train $X(t)$ [4] is the response to $Z(t)$ through the linear time-varying filter $h(t,s)$, defined in the sense of the distributions, such that:

$$h(t,s) = \sum_{k=-\infty}^{+\infty} q(t - t_0 - kT)\delta(s - kT) \tag{21}$$

For any $q(t - t_0)$, we see that a PAM is a sum of PCC with:

$$f_n(t) = t[T] + nT \quad \text{and} \quad g_n(t) = q(t[T] + nT - t_0) \tag{22}$$

where $[]$ is the modulus function and the number of elements in the sum is the integer ratio of the $q(t)$ support and T. Pulse-amplitude modulation is therefore a particular case of LPTV filtering.

5 Linear reconstruction after an LPTV filtering

In this section, we suppose that the output process, $X(t)$, is observed and that the filter, $h(t,s)$, is known. We show that the linear reconstruction of the input process $Z(t)$ is possible in theory without error. We apply the resulting reconstruction formulae in particular cases which correspond to a new form of spread spectrum [8].

5.1 Theory

Let $\Theta_X(\omega)$ and $\Theta_{G_k}(\omega)$ be the Cramér-Loève spectral representations of $X(t)$ and $G_k(t)$ respectively, where $G_k(t)$ is the response to $Z(t)$ through the time-invariant linear filter of frequency response $\psi_k(\omega)$. The continuous-series decomposition defined in section 3.1 yields the Cramér-Loève spectral representation of $X(t)$:

$$d\Theta_X(\omega) = \sum_{k=-\infty}^{+\infty} d\Theta_{G_k}(\omega - k\omega_0) = \sum_{k=-\infty}^{+\infty} \psi_k(\omega - k\omega_0)d\Theta_Z(\omega - k\omega_0) \quad (23)$$

$d\Theta_X(\omega)$ is an infinite sum of weighted shifted versions of $d\Theta_Z(\omega)$. The individual versions are centered around multiples of the LPTV filter frequency and the weights depend on the filter. For example, if the support of $d\Theta_Z(\omega)$ is included in $[-3\omega_0, 3\omega_0[$, the modulus of (23) can be represented by:

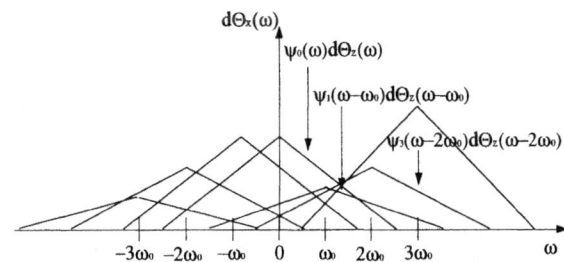

If we suppose that $d\Theta_X(\omega)$ and the functions $\{\psi_k(\omega)\}_{k \in Z}$ (i.e. the LPTV filter) are known, the inversion of the equation (23) allows the identification of $d\Theta_Z(\omega)$. In practice, we take $Z(t)$ with a band-limited spectral support. Under certain conditions of inversibility, the linear reconstruction of $Z(t)$ without error is then possible.

5.2 Application to band-limited signals

Reconstruction without overlapping spectra

For this application, we will consider that the support of $d\Theta_Z(\omega)$ is included in

$[-\omega_0/2, \omega_0/2[$. In equation (23) the shifted versions of $d\Theta_Z(\omega)$ do not overlap. Let Δ be the integer set such that the functions $\{\psi_k(\omega)\}_{k\in\Delta}$ do not take zero values on the spectral support of $Z(t)$. Then equation (23) becomes:

$$\forall k \in \Delta, \quad d\Theta_Z(\omega) = \psi_k^{-1}(\omega)d\Theta_X(\omega + \omega_0)\prod_{[-\omega_0/2,\omega_0/2[}(\omega) \qquad (24)$$

where $\prod_{[a,b[}(\omega)$ is the indicator function of the interval $[a,b[$. This expression shows that $Z(t)$ can be reconstructed using any band of $d\Theta_X(\omega)$ around $k\omega_0$ with $k \in \Delta$. In real scrambling cases, these multiple reconstructions are useful. When we are in presence of a frequency selective noise or when we want to improve the quality of the reconstruction, the mean of several reconstructions therefore permits better identification of the original process.

Reconstruction with overlapping spectra

For this example, we will consider that the support of $d\Theta_Z(\omega)$ is included in $[-\omega_0, \omega_0[$. Then equation (23) shows that the weighted shifted versions of the Cramér-Loève spectral representation of $Z(t)$ are overlapped in $d\Theta_X(\omega)$. To find $d\Theta_Z(\omega)$, we use the redondant system:

$$\forall l \in \mathbb{Z}, \quad \forall \omega \in [0, \omega_0[,$$
$$\begin{cases} \psi_l(\omega)d\Theta_Z(\omega) + \psi_{l+1}(\omega - \omega_0)d\Theta_Z(\omega - \omega_0) = d\Theta_X(\omega + l\omega_0) & (25) \\ \psi_{l-1}(\omega)d\Theta_Z(\omega) + \psi_l(\omega - \omega_0)d\Theta_Z(\omega - \omega_0) = d\Theta_X(\omega + (l-1)\omega_0) \end{cases}$$

The solution to this system exists if there exists an integer l such that:

$$\forall \omega \in [0, \omega_0[, \quad \psi_l(\omega)\psi_l(\omega - \omega_0) - \psi_{l-1}(\omega)\psi_{l+1}(\omega - \omega_0) \neq 0 \qquad (26)$$

Then the spectral representation of $Z(t)$ is given by:

$$\begin{cases} \forall \omega \in [-\omega_0, 0[, \\ \quad d\Theta_Z(\omega) = \dfrac{\psi_l(\omega + \omega_0)d\Theta_X(\omega + l\omega_0) - \psi_{l-1}(\omega + \omega_0)d\Theta_X(\omega + (l+1)\omega_0)}{\psi_l(\omega)\psi_l(\omega + \omega_0) - \psi_{l-1}(\omega + \omega_0)\psi_{l+1}(\omega)} \\ \forall \omega \in [0, \omega_0[, \\ \quad d\Theta_Z(\omega) = \dfrac{\psi_l(\omega - \omega_0)d\Theta_X(\omega + l\omega_0) - \psi_{l+1}(\omega - \omega_0)d\Theta_X(\omega + (l-1)\omega_0)}{\psi_l(\omega)\psi_l(\omega - \omega_0) - \psi_{l-1}(\omega)\psi_{l+1}(\omega - \omega_0)} \end{cases}$$

$$(27)$$

Perfect linear reconstruction is then in theory possible by means of a time-varying filtering given by (27). Like in the previous case, we may obtain multiple reconstructions. The mean of several reconstructions will allow a better identification of the original signal.

Spread spectrum

We have here a kind of spread spectrum transmission tool. Spread spectrum has been defined in [8] as "a means of transmission in which the signal occupies a bandwidth in excess of the minimum necessary" such that "the band spread is accomplished by means of a code". The use of multiple spectral reproductions of the initial signal defines then an unconventional case of spread spectrum, where the analogue scrambling introduced by the linear periodic filter can be seen as a code. The principal advantage of this method compared to digital methods is that it does not need quantization which would significantly increase the bandwidth of the initial signal. Our scrambling method is moreover simpler than classical analogue methods of scrambling [1] [2]. Its implementation is just a generalization of the method proposed in [10].

6 Simulations

6.1 Example 1

Let $Z(t)$ be an N.R.Z. signal [12] approximately band-limited on $[-\omega_0/2, \omega_0/2[$ and $\tilde{h}$ be an LPTV filter defined as a sum of two PCC (see 4.2) such that:

$$\begin{cases} f_1(t) = -\alpha \sin \omega_0 t \\ g_1(t) = 1 \end{cases} \quad \text{and} \quad \begin{cases} f_2(t) = -\beta \sin (\omega_0 t/2) \\ g_2(t) = \exp (\omega_0 t/2) \end{cases} \tag{28}$$

These functions are periodic with common circular frequency $\lambda = \omega_0/2$ and the coefficients $\psi_k(\omega)$ are given by:

$$\psi_k(\omega) = \frac{1 + (-1)^k}{2} J_{k/2} (\alpha\omega) + J_{k-1} (\beta\omega) \tag{29}$$

where $J_k(\omega)$ is the k'th order Bessel function. We assume that the observed signal $X(t)$ contains a low frequency noise, chosen as sinusoidal interference, $n(t) = 2\sin(\omega_1 t)$. We take $\alpha = 0.5$, $\beta = 0.5$, $\omega_0 = 40\pi$ and $\omega_1 = 0.4\pi$. In Figure 1, we represent the initial N.R.Z. signal.

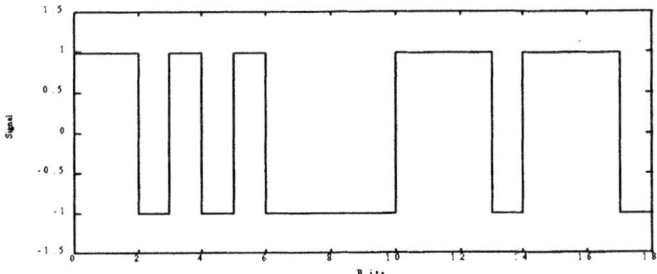

Figure 1, Initial signal

Figure 2 shows the initial signal filtered by the sum of PCC and affected by the additive frequency selective interference.

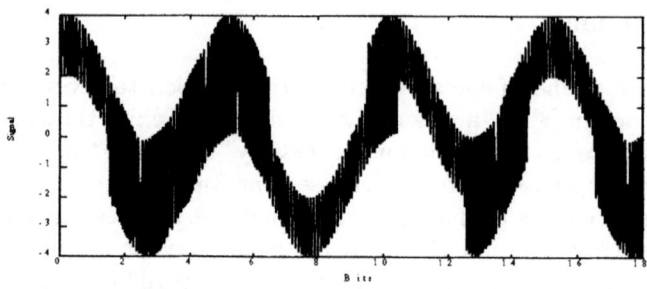

Figure 2, Observed signal

The frequency of the sinusoidal interference is in the low band of the observed signal. Then, if we try to reconstruct the process around $\omega = 0$ with $\psi_0^{-1}(\omega)$, the reconstructed signal is dominated by the interference and the information is lost (see Figure 3).

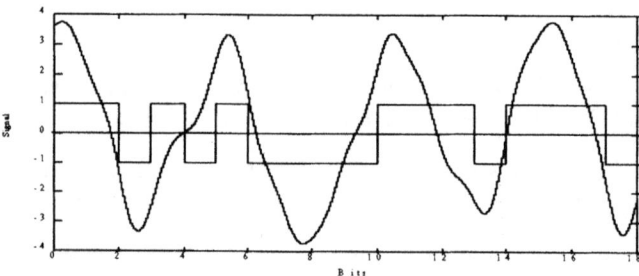

Figure 3, Low band reconstruction

In contrast, in another band where no noise is present, the filtering operation yields the desired response. For example, Figure 4 shows the result obtained around $\omega_0/2$.

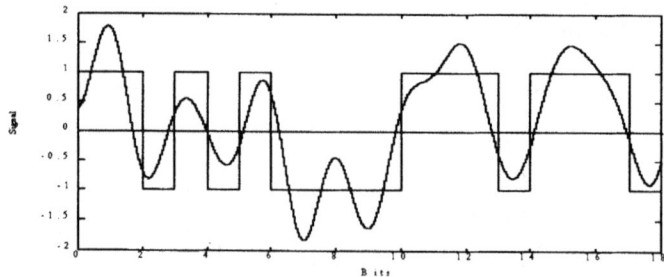

Figure 4, Reconstruction around $\omega_0/2$

Then, any method of bit detection permits the exact knowledge of the transmitted bits without error. Because the frequency of the interference is unknown, we are normally not able to choose a noise free frequency band for the reconstruction. One simple remedy, which does not require this knowledge, is to take the average of reconstructions based on several bands. Figure 5 presents the mean of reconstructions around $-\omega_0/2, 0, \omega_0/2, \omega_0$. After bit detection, the signal is well reconstructed.

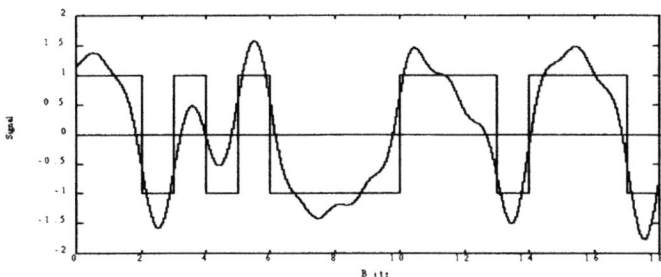

Figure 5, Final reconstructed signal

6.2   Example 2

We consider now that $Z(t)$ is a continuous valued stationary signal. Its spectral representation is always assumed to be band limited on the interval $[-\omega_0/2, \omega_0/2[$ and the LPTV filter be the same as in example 1. $Z(t)$ is given by the signal in Figure 6.

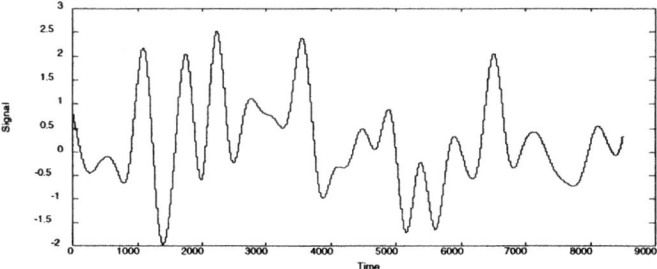

Figure 6, Original signal

We chose $\omega_0 = 0.04\pi$. For $\alpha = \beta = 250$, at the output of the filter we obtain a scrambled signal (see Figure 7).

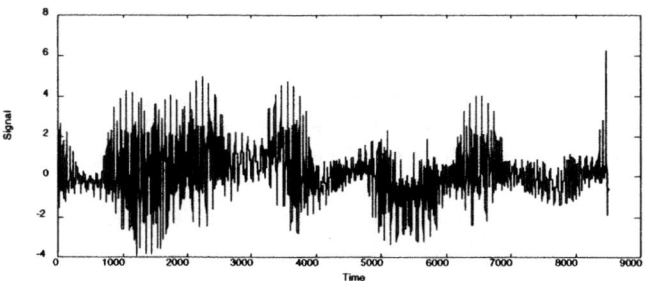

Figure 7, Observed signal

The quality of the reconstruction can be measured by the variance between the original signal and the reconstructed signal. For $\alpha = \beta$, Figure 8 compares the error performance of three different reconstruction schemes: simple low pass filter (———), ideal reconstruction filter at baseband (- - -) and mean of the reconstructions for the bands corresponding to $k = 0$ and $k = 2$ (— —).

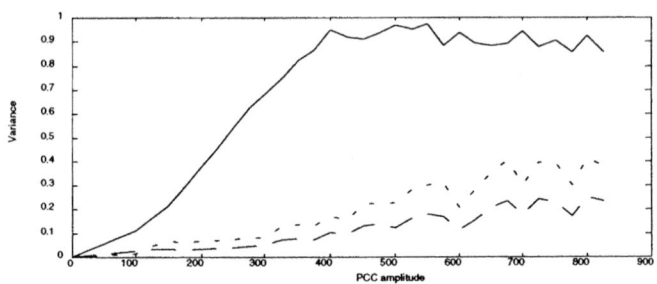

Figure 8, Error reconstruction

Figure 8 shows that taking the average of several reconstructions reduces the error when the amplitudes of the PCC increase ($\alpha = \beta > 120$), i.e. when the power of the signal in the reproduced bands is non-negligeable.

7 Conclusion

In this paper we have studied linear time-varying periodic filtering of stationary processes. We have used their cyclostationary properties to introduce a method of perfect reconstruction. We have shown, in particular, that this method can be applied in the case of periodic clock changes. In fact, these are linear time-varying periodic filters. As examples, we presented a scrambling application giving rise to an unconventional form of spread spectrum. In simulations, we verified that our novel spread spectrum method can correct errors due to frequency selective noise for a binary signal and improve the reconstruction for a continuous valued signal.

References

1. A.D. Wyner, "An Analog Scrambling Scheme which Does Not Expand Bandwidth, Part I: Discrete Time", *IEEE Trans. Inform. Theory*, pp. 261-274, 1979
2. A.D. Wyner, "An Analog Scrambling Scheme which Does Not Expand Bandwidth, Part II: Continuous Time", *IEEE Trans. Inform. Theory*, pp. 415-425, 1979
3. B. Goldburg and S. Sridharan, "Design and Cryptanalysis of Transform-Based Analog Speech Scramblers", *IEEE Journ. Select. Areas in Comm.*, pp. 735-744, 1993
4. W.A. Gardner, *Introduction to Random Processes with Applications to Signals and Systems*, 2nd ed., McGraw-Hill, 1990
5. T.H.E. Ericson, "Modulation by Means of Linear Periodic Filtering", *IEEE Trans. Inform. Theory*, pp. 322-327, 1981

6. P. Bolzern, P. Colaneri and R. Scattolini, "Zeros of discrete-time linear periodic systems", *IEEE Trans. Autom. Contr.*, pp. 1057-1058, 1986

7. R. Ishii and M. Kakishita, "A Design Method for a Periodically Time-Varying Digital Filter for Spectrum Scrambling", *IEEE Trans. Acous., Speech and Signal Proc.*, pp. 1219-1222, 1990

8. R.L. Pickholtz, D.L. Shilling and L.B. Milstein, "Theory of Spread-Spectrum Communications - A Tutorial", *IEEE Trans. Commun.*, pp. 855-884, 1982

9. W.A. Gardner and L.E. Franks, "Characterization of cyclostationary random signal processes", *IEEE Trans. Inform. Theory*, pp. 4-14, 1975

10. N.D. Aakvaag, B. Lacaze and A. Duverdier, "On the use of periodic timebase companding in transmission systems: Scrambling by Means of Periodic Clock Changes", *Lecture Notes in Computer Science*, 1996

11. H. Cramér and M.R. Leadbetter, *Stationary and Related Stochastic Proc.*, Wiley, 1967

12. J.G. Proakis, *Digital Communications*, Mc-Graw-Hill, 1989

On Random-Like Codes

Gérard Battail

Ecole Nationale Supérieure des Télécommunications
Département Communications and URA 820 of CNRS,
46, rue Barrault,
F-75634 Paris Cedex 13, France

Abstract. A random-like code is designed so as to make its normalized distance distribution (its weight distribution if it is linear) close to that obtained in the average by random coding. The iterated product of single-parity-check codes, for testing the use of several successive weighted-output decodings in order to reduce the overall decoding complexity, was the first example of a random-like code with a poor minimum distance but a good bit-error rate.

A distinction is introduced between strongly and weakly random-like codes. Strongly random-like codes are such that each term of their weight distribution is close to that of random coding, in contrast with weakly random-like ones which exhibit only a global shape similarity (as measured e.g., by the cross-entropy). Strongly random-like codes are good for the minimum distance criterion as meeting the Gilbert-Varshamov bound and their word-error rate (WER) is good. In contrast, weakly random-like codes may have small bit-error rate (BER) but bad WER. From an engineering point of view, however, the BER is the most significant parameter so weakly random-like codes are actually interesting. Moreover, their design is easier.

We briefly review some codes which may be considered as random-like, with special emphasis on so-called "pseudo-random" systematic convolutional codes which are recursive and such that the recursive part of the encoder corresponds to a maximum-length generator. Even if such a code has a small free distance and thus does not result in a low enough BER, combining several such codes according to the turbo-code scheme enables control of the weight distribution tails, hence the BER, by increasing the number of combined codes, so as to meet any specified performance at a rate lower than the channel capacity.

1 Introduction

The author is grateful to the organizers of this workshop for the opportunity he is given to express some unconventional ideas about channel coding. He believes that the (heretic?) point of view discussed here is closer to both the spirit of Information Theory and the engineering needs than the conventional theory of error correcting codes. This talk will not discuss a well rounded topic and maybe it raises more questions than it provides answers, but hopefully contains seeds for future works.

The title "Random-Like Codes" does not refer to a single family of codes but rather to a *way of designing codes* not relying on the usual minimum distance criterion. One of the main conclusions of this talk will be that the newly discovered "turbo-codes" of Berrou, Glavieux and Thitimajshima [1] are actually random-like, which explains some of their properties, including their outstanding performance.

Often, discussing some topic does not follow the initial path which led to its elaboration. Here, however, we wish to begin this talk with the actual starting point of our research, namely the concept of *soft-output decoding*, as providing both a motivation for it and an example of some important features of random-like codes.

2 Why Search for Random-Like Codes?

2.1 Soft-Output Decoding and its Use for the Iterated Product of Single-Parity-Check Codes

In the presence of real-valued noise (as in most channels of the real world), soft-input decoding takes into account the probabilities of the input symbols as measured by the decision margins, thus avoiding the loss of an information available at the channel output. Extending the decoding role to reestimate the codeword probabilities to take account of the coding constraints, as proposed in [2], leads to *soft-output* decoding which still avoids the loss of an available information. Earlier decoding algorithms were actually of this kind e.g., [3]-[5], and soft-output decoding is now a fast extending field [6, 7]. Ideally, such decoding does not incur information loss. Decoding a long code constructed by product or concatenation of simple component codes can thus involve cascading a number of decoders. Instead of a large, necessarily complicated single decoder, one can use a chain of simple decoders, each of them taking into account only some partial coding constraint, but the whole set of decoders taking account of all constraints.

In order to illustrate the potential of soft-output decoding in a system combining several codes, we tested it in the case where the *simplest codes* are combined in the *simplest way*. The binary single-parity-check (SPC) codes are undoubtly the simplest non-trivial codes [8], and Elias' iterated product is probably the simplest way of combining codes [9]. A single-parity-check code $(k+1, k)$ results from appending to the k information bits a single parity check bit equal to their sum modulo 2. Soft-output decoding of an SPC code is easily performed or simulated in the presence of additive white Gaussian noise. Moreover, this code is so simple that its performance can be accurately predicted by computation with a very good agreement with simulation results [10].

The product of two systematic codes of parameters (n_1, k_1) and (n_2, k_2), denoted by $(n_1, k_1) \otimes (n_2, k_2)$, is best described by a bidimensional array: k_2 codewords of the first code (n_1, k_1) are written as its rows, and they are encoded column-wise according to the second code (n_2, k_2), thus resulting in an $n_1 \times n_2$

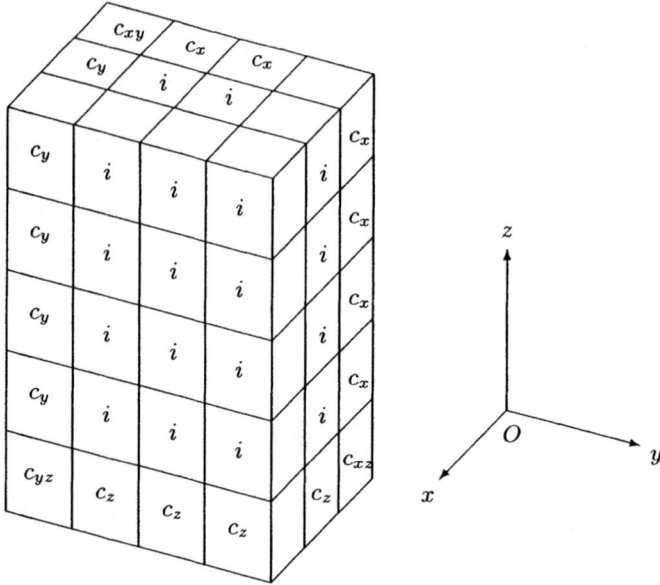

Fig. 1. Example of a product of 3 SPC codes: $(3,2) \otimes (4,3) \otimes (5,4)$. The 3-dimensional array is represented as a packing of cubes, each associated with a bit; i denotes an information bit, c_x denotes a check bit on information bits in the x direction, c_{xy} denotes a check bit in the y direction on checks in the x direction, etc.

array which is a word of an $(n_1 n_2, k_1 k_2)$ code. Extension to an array of arbitrary dimension is straightforward. As an example, the product of three SPC codes is depicted in Fig. 1.

The iterated product in Elias' sense consists of a product of the codes of an infinite family such that its rate $R_j = \prod_{i=0}^{j-1} k_i / \prod_{i=0}^{j-1} n_i$ approaches a finite limit as the number j of codes in the product approaches infinity. The parameters of the SPC code family initially used in [10] were $k_i = 2k_{i-1} = 2^i k_0$, where k_i is the dimension of the i-th code in the product, $i = 0, 1, \ldots$; k_0 is a given integer. This choice ensures that the rate of the product of j codes, $R_j = \prod_{i=0}^{j-1} k_i / \prod_{i=0}^{j-1} (k_i + 1)$, approaches a non-zero limit R as $j \to \infty$. For $k_0 = 2$, this limit is of about $0.4194\ldots$

It was later shown by Caire, Taricco and Battail [11] that a better choice is the product of m SPC codes of same length n, namely $(n, n-1)^{m\otimes}$, with $n = \lfloor m/\ell \rfloor$, where m, an integer, and ℓ are given. Then $R = \lim_{m \to \infty} (\frac{n-1}{n})^m = \exp(-\ell)$. These codes will be referred to in the sequel as "hypercube SPC codes." Their decoding is somewhat more difficult than that of the previously mentioned ones.

Surprisingly, the performance of the product of unequal-dimension codes was close to the best performance reported at that time (circa 1987), which was achieved by far more complicated means. Using this iterated product over the

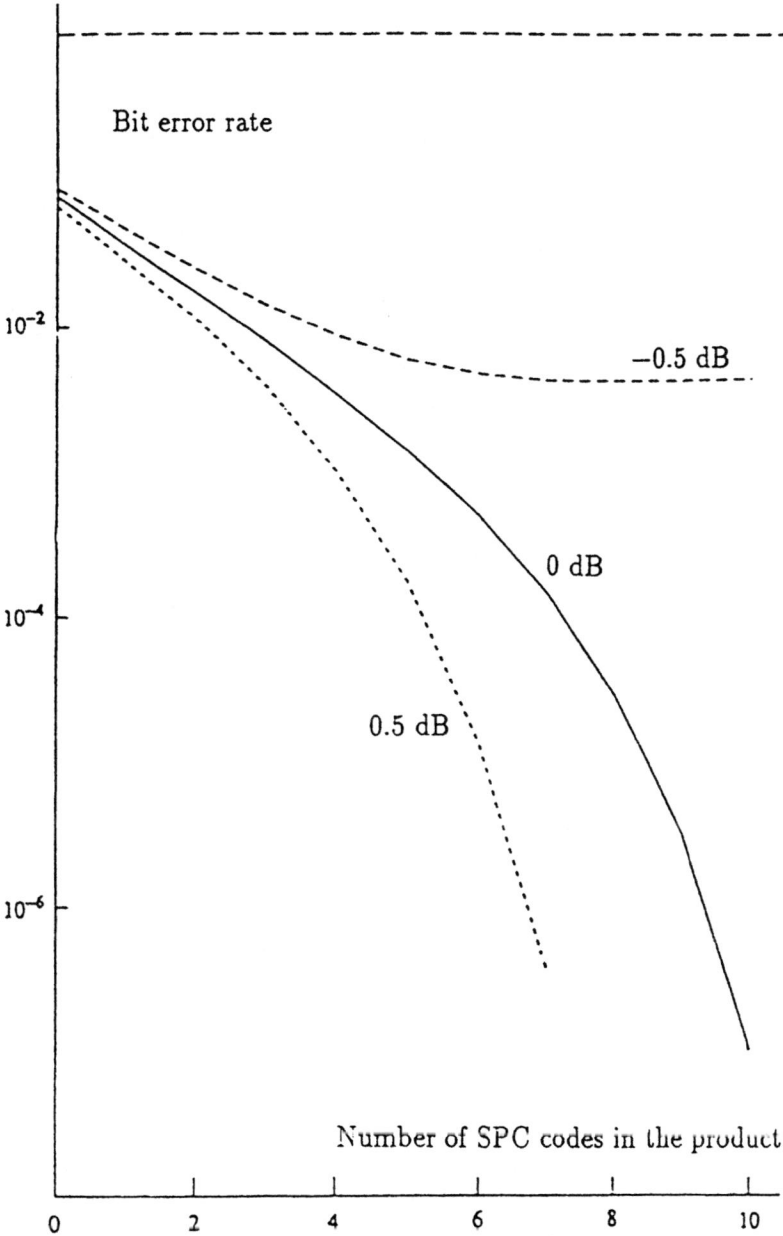

Fig. 2. BER of the product $(k_0 + 1, k_0) \otimes (k_1 + 1, k_1) \otimes \cdots \otimes (k_{j-1} + 1, k_{j-1})$, with $k_0 = 2$ and $k_i = 2k_{i-1} = 2^i k_0$, $i = 0, 1, \ldots, j - 1$, in terms of the number j of codes in the product. The channel signal-to-noise ratio (SNR) is indicated as a parameter. If the SNR is above a threshold slightly below $0\,$dB, the BER vanishes as the number of codes in the product increases.

additive Gaussian channel resulted in a vanishingly small bit error rate near the channel cutoff rate, as the number of codes in the product increased. As an example, the performance of the iterated product of SPC codes with $k_i = 2k_{i-1} = 2^i k_0$ and $k_0 = 2$ is shown in Fig. 2. As another example, the hypercube SPC code with $n = 6$ and $m = 5$ is a $(7776, 3125)$ code of rate about 0.4019. Its minimum distance is $d_{\min} = 32$. The proportion of codewords of minimum weight is amazingly small: $1.45\,10^{-935}$. At a signal-to-noise ratio of $-2\,\mathrm{dB}$ ($2.29\,\mathrm{dB}$ better than the channel capacity), a bit error rate of $2.17\,10^{-3}$ was obtained; an error pattern had then an average weight of about 53.

This good performance was quite unexpected, because such codes are bad according to the conventional minimum distance criterion since the ratio d/n of their minimum distance d to their length n approaches 0 as n tends to infinity. Obtaining *good* performance by the use of *bad* codes leads obviously to the suspicion that it is the criterion rather than the codes that is bad!

2.2   Random-Like Criterion

The minimum distance criterion cannot indeed be entirely justified, although there exist good reasons in favour of it. Criticizing it would have been too negative without proposing a hopefully better alternative. We proposed to look for *random-like* codes i.e., for codes with a distance distribution close to that which is obtained in the average by random coding [10]. For the sake of brevity (with some abuse of language), this criterion will be referred to as random-like, too.

The motivation for proposing the random-like criterion was twofold. First, a heuristic analysis of the distance distribution of the iterated product of SPC codes, based on a decimation argument, showed that it was indeed close to that of random coding, a statement later shown to be true by Caire, Taricco and Battail [11], and Biglieri and Volski [12], using more refined means. Second and more important, the role of random coding in the proof of the channel coding theorem of Information Theory strongly suggests that a random-like code should be "good," at least in a more comprehensive sense than according to the conventional minimum distance criterion.

2.3   Strongly Random-Like Codes

Let us consider a linear binary (n, k) code. Assuming linearity enables considering the weight distribution of all its words instead of the distance distribution of all its pairs of words, for the sake of simplicity. Let $A(w), w = 0, 1, \ldots, n$, denote the number of its words of weight w, and $a(w) \overset{\triangle}{=} 2^{-k} A(w)$ denote its normalized weight distribution.

There are several manners to specify that the weight distribution of this code is "close" to that of random coding. The most stringent one consists of assuming that each $A(w)$ is close to the average number of words of weight w obtained in the average by random coding namely, for an (n, k) binary code, $R(w) = 2^{-(n-k)} \binom{n}{w}$. For instance, $A(w)$ may be the nearest integer to this average number. One should then have $A(w) = 0$ if $2^{-(n-k)}\binom{n}{w} < \lambda$, where λ is some positive constant e.g., $1/2$, for an approximation to the nearest integer. The first non-zero coefficient $A(d)$ thus corresponds to the smallest weight d such that $2^{-(n-k)}\binom{n}{d} \geq \lambda$. Assuming that n and d are as large as to enable using the Stirling approximation, d is the smallest integer such that

$$\frac{n^{n+1/2}}{\sqrt{2\pi}} d^{d+1/2}(n-d)^{n-d+1/2} \geq \lambda 2^{n-k} \ .$$

Taking logarithms to the base 2 and dividing by n results in

$$(1 + 1/2n)\log_2 n - (1/2n)\log_2(2\pi) - (d/n + 1/2n)\log_2 d$$
$$- (1 - d/n + 1/2n)\log_2(n-d) \geq 1 - k/n + (1/n)\log_2 \lambda \ .$$

Letting $d/n = \delta$, $k/n = R$ and neglecting the terms in $1/n$ results in the inequality:

$$\mathcal{H}_2(\delta) \geq 1 - R \ , \tag{1}$$

where $\mathcal{H}_2(x) \overset{\triangle}{=} -x\log_2(x) - (1-x)\log_2(1-x)$ is the entropy of a binary random variable one of the outcomes of which occurs with probability x, expressed in binary unit. Notice that λ no longer appears in (1). This inequality turns out to be the Gilbert-Varshamov bound i.e., a lower bound on the largest possible minimum distance. (This result has been known for many years e.g., [13], but we believe it was interesting to give this very simple proof of it.) Therefore, a code which is random-like in this sense, to be referred to as *strongly random-like*, is good also for the minimum distance criterion as meeting the Gilbert-Varshamov bound. We may define the Gilbert-Varshamov distance of an (n, Rn) code as $d_{\text{GV}} \overset{\triangle}{=} n\mathcal{H}_2^{-1}(1 - R)$ where $y = \mathcal{H}_2^{-1}(x)$, $0 < y \leq 1/2$, is such that $x = \mathcal{H}_2(y)$. For a q-ary linear code, the same argument results in the Gilbert-Varshamov bound in the form

$$\mathcal{H}_q(\delta) + \delta\log_q(q-1) \geq 1 - R \ , \tag{2}$$

with $\mathcal{H}_q(x) \overset{\triangle}{=} -x\log_q(x) - (1-x)\log_q(1-x)$. Now $d_{\text{GV}} = n\delta_0$, where δ_0 is the smallest positive solution of the implicit equation which results from (2) by substituting $=$ for $\geq$.

A strongly random-like code can be represented as a constellation in an n-dimensional space with the Hamming metric, as schematically shown in Fig. 3-A: the closest point to the transmitted one is at least at the Gilbert-Varshamov distance d_{GV}. This situation appears as ideal since this code would reach the channel capacity when the code length n approaches infinity. No explicit construction of a strongly random-like code is known in general for some given length n and rate R. Therefore, an actual implementation will most often lead to the consideration of means for relaxing the constraints on the weight distribution with respect to the strongly random-like case.

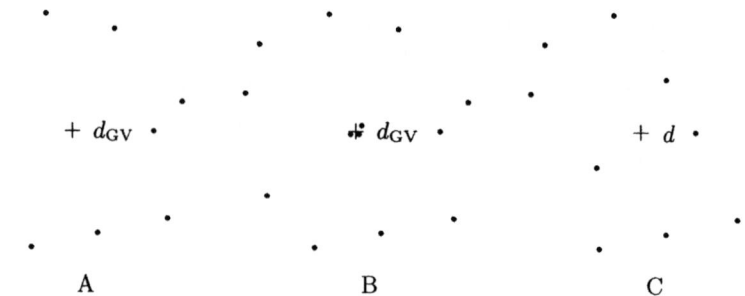

Fig. 3. Three constellations schematically representing codes in the n-dimensional space. The transmitted codeword is represented by +, other codewords by dots. (Distances as drawn here are meaningful only with respect to the transmitted codeword.)
A: strongly random-like code. The distance distribution of points with respect to the transmitted one is the same as that of random coding in the average. The nearest point to the transmitted one is at least at the Gilbert-Varshamov distance, d_{GV}, from it.
B: weakly random-like code. Almost the same as the previous distribution. Almost all points are farther from the transmitted one than the Gilbert-Varshamov distance d_{GV} (as in A) except for a few close neighbours.
C: code designed according to the conventional minimum distance criterion. The points are separated by at least a given distance d, but the remainder of the distribution is not controlled.

2.4 Weakly Random-Like Codes

Rather than the very stringent strongly random-like criterion, we suggested in [10] to use a proximity measure of the *normalized* weight distribution with respect to that of random coding like the cross-entropy (also referred to as Kullback divergence) which guarantees only a global shape similarity, especially in the central part of the distribution. Let us recall that if p and q are two discrete distributions $p = \{p_1, p_2, \ldots, p_n\}$ and $q = \{q_1, q_2, \ldots, q_n\}$ with $\sum_{i=1}^{n} p_i = \sum_{i=1}^{n} q_i = 1$ and q_i being assumed to be non-zero for any i, then the cross-entropy of p and q is defined as $H(p,q) \triangleq \sum_{i=1}^{n} p_i \ln(p_i/q_i)$. Then, a code C is random-like according to this proximity measure for a given positive constant ε if $H(a, r) < \varepsilon$, where $a = \{a(0),\ a(1),\ \ldots, a(n)\}$ is the normalized weight distribution of code C and $r = \{r(0),\ r(1),\ \ldots, r(n)\}$, with $r(w) = 2^{-n}\binom{n}{w}$, is the normalized average distance distribution of random coding.

Due to the factor p_i in the i-th term of the sum which defines the cross-entropy $H(p,q)$, a small value of the cross-entropy implies a much better fit of the distributions p and q where p_i is large than where it is small i.e., for "bell shaped" ones, in the central part rather than in the tails. Therefore, a code may be good in this sense for the random-like criterion but bad according to the conventional minimum distance one, which precisely concerns the low-side tail of a weight distribution.

Such a code will be referred to as *weakly random-like*. The product of SPC codes is an example of this case where $d = 2^m$ is the minimum distance, with m the number of codes in the product. It is smaller than the Gilbert-Varshamov distance d_{GV} if the code is long enough. For instance, the hypercube product $(6, 5)^{5\otimes}$ already mentioned has a minimum distance of $2^5 = 32$, while the Gilbert-Varshamov distance for the same parameters, $(7776, 3125)$, is $d_{GV} = 1,130$.

Cross-entropy was proposed because it is a meaningful quantity of Information Theory, but it needs an adaptation to local details like the possible cancellation of certain weights e.g., of all odd weights [14]. Other proximity measures may be used, especially those involving cumulative distributions which are simpler as avoiding the previously mentioned difficulty.

2.5 Relaxing the Constraints of a Strongly Random-Like Distribution

The minimum distance is the most sensitive parameter of the weight distribution of a linear code as regards the *word error rate* [15], and the conventional criterion aims at obtaining it as large as possible. The only feature of the random coding average weight distribution which is retained is thus its large minimum distance. Instead of the Gilbert-Varshamov distance d_{GV}, however, the closest point to the transmitted one is generally at some smaller minimum distance $d < d_{GV}$ since most code families do not reach the Gilbert-Varshamov bound. Moreover, the number of points at minimum distance may be large if the minimum distance criterion is used without care. The constellation in this case is depicted in Fig. 3-C.

On the other hand, the weight distribution of a weakly random-like code is close to that of random coding in its central region, but in the tails it is the more different, the normalized distribution is the smaller. Can codes be good with a weakly random-like distance distribution i.e., with a low-distance tail markedly different from that of random coding, possibly resulting in a poor minimum distance? The performance of the iterated product of SPC codes suggests a positive answer *provided* that one uses the bit error rate (BER) as a quality criterion rather than the word error rate: at a low signal-to-noise ratio, these codes have a small BER but a large word error rate[1]. The BER appears indeed as a more natural measure of performance, especially for long codes. For instance, it is the standard performance measure used in the field of convolutional coding. It actually specifies the performance in almost all engineering applications.

In the weakly random-like case, the constellation is as in Fig. 3-B. Almost all points are located with respect to the transmitted word as shown in Fig. 3-A, except that very few points are close to the transmitted one. At a small signal-to-noise ratio, an error in favour of such points is *likely*. It is however *slight*, in the sense that it results in a small bit error rate[2].

---

[1] A more general quality criterion would be that decoding as extended to reevaluate the word probabilities (as discussed in Sect. 2.1) reduces the uncertainty concerning the transmitted word by a large enough amount.

[2] This conclusion holds regardless of the channel metric in the binary case only. No

Decoding will in this case result in a large word error probability. However, one may interpret decoding, in accordance with the extension of its role which led to start the study of the iterated product of SPC codes (as discussed in Sect. 2.1), as intended to diminish the uncertainty regarding the knowledge of the transmitted word (instead of the conventional point of view i.e., almost always suppressing it except if an error occurs). Then almost surely knowing that the transmitted word belongs to a small subset of all codewords may be considered as a satisfying result. Moreover, at least in the binary case, the words of this subset are at a small distance of the transmitted word. Therefore, a word error results in some small BER which may be, or not, acceptable for the user depending on its own specification.

We already noticed that the engineering problem consists most often of achieving a bit error rate lower than a prescribed value. A fast improvement of the bit error rate as the channel signal-to-noise ratio increases is seldom an advantage, and the very small error probabilities eventually achieved become deprived of any physical meaning. Moreover, it has as a counterpart a steep increase of the bit error rate if the signal-to-noise ratio falls below some threshold. In many problems, for instance in the case of speech communication, it is rather a "graceful degradation" which is sought. The concept of "asymptotic gain" has in this context very little meaning, since it enables comparing coding schemes at large signal-to-noise ratios i.e., when they are of no use.

Even if the user's BER specification is not met for a given weakly random-like code, the knowledge of a small subset which contains almost certainly the transmitted word, instead of this transmitted word alone provided no error occurred, can easily be used to achieve a tighter performance. The key word here is *concatenation*. Instead of an arbitrary information sequence, one may choose a sequence which belongs to some known subset of all sequences e.g., which results from some outer coding. Then, it becomes much less likely that an erroneous sequence belongs to both this subset and the vicinity of the transmitted word which is known after decoding, so the uncertainty about the transmitted word is widely diminished. This process can be extended to an arbitrary number of codes while the rate is kept constant by puncturing, as will be discussed with more detail in Sect. 4.3.

Incidentally, the cases where a fairly large BER may be specified e.g., speech transmission, can generally be interpreted as actually involving concatenation. In order to describe any speech transmission system, one may indeed include in it the speaker and the listener. A complex concatenated scheme results where the technical link implements only an inner code, while outer linguistic codes involve redundancy at the phonetic, grammatical and semantic levels. Almost error-free transmission is eventually obtained despite the relatively high BER in the inner link.

---

such simple relation exists, for instance, if the channel metric is Euclidean and the alphabet q-ary, with a one-to-one mapping of the alphabet symbols into the points of some constellation.

3 Qualitative Behaviour of Random-Like Codes

We shall now discuss the qualitative behaviour of random-like codes. To this end, we shall use union bounds. Such bounds are poor at a low signal-to-noise ratio and when the codewords are very many i.e., precisely in the conditions of interest to us. They are nevertheless helpful for a qualitative understanding of the phenomena, although they are not for a precise performance prediction nor for designing code parameters.

3.1 Union Bounds

The union bound on the word error rate (WER), P_W, is:

$$P_W \leq \sum_w A(w) P(C_w|C_0) \ , \tag{3}$$

where $A(w)$ is the number of words of weight w and $P(C_w|C_0)$ is the probability of receiving a particular codeword C_w of weight w when the all-zero word C_0 is transmitted. For the additive Gaussian channel, we have:

$$P(C_w|C_0) = \frac{1}{2} \mathrm{erfc}(\sqrt{wE_s/N_0}) \ , \tag{4}$$

where E_s is the received energy per channel symbol and N_0 the one-sided spectral noise density. The "complementary error function" is defined as

$$\mathrm{erfc}(x) \stackrel{\triangle}{=} \frac{2}{\sqrt{\pi}} \int_x^\infty \exp(-t^2) \mathrm{d}t \ .$$

The bound on the bit error rate, P_B, corresponding to (3) is:

$$P_B \leq \sum_w \frac{w}{n} A(w) P(C_w|C_0) \ , \tag{5}$$

since assuming without loss of generality the code to be systematic an information bit is as likely to be in error as any other bit in the word.

3.2 Bounds on the Word and Bit Error Rates in Terms of the Signal-to-Noise Ratio: an Example

Word error rate. A rough but very convenient approximation of the complementary error function is well known to be:

$$\mathrm{erfc}(x) \approx \exp(-x^2) \ .$$

Using this approximation, we may rewrite (3) as:

$$P_W \lesssim \frac{1}{2} \sum_w A(w) \exp(wE_s/N_0) = \sum_w \exp(a_w - \gamma w) \ , \tag{6}$$

the equality being obtained by letting $a_w = \ln[A(w)/2]$ and $\gamma \triangleq E_s/N_0$. Let p_w be the term in w in the right hand side of (6). In a Cartesian diagram with

$$y_w \triangleq \ln p_w = a_w - \gamma w$$

plotted in terms of γ, each term in the right hand side of (6) may thus be represented by a straight line, as in Fig. 4-A. The code considered as an example for drawing this figure was the product $(28,18)$ of the $(4,3)$ and $(7,6)$ SPC codes. The weight enumerator polynomial of this code is [10]: $1 + 126z^4 + 840z^6 + 6{,}531z^8 + 23{,}520z^{10} + 59{,}892z^{12} + 80{,}304z^{14} + 59{,}927z^{16} + 23{,}520z^{18} + 6{,}510z^{20} + 840z^{22} + 133z^{24}$, so we have $a_4 = 4.14$, $a_6 = 6.04$, $a_8 = 8.09$, $a_{10} = 9.37$, $a_{12} = 10.31$, etc. The actual bound is represented by a curve above these lines, close to the highest one if the others are low enough. Of course, 1 is an upper bound on any probability, so the plots are significant, and have been drawn, only under the γ-axis. It is clear in this figure that the WER curve is dominated by the lowest term y_4, regardless of the signal-to-noise ratio γ.

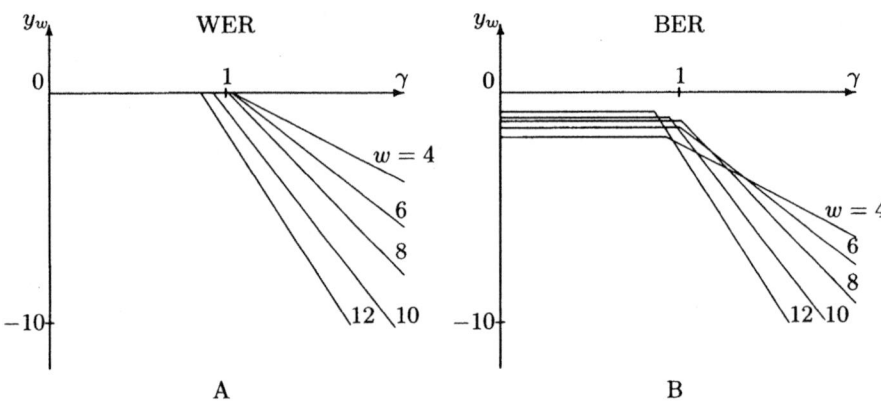

Fig. 4. Straight line representation of union bounds.
A: Representation of terms of (6), for the $(28,18)$ product of SPC codes $(4,3)$ and $(7,6)$. Each line corresponds to a term in the right hand side.
B: Representation of terms of (7) for the same code. Each oblique line corresponds to a term in the right hand side. Horizontal lines representing the obvious upper bound w/n on this term have also been drawn.

Bit error rate. Similarly, we may approximately rewrite (5) as:

$$P_B \overset{<}{\sim} \sum_w \exp[a_w - \gamma w + \ln(w) - \ln(n)] \ . \tag{7}$$

Now, we represent each term of the right hand side of (7) by a line of equation

$$y_w \stackrel{\triangle}{=} a_w - \gamma w + \ln w - \ln n = b_w - \gamma w \ ,$$

where $b_4 = 2.19$, $b_6 = 4.50$, $b_8 = 6.84$, $b_{10} = 8.34$, $b_{12} = 9.46$, etc for the same code as above. An obvious upper bound on this term is also w/n, so we obtain Fig. 4-B. We see now that y_4 dominates above some value of γ, but other terms are more significant at a smaller signal-to-noise ratio.

One may notice also that the coefficients $\{A(w)\}$ are exactly or approximately symmetrical with respect to $n/2$. Therefore, the corresponding $\{a(w)\}$ are nearly the same for symmetrical weights. Since $b(w)$ differs from $a(w)$ by the relatively small amount $\ln(n) - \ln(w)$, the straight line of slope $-(n-w)$ will generally be well below the line of slope $-w$ if $w < n/2$ (below the γ-axis), which shows that only the weights smaller than $n/2$ are actually significant.

3.3 A Crude Model of a Weakly Random-Like Code

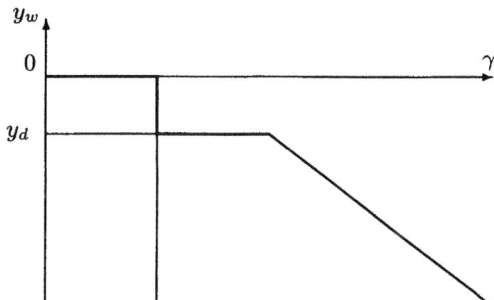

Fig. 5. Straight line representation of the bit error rate for the model of Sect. 3.3. The actual curve is close to the thick broken line, where the leftmost very steep line (actually drawn as vertical) represents the performance of random coding, the rightmost line has slope $-d$, and a horizontal line in the middle at $y_d = \ln(d/n)$ corresponds to the limiting BER d/n.

As a very crude model of the weakly random-like case, we may consider a code with a weight distribution equal to the true normalized distribution of random coding, multiplied by a coefficient close to 1, say $1 - \varepsilon$, plus a single coefficient $\varepsilon \ll 1$ at a weight $d \ll d_{GV}$. This code has thus an almost random-like weight distribution, except for a fraction ε of all codewords which has the small weight d. Then, as shown in Fig. 5, the figure representing the terms y_w as a function of γ reduces to a very steep curve at a low signal-to-noise ratio, which corresponds to the performance of random coding, and to a much less steep line, of slope $-d$, at a higher signal-to-noise ratio, which corresponds to the few words at distance d. For a small enough value of ε, the dominating term at an intermediate signal-to-noise ratio will be the horizontal line at $y_d = \ln(d/n)$, corresponding to the obvious bound d/n on the bit error rate. The overall BER curve will thus be very steeply descending at a low signal-to-noise ratio, slowly descending with slope $-d$ at a high signal-to-noise ratio and, if the lines are located as in the figure, horizontal at y_d in the middle. Similar shapes were observed in the simulation of turbo-codes, especially by Robertson [16] and Svirid [17]. Although oversimplified, this model of weakly random-like codes thus seems to adequately describe the qualitative behaviour of certain actual codes.

4 A Review of Some Random-Like Codes

We shall now review some code families which can be considered as random-like in the sense we just defined.

4.1 Random-Like Block Codes

We already mentioned the iterated product of SPC codes which was the starting point of our research, used above as an example of weakly random-like block codes.

Let us also mention among binary block codes the family of *strongly* random-like autodual codes recently investigated by Kalouti, Lazić and Beth [14].

As another block code family, the *Maximum Distance Separable* codes, which have the largest possible distance d according to the Singleton bound namely, $d = n - k + 1$, where n is the code length and k its dimension, are other examples of random-like block codes since their Hamming weight distribution approaches that of random coding for n and k large enough [18]. The interesting codes of this family are non-binary, the best known ones being the Reed-Solomon codes. We tried to combine such a code with a q-PSK modulation, q being the alphabet size, which results in a random-like code for the Euclidean metric fitted to the Gaussian channel. We shall not here discuss this topic in spite of its interest (it has already been discussed in [19, 20] and more recently in [21]).

4.2 Pseudo-Random Recursive Convolutional Codes

May be the most important applications of the concept of random-like codes are found in the field of convolutional codes. A convolutional encoder is just considered here as a simple means for generating coded blocks of large but finite length, as in any practical situation. We shall first consider the case of a single convolutional code and show that only *recursive* convolutional codes can be good for the random-like criterion.

Let us consider a systematic convolutional encoder of rate $1/2$ and memory K which encodes a block of L information bits, with $K \ll L$. We may thus think of coding as a means for associating a redundancy vector of length L with any information vector of same length (we neglect the termination). Graphically, let us

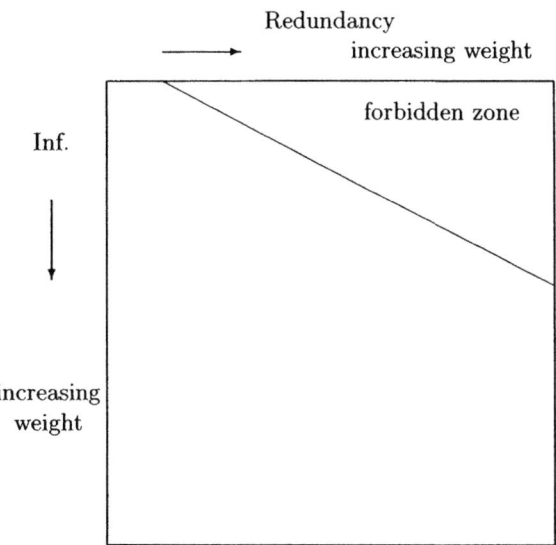

Fig. 6. Array representing non-recursive systematic coding. The border of the forbidden zone at the right top of the array is much more complicated than the straight line drawn.

consider an array of $2^L \times 2^L$ points, with its rows indexed by all information vectors and its columns by all redundancy vectors, both ordered by non-decreasing weight. Coding then consists of choosing a point in each row so as to tell what redundancy vector is associated with the corresponding information vector. If the encoder is non-recursive, the weight of the redundancy vector is at most equal to the length of the information vector plus K. Therefore, *small weight* redundancy vectors are associated with *small weight* information vectors, and there exists in the array a forbidden zone (near the upper right corner) where no point can be found (see Fig. 6). This is not consistent with random-like coding, where the points should be evenly distributed in the array. On the other hand, if a recursive encoder is used, infinite-weight redundancy sequences (to be truncated to the length L) are associated with small-weight information vectors, so recursive convolutional codes can be random-like.

We defined *pseudo-random* codes which are systematic recursive convolutional codes with a weight distribution very close to that of random coding, except for a set of low weight sequences which can be made very small by increasing the encoder memory K (these sequences are a fraction of all sequences less than an exponentially decreasing function of K) [22, 23]. Such codes are thus weakly random-like in the above sense. As such, when used to build coded blocks of arbitrary large length, they exhibit non-zero block error rate but small bit error rate at low signal-to-noise ratios.

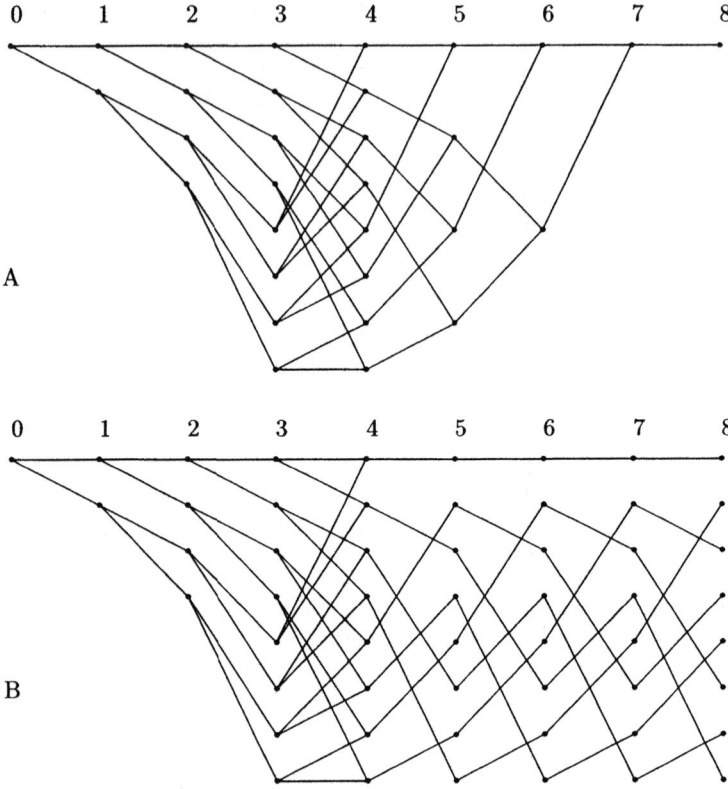

Fig. 7. Trellises representing the set of coded sequences corresponding to all informa-
tion sequences of length 4 for convolutional encoders of memory K=4.
A: non-recursive encoder.
B: recursive encoder.

These codes are systematic recursive with a generator matrix of the form

$$\mathbf{G} = [1 \quad N(D)/P(D)] \; , \qquad (8)$$

where D is the unit-delay operator, $P(D)$ is a primitive polynomial of degree K
and $N(D)$ is a polynomial of degree at most equal to K (the constraint length is
thus $K + 1$). The encoder of such a code constitutes, in the absence of an input
information sequence but assuming the encoder initial state to be different from
0, a generator of "maximum length sequences" of period $2^K - 1$, often referred to
as *pseudo-random*, so we shall refer to this encoder (and to the code it generates)
as pseudo-random, too.

Then, infinite-weight sequences are generated in response to all information sequences, whatever their weight, provided their polynomial representation in terms of the unit delay operator D is not a multiple of $P(D)$. Almost all generated sequences have thus infinite weight for K large enough.

The choice of K large, which is necessary in order to obtain a reasonable performance, entails that the Viterbi algorithm is far too complex for decoding such codes. We therefore used the *replication decoding* algorithm, the complexity of which is independent of the memory size K [4, 24]. It depends only on the chosen number of replicas, r, equal to the degree of $P(D)$ and $N(D)$ (assumed to be the same for keeping full symmetry between the information and redundancy bits, which are both decoded). We thus have $r \geq 3$ since no irreducible polynomial of weight 2 exists for $K > 1$. The operation of this algorithm can be represented using a trellis with 2^r states. Limiting the complexity leads to the use of small values of r, which results in a low free distance.

With a limited number of replicas, this algorithm takes into account a fairly small set of received bits, or context, while each bit actually depends on an infinite context. This problem may be solved by the *iteration* of decoding, made possible by a normalization after each stage which essentially removes the dependence due to previous decodings.

Still another problem is that a recursive code shares the same infinite trellis with the non-recursive equivalent one, of generator matrix resulting from (8) by multiplication by the scalar $P(D)$ [25]:

$$\mathbf{G} = [P(D) \quad N(D)] \ .$$

The difference lies in the way of associating the finitely many information vectors to certain paths of the trellis, as illustrated in Fig 7. It is not clear as yet if decoders can be designed to take into account the specific distance distribution properties of recursive codes, which are described only by statistical means.

4.3 Turbo-Codes

The BER of a single pseudo-random code may thus be poor if the code has been chosen such that the number of available replicas r is small. Its performance can however be improved by the use of concatenation according to the *turbo-code* scheme, as illustrated in Fig. 8. The weight distribution remains almost the same in its central part (where it is already very close to that of random coding), but it becomes also much closer to it in its tails. In effect, if we interpret interleaving as performing a random permutation, the weight distribution of the redundancy sequence (before puncturing) is the *convolution product* of those of the component codes, analogously to the probability distribution of the sum of independent random variables. Therefore, it tends to the Gaussian distribution as the number $N + 1$ of component codes approaches infinity as a consequence of the central limit theorem. The Gaussian distribution is itself the limiting distribution of a large random sequence. Choosing the number N thus enables to adjust the BER to any given specification consistent with the channel capacity.

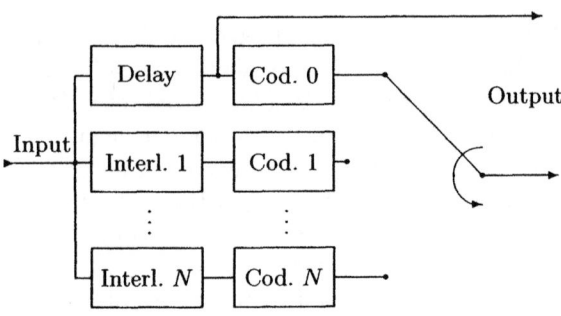

Fig. 8. Encoder of a pseudo-random turbo-code of rate 1/2. The boxes labelled "Cod. 0, Cod. 1, . . . , Cod. N" generate the redundancy sequences associated with the input sequences according to a pseudo-random code of rate 1/2. The boxes labelled "Interl. 1, . . . , Interl. N" represent interleavers, the delay of which is compensated for in the box labelled "Delay". Periodic puncturing keeps the rate to its initial value 1/2.

Inasmuch as the redundancy sequences are random-like, periodic puncturing clearly keeps their random-like property.

The properties of recursive convolutional codes as well as the convolution of the weight distribution which results from the "parallel concatenation" and interleaving seem to be the main keys for understanding the performance of turbo-codes. Incidentally, turbo-codes also provide an outstanding illustration of the benefit of soft-output decoding, as discussed in Sect. 2.1. One may notice also that the turbo-code scheme is somewhat analogous to the product of SPC codes also discussed in this section. The latter scheme combines coding and interleaving several times, too, the main difference being that both functions are implemented in a more refined fashion in the former.

Of course, using recursive pseudo-random convolutional codes as components in the turbo-code scheme will result in a weight distribution very close to that of random coding for small values of the number $N + 1$ of component codes, even for $N = 1$, since the initial weight distribution of each of these codes will itself be close to that of random coding except in the distribution tails. However, the central limit theorem results in an asymptotically random weight distribution regardless of the initial one, as N tends to infinity, so the turbo-code scheme with many component codes may be considered as *intrinsically random-like*.

5   Conclusion

As claimed in the introduction, random-like coding appears as an unusual way for designing codes. Designing strongly random-like codes, hence with very small word error rate, seems very difficult although results have already been reported [14]. Weakly random-like coding appears as easier, but results at low signal-to-noise ratios in a small bit error rate rather than a small word error rate.

Schemes which combine concatenated coding and interleaving, like the turbo-codes, can be used for improving the bit-error-rate performance, as tightening the random-like properties in the tails of the weight distribution. Besides splitting decoding into a few simple operations, to the benefit of an overall low complexity, concatenation associated with interleaving thus performs the important task of random-like shaping the weight distribution.

Needless to say that much work remains to be done to refine the proximity criterion with respect to random coding, find good code families for this criterion, analyse the properties of random-like codes, assess their performance, and design systems for their implementation, to cite only a few points worth further study. This may initiate a renaissance of coding techniques with criteria and specifications closer to both the spirit of Information Theory and the engineering needs.

Acknowledgements. The paper was improved thanks to helpful suggestions of anonymous referees.

References

1. Berrou, C., Glavieux, A., Thitimajshima, P.: Near Shannon Limit Error-Correcting Coding and Decoding : Turbo-Codes. Proc. ICC'93, Geneva, Switzerland (1993) 1064–1070
2. Battail, G.: Le décodage pondéré en tant que procédé de réévaluation d'une distribution de probabilité. Ann. Télécommunic. **42** (1987) 499–509
3. Bahl, L. R., Cocke, J., Jelinek, F., Raviv, J., Optimal Decoding of Linear Codes for Minimizing Symbol Error Rate. IEEE Trans. Inf. Th. **IT-20** (1974) 284–287
4. Battail, G., Decouvelaere, M.: Décodage par répliques. Ann. Télécommunic. **31** (1976) 387–404
5. Battail, G.: Pondération des symboles décodés par l'algorithme de Viterbi. Ann. Télécommunic. **42** (1987) 31–38
6. Hagenauer, J., Hoeher, P.: A Viterbi algorithm with soft-decision outputs and its applications. Proc. GLOBECOM'89, Dallas, U.S.A. (1989) 47.1.1–47.1.7
7. Hoeher, P.: Advances in soft-output decoding. Proc. GLOBECOM'93, Houston, U.S.A. (1993) 793–797
8. Silverman, R. A. Balser, M.: Coding for constant data rate systems. Part I. A new error-correcting code. Proc. IRE (1954) 1428–1435
9. Elias, P.: Error-Free Coding. IRE Trans. Inf. Th. (1954) 29–37
10. Battail, G.: Construction explicite de bons codes longs. Ann. Télécommunic. **44** (1989) 392–404
11. Caire, G., Taricco, G., Battail, G.: Weight distribution and performance of the iterated product of single-parity-check codes. Proc. GLOBECOM'94, Communication Theory Mini-Conference, San Francisco, U.S.A. (1994) 206–211, and Ann. Télécommunic. **50** (1996) 752–761
12. Biglieri, E., Volski, V.: The weight distribution of the iterated product of single-parity-check codes is approximately Gaussian. Electronics Letters **30**, 9th June 1994, 923–924
13. Pierce, J. N. Limit distribution of the minimum distance of random linear codes, IEEE Trans. Inf. Th. **IT-13** (1967) 595–599

14. Kalouti, H., Lazić, D. E., Beth, T.: On the relation between distance distributions of block codes and the binomial distribution. Ann. Télécommunic. **50** (1996) 762–778

15. Lazić, D. E., Šenk, V.: A direct geometrical method for bounding the error exponent for any specific family of channel codes – Part I: Cutoff rate lower bound for block codes. IEEE Trans. Inf. Th. **38** (1992) 1548–1559

16. Robertson, P.: Illuminating the structure of code and decoder of parallel concatenated recursive systematic (turbo) codes. Proc. GLOBECOM'94, San Francisco, U.S.A. (1994) 1298–1303

17. Svirid, Y. V.: Weight distributions and bounds for turbo-codes. Europ. Trans. on Telecommunic. **6** (1995) 543–556

18. Cheung, K.-M.: More on the decoder error probability for Reed-Solomon codes. IEEE Trans. Inf. Th. **35** (1989) 895–900

19. Battail, G., Magalhães de Oliveira, H., Zhang, W.: Codage déterministe imitant le codage aléatoire pour le canal à bruit gaussien additif. Ann. Télécommunic. **47** (1992) 443–447

20. Battail, G.: We can think of good codes, and even decode them. Springer, CISM Courses and Lectures **339** (1993) 353–368

21. Battail, G.: Direct combination of Reed-Solomon encoding over $GF(q)$ and q-PSK modulation. Proc. of IEEE Inf. Th. Workshop, Rydzyna, Poland (1995) 4.1

22. Battail, G., Berrou, C., Glavieux, A.: Pseudo-random recursive convolutional coding for near-capacity performance. Proc. GLOBECOM'93, Communication Theory Mini-Conference, Houston, U.S.A. (1993) 23–27

23. Battail, G.: Codage convolutif récursif pseudo-aléatoire. Ann. Télécommunic. **50** (1996) 779–789

24. Battail, G., Decouvelaere, M., Godlewski, P.: Replication Decoding. IEEE Trans. Inf. Th. **IT-25** (1979) 332–345

25. Forney G. D. Jr: Convolutional Codes I: Algebraic Structure. IEEE Trans. Inf. Th. **IT-16** (1970) 720–738

An Alternative Approach to the Design of Interleavers for Block "Turbo" Codes

Paul Guinand[1] , John Lodge[1] and Lutz Papke[2]

[1] Communications Research Centre
3701 Carling Avenue, Ottawa Canada K2H 8S2
(tel) 613-998-2284, (fax) 613-990-6339
[2] Institute for Communications Technology German Aerospace Research
Establishment D-82230 Wesling P.O. Box 1116, Germany

Abstract. In this paper we describe two methods of interleaving for use with generalized product block codes that can be decoded using a "turbo" decoding strategy. These methods are designed to avoid certain specific error patterns. One method has a number theoretic basis while the other uses finite projective planes. Simulation results are presented.

1 Introduction

Recently [1-3] there has been considerable development in the use of iterative decoding techniques for complex codes which are constructed in some fashion from simpler codes. The rationale for the use of these constructions is that the resultant code although very powerful still allows for efficient decoding. For the most part these codes involve coding the data bits using the simple constituent code (almost always systematic) and then interleaving the data bits and applying the same code again to the interleaved bits. Typically the method uses some form of a maximum a posteriori (MAP) algorithm to do the decoding. The resulting decoding method is suboptimal but because the codes which are produced are usually more powerful than any codes that one could optimally decode the results are very good. The term "turbo" code is being used by a number of authors to describe codes to which such an iterative MAP decoding strategy can be applied. In this paper we describe an approach to constructing generalizations of product block codes which admit such a strategy.

2 Description of the Problem

The most obvious way to construct, from simple block codes, a more powerful code to which a turbo decoding strategy may be applied is to form a product code. That is take $k_1 \mathrm{x} k_2$ information bits, form them into a rectangular $k_1 \mathrm{x} k_2$ array and apply a (n_1, k_1) code to the rows and a (n_2, k_2) code to the columns. Fig. 1 illustrates this construction. The most classical version of a product code includes the parity of parity bits (the parity bits in the dotted box). The MAP algorithm is somewhat simpler if one omits these parity bits and it is this version of a product code which we will consider and generalize. Thus the product code we have in this case is a (40,16) code. The problem that arises with this approach is that the interleaving – the division into rows and columns – allows certain error patterns to have too localized an effect. In particular a rectangular error pattern of four errors, as shown in Fig. 1, introduces two errors into two rows and into two columns. If one is using a Hamming code, which will only correct single errors, this error pattern is not correctable. Such error patterns are the primary source of errors for product Hamming codes at high signal to noise ratio.

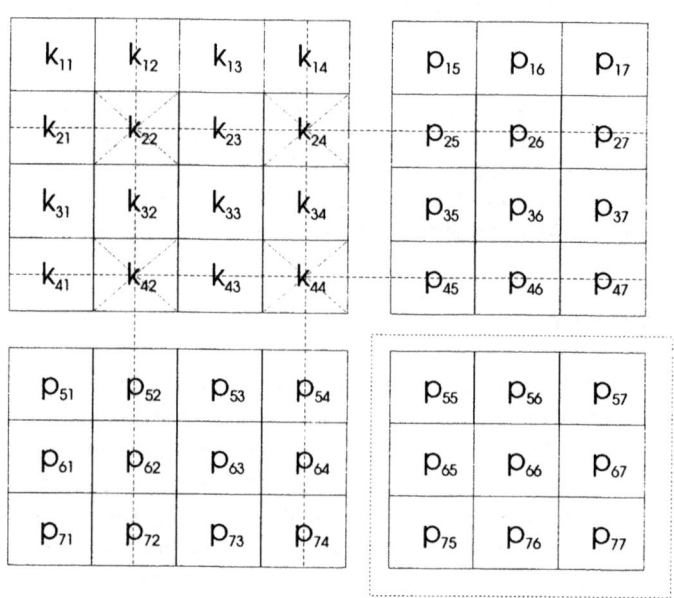

Fig. 1. Construction of product code showing rectangular error pattern

3   A Graphical Representation of the Problem

For reasons of expository clarity a graphical representation of the situation is useful. Given a set of m elements (corresponding to the data block) and two partitions P_1 and P_2 of it into k_1 element sets and k_2 element sets we define an "Incidence Structure" as follows. For each of the m/k_1 sets in the partition P_1 draw a point and associate to that point one of the sets in the partition P_2. Thus we have m/k_1 points each labeled with a set from P_1. For each of the m/k_2 sets in P_2 draw a line joining the points whose labels include the elements corresponding to this set from P_2. The resulting set of points and lines we call an incidence structure. In general, given arbitrary partitions this construction would give rise to lines that have the same point on them several times. This however corresponds to the case where two sets, one from each partition, overlap in several elements. This situation is obviously to be avoided from the coding point of view so we will restrict attention to the situation where each line has k_2 distinct points on it. Note each point has k_1 lines through it. Thus we will be interested in structures which have the following properties

1. given two points there is at most one line they are both on
2. each point lies on k_1 lines
3. each line has k_2 points on it

Given such a structure it is simple to recover a pair of partitions. The first partition amounts to assigning k_1 element subsets of distinct elements to each point in the incidence structure. Then to obtain the second partition you work your way methodically through the lines, constructing subsets from elements you have not already used in the second partition. If the constituent codes that you are using is are (n, k) codes the resulting code will be an $(m(1 + 2(n - k)/k), m)$ code. Note that the rate of the code is independent of the size of the set one is partitioning.

As a first example consider what happens if you consider the partitions corresponding to a normal $k \times k$ block code. The resulting incidence structure consists of k points with k lines going through all k points (cf. Fig. 2).

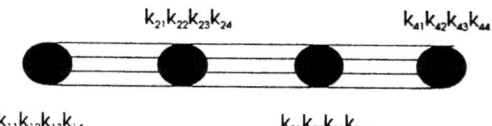

$$k_{21}k_{22}k_{23}k_{24} \qquad\qquad k_{41}k_{42}k_{43}k_{44}$$

$$k_{11}k_{12}k_{13}k_{14} \qquad\qquad k_3,k_{32}k_{33}k_{34}$$

Fig. 2. Incidence structure for product code

In terms of the incidence structure the condition which we wish to avoid (a "rectangular" error pattern) corresponds to the existence of two lines which intersect in more than one point. Higher order error patterns which one might wish to avoid also have natural expressions in terms of the incidence structure. For instance, hexagonal error patterns correspond to the existence of triangles in the incidence structure. For the purposes of the current study we have only considered the "rectangular" error patterns. Also we are, in the present work, primarily concerned with the case where k_1 and k_2 are the same.

4 Some Solutions

The first approach to the problem which we will consider has a number theoretic basis. We will initially consider the case where the constituent codes are (n, k) codes where k is a prime. In this approach we have $k \times k$ points in our configuration. Arrange these points as a $k \times k$ square array. From each point along the top of the array extend lines with slopes: vertical, -1, -1/2 etc. up to -1/(k-1), wrapping around as required from left to right of the configuration (cf. Fig. 3a). The fact that any two of these lines intersect in only a single point is a consequence of the fact that the congruence $a + bx = c \bmod p$, p **prime**, $b \neq 0$ has a unique solution. This construction was employed in [2]. In that paper the construction was also used for the case where k is not prime (in particular k=4). As can be seen from Fig. 3b this will substantially reduce the number of possible rectangular error patterns but it does not eliminate them. Probably the simplest way to extend this approach to eliminate rectangular error patterns for k non-prime is to go to the first prime larger than k construct the configuration, in the fashion described, and then remove some of the points from the lines. One has to remove the points in such a fashion as to end up with the same number of lines through each point. One way to do this is to construct a rectangular pxp array of points where p is the first prime larger than k. Construct the lines as above emerging from the first point in the first row. Then remove the first $p - k$ points from the vertical line – just remove them from the line not from the configuration. For the line of slope -1 leave the first point and then remove the next $p - k$ points. Continue this procedure wrapping around the omitted sets top to bottom as required. The lines emerging from the other points in the top row are then simply obtained by taking translates of the points coming from the first point (equivalently repeating the process). The result is a configuration with k points on each line, k lines through each point and no two lines meeting in more than one point.

The second method that we consider for the construction of the interleavers employs finite projective planes. For the fundamental definitions and results about finite projective planes see [4, 5]. That finite projective planes lead to interleavers with the desired property is an immediate consequence of the definition. A finite projective plane consists of a finite set P ("points") and a collection L ("lines") of subsets of P which satisfy

1. two points are on precisely one line
2. any two lines meet in precisely one point
3. every line has at least 2 points
4. there exist 4 points no 3 of which are collinear.

This construction is not possible for all sizes of codes – there do not exist projective planes of all orders. The order of a finite projective plane is the number of points on a line in the plane minus 1. An example of a projective plane of order 3 is shown in Fig. 4. This is a configuration of 13 points with 4 points on each line. In general a projective plane of order n has $n^2 + n + 1$ points. Thus a projective plane interleaver appropriate for an (n, k) code is generated by a configuration with $k^2 - k + 1$ points. This construction is notable in that the resulting interleaver will be the smallest possible interleaver that avoids rectangular error patterns. This is apparent by considering the associated configuration. A point in the configuration must have k lines through it each of which have $k - 1$ other points on them and none of these points can coincide. The only known projective planes are of prime power order so the codes for which these interleavers can be used directly have $k = p^s + 1$. One can of course play the game described

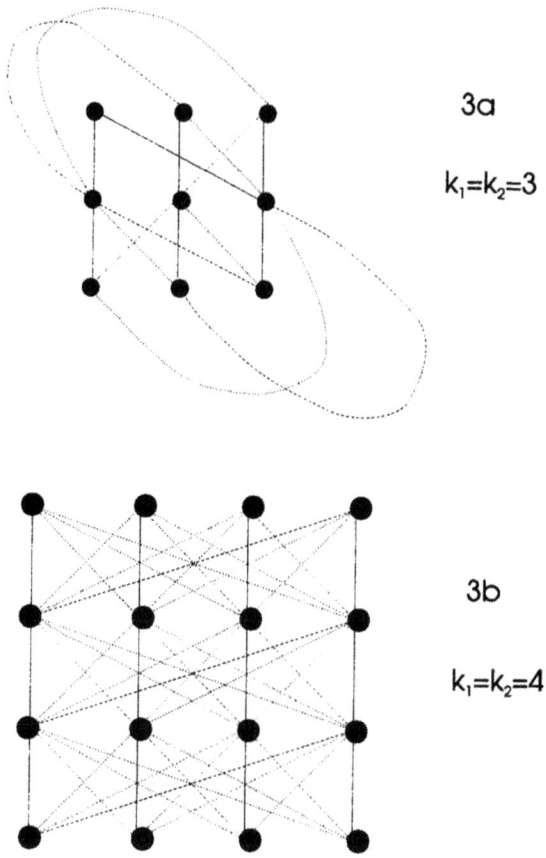

3a

$k_1=k_2=3$

3b

$k_1=k_2=4$

Fig. 3. Number theoretic interleaver construction

above of constructing a configuration with more points on the lines than desired and then remove points from the lines until each line has the desired number of points on it. The problem is that one has to have the same number of lines through each point. One way to do this, starting with a projective plane of order k, is to remove a line and all the points on it. The result is an affine plane with k points on each line and $k+1$ lines through each point. One can further reduce this configuration by removing sets of parallel lines. Each such removal decreases the number of lines through each point by one. The resulting interleavers are no longer necessarily the smallest which avoids rectangular errors. The first removal leads to a configuration which can be used in the case $k = p^s$. If $s = 1$ this is just the configuration that was produced using the first number theoretic construction.

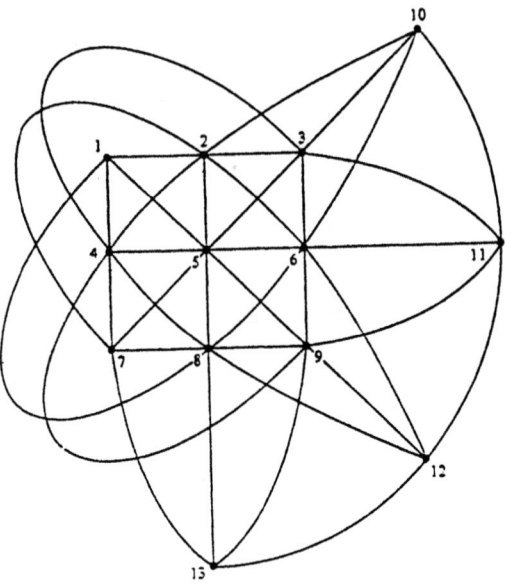

Fig. 4. Projective plane of order 3

5 Simulation Results

Simulation results were generated for (7,4), (15,11) and (31,26) Hamming codes (cf. Figs. 5-7). For the (7,4) Hamming code (Fig. 5) results were generated for the product code (a (40,16) code), for the interleaver produced using the straight number theoretic construction (a (160,64) code), for the interleaver produced by using a projective plane of order 3 (130,52) and for the modified number theoretic interleaver constructed by extending to a 5x5 array and then removing points from lines (250,100). The extended interleavers produce slightly better results ($\sim$ .15 dB).

For the (15,11) Hamming code results (Fig. 6) were produced for the product code (209,121) and the number theoretic interleaver (2299,1331). In this case there is no interleaver arising from a projective plane as there is no projective plane of order 10 (a fairly recent result obtained through a massive computational effort). The improvement in performance is $\sim$ .5 dB. at an E_b/N_o of 2.5 dB.

For the (31,26) Hamming code simulations (Fig. 7) were run using the product code (936,676), the straight number theoretic interleaver (24336,17576) and the projective plane interleaver obtained from a projective plane of order 25 (23436,16926). In this case the extended interleavers gave an improvement of $\sim$ .7 dB. at an E_b/N_o of 2.5 dB. The simulations were typically run for sufficient time to produce at least 100 blocks with errors in them. For some of the large block sizes and low error rates this requirement was reduced to 30 blocks.

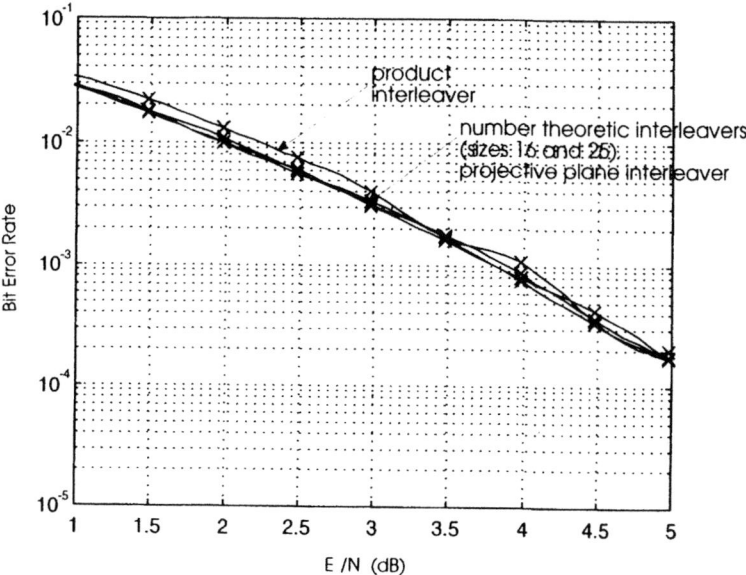

Fig. 5. Simulation results arising from (7,4) Hamming code using various interleavers

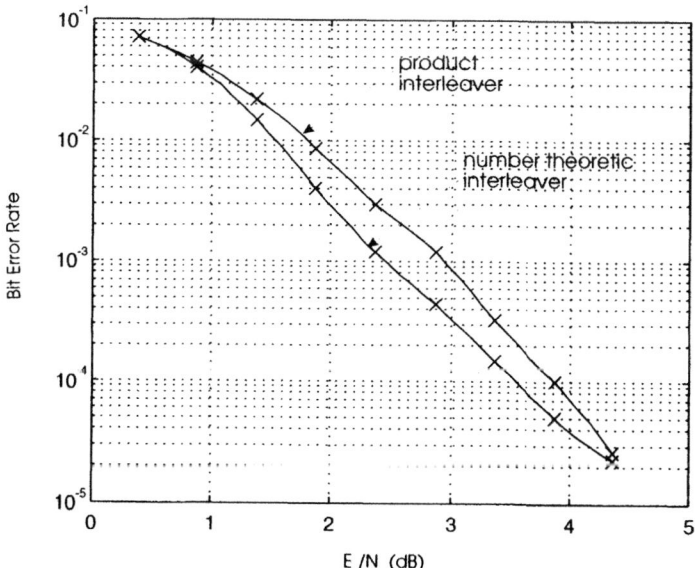

Fig. 6. Simulation results arising from (15,11) Hamming code for product interleaver and number theoretic interleaver

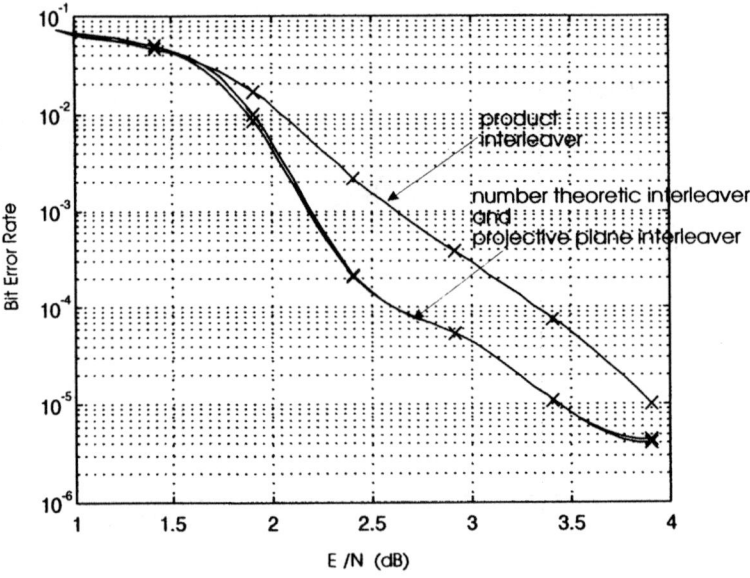

Fig. 7. Simulation results arising from (31,26) Hamming code using various interleavers

6 Discussion

We have constructed generalizations of product block codes which can be decoded using an iterative MAP technique. The codes provide better performance than standard product codes by removing the possible occurrence of certain error patterns. The performance margin of these codes over straight product codes increases as the size of the constituent codes gets larger. In effect, by using larger interleavers we are giving ourselves enough room to construct interleavers which "spread out" the error correction. This is somewhat analagous to interleavers designed to improve burst error performance. We are removing the possiblility of certain error patterns whereas a burst error interleaver is diminishing the probablility of burst type errors for codes which are sensitive to burst errors. Thus the increased size of interleaver we are employing is analagous to the improved performance one can get using a larger interleaver to improve burst error performance. There are two directions in which one can extend the ideas presented in this paper. One can try to get rid of other problematical error patterns. For instance the existence of a "hexagonal" error pattern corresponds to the nonexistence of triangles in the incidence structure describing a pair of partitions. More generally the existence of a "2n-gon" error pattern corresponds to the nonexistence of a n sided closed path in the incidence structure. The other direction in which the concerns of the paper can be extended is to product codes of higher dimension. For an n-dimensional product code the problem at hand corresponds to the partitioning of a set in n different ways such that no two distinct sets taken from any of the sets in the partitions intersect in more than one point. To extend the finite projective plane construction of partitions to this setting one needs to consider block designs. A block design [5] is an arrangement of v distinct objects into b blocks such that each block contains exactly k distinct objects, each object occurs in exactly r different blocks, and every pair of distinct objects occurs together in exactly λ blocks. To generalize the construction one considers block designs with k being the number of information bits in the constituent code which one is employing, $r = (n-1)k$ and $\lambda = 1$.

References

1. J. Lodge, P. Hoeher, and J. Hagenauer, "The decoding of multidimensional codes using separable MAP 'filters'," in Proc. of Queen's University 16th Biennial Symp. on Communications, pp. 343-346, May 1992.
2. J. Hagenauer, E. Offer, and L. Papke, "Iterative ('turbo')-decoding of binary block and convolutional codes," IEEE Trans. in Info. Thy. Vol.42, No.3, pp. 429-445, Mar. 1996.
3. C. Berrou, A. Glavieux, and P. Thitimajshima, "Near Shannon limit error- correcting coding and decoding: Turbo-codes (1)," Proc. of IEEE Int. Conf. on Comm., Geneva, pp. 1064-1070, May 1993.
4. L.M. Batten, "Combinatorics of Finite Geometries," Cambridge, Cambridge University Press, 1986.
5. M. Hall, Jr., "Combinatorial Theory," New York, John Wiley & Sons, 1986.

Improved VLSI Design for Decoding Concatenated Codes Comprising an Irreducible Cyclic Code and a Reed-Solomon Code

D.B. Gravel, G. Drolet, C.N. Rozon

Electrical and Computer Engineering Department, Royal Military College of Canada, Kingston, Ontario, Canada, K7K 5L0

Abstract. The concatenation of a Reed-Solomon code with an irreducible cyclic code is used to increase the error correction capabilities on a bursty digital communication channel, in addition to both reduce the complexity and increase the throughput of a VLSI implementation of the decoder. The concept is illustrated with the $\left(2^{10} - 1, k \right)$ Reed-Solomon code and the (11, 10, 2) irreducible binary cyclic code (even parity code). We compare the Berlekamp-Massey processors for the $\left(2^{10} - 1, k \right)$ Reed-Solomon code alone, and for the concatenation on the other hand. Both Berlekamp-Massey algorithms in the frequency and in the time domain are used for comparisons. The VLSI structures of the data processing units for each algorithm reveals that the concatenation with irreducible cyclic code results in a significant decrease in the physical size of the decoding algorithm's processor for both the time and frequency domain algorithms, and in an increased operating clock frequency with the frequency domain algorithm.

1. Introduction

Consider a Reed-Solomon code $\left(2^m - 1, k \right)$ defined over the finite field GF (2^m), and a bijective mapping from the set of binary m-tuples (called symbols) onto GF (2^m). When the information from a binary source is to be transmitted over a bursty binary communication channel, the stream of information bits is divided into information blocks of k symbols which are encoded into codewords of $2^m - 1$ symbols. Due to the algebraic structure of Reed-Solomon codes, the encoder and decoder must contain finite field multipliers, adders and inverters designed to operate on the m-tuples representing (under the bijective map) the elements of GF (2^m). Encoders and decoders for the correction and/or detection of errors are often implemented using VLSI technology to achieve a higher throughput under severe size and power constraints. Several Galois field multiplier designs have been developed over time for use in error-correcting code applications. However, circuit designs have typically not been well suited for VLSI design due to their irregular wire routing and non-modularity. Various designs have been proposed to render the design of multipliers much simpler. Yeh, Reed and Truong [1] developed designs for both a serial and a parallel systolic Galois field multiplier. Their designs offered the advantage of being modular in nature thus simplifying VLSI construction. In addition, both the serial an parallel implementations are programmable for the selection of the irreducible polynomial defining the Galois field. All such irreducible polynomials define the same Galois field, but the programmable selection offers versatility which is appealing from the manufacturer's point of view. The cost of this versatility is an increase in the number of gates required and a longer delay path as compared with other multipliers. The Massey-Omura multiplier was designed to perform multiplications when field elements are represented with respect to a normal basis [2]. The multiplier provides a certain modularity for VLSI design as a common boolean function f is used to calculate all coefficients of the product $P = A \cdot B$ where A and B are two elements of the field GF (2^m) . In addition, it offers the advantage of simplified squaring of an element, leading to faster calculation of the element's inverse and to discrete exponentiation. In a paper by Wang et als [3], the VLSI design of the parallel type Massey-Omura multiplier is discussed. However, the parallel type Massey-Omura multiplier also requires a large number of gates as compared with other multipliers.

In order to both increase the error detection and/or correction capabilities of a Reed-Solomon code as well as to simplify the implementation of the finite field arithmetic circuits within the encoder and decoder, it is proposed to use a concatenation of a Reed-Solomon code with an irreducible cyclic code. The effect of the concatenation is to represent the elements of GF (2^m) by some blocks of n bits, $n > m$. This field element representation uses $n - m$ more bits than conventional representations using a basis and specific multiplier circuits can be designed for the novel representation. It has been shown that for some values of m , the novel representation leads to multiplier structures that are less complex than conventional finite field multiplier structures [4]. While the burden of carrying extra bits pays off for the implementation of multipliers, it is unclear whether it will pay off for the implementation of a complete decoder system.

After a review of the multiplier circuit presented in [4], we compare the VLSI structures of the data processing units of Reed-Solomon decoders using first a Massey-Omura multiplier circuit with normal basis representation of the field elements (conventional approach), and second using the multiplier circuit of [4] with the novel representation of the field elements (approach based on the concatenation of the Reed-Solomon code with an irreducible cyclic code). Moreover the comparison will be carried out on two decoding algorithms: Berlekamp-Massey in the time domain and in the frequency domain [5].

2. Finite Field Arithmetic Circuit

Let $h(x)$ be an irreducible polynomial of degree m, n be the smallest positive integer such that $h(X) | (X^n + 1)$, and $g(X) = (X^n + 1) / h(X)$. $n > m$ is odd, and the binary irreducible cyclic code $\langle g(X) \rangle / \langle X^n + 1 \rangle$ of block length n and dimension $m = n - \deg(g(X))$ over GF (2) is isomorphic to the finite field GF (2^m) . Under the isomorphism, elements of GF (2^m) are represented by symbols of $n > m$ bits and the representation is thus termed *redundant*; the elements of GF (2^m) are not represented with respect to a basis of GF (2^m) over GF (2) .

While performing arithmetic operations such as multiplication with conventional multiplier circuits, intermediate polynomials are obtained which contain cross-product terms that must be expanded in the basis used; this is done by reducing modulo an irreducible polynomial when a canonical basis is used [1], or by the boolean function f when a normal basis is used with the Massey-Omura multiplier [2], [3]. In the ideal $\langle g(X)\rangle / \langle X^n + 1\rangle$ however, all polynomials are reduced modulo the polynomial $X^n + 1$ which contains fewer coefficients than a binary irreducible polynomial and may yield simpler circuitry. The following result is well known.

Lemma: Let $a(X) = \sum\limits_{i=0}^{n-1} a_i X^i$, $b(X) = \sum\limits_{i=0}^{n-1} b_i X^i$ be polynomials in $\langle g(X)\rangle / \langle X^n + 1\rangle$. Then the polynomial $c(X) = a(X)b(X)$ lying in $\langle g(X)\rangle / \langle X^n + 1\rangle$ is given by:

$$c(X) = \sum_{k=0}^{n-1} \left(\sum_{i+j \equiv k \,(\mathrm{mod}\, n)} a_i b_j \right) X^k .$$

The novel multiplier performs the multiplication according to the double summation of the lemma [4]. It has a structure which is similar to the first stage of the multiplier developed in [6]. There, the elements of the field GF (2^m) are represented with respect to a canonical basis. Elements to be multiplied together are first mapped into the ring GF $(2)[X]/\langle X^{m+1} + 1\rangle$ (but not necessarily in the ideal $\langle g(X)\rangle / \langle X^{m+1} + 1\rangle$) in order to exploit the simplicity of reduction modulo $X^{m+1} + 1$, and an intermediate polynomial is obtained. This polynomial is then mapped back to the product element of GF (2^m) . The findings of [6] only apply for values of m such that $\sum\limits_{i=0}^{m} X^i$ is irreducible in the ring GF $(2)[X]$. With the novel multiplier, the field element representation is the isomorphic image of the elements of GF (2^m) in $\langle g(X)\rangle / \langle X^n + 1\rangle$; it consequently need not map back and forth between GF (2^m) and GF $(2)[X]/\langle X^n + 1\rangle$. The novel representation works for any value of m and offers the advantage that the square of an element is obtained by permuting the coefficients of its polynomial. The implementation of inverter and discrete exponentiation circuits then becomes very efficient.

The cost of using the novel multiplier structure however, is that n bits are now required to represent each element of GF (2^m) instead of only m bits. For specific values of m however, n is not too much bigger than m and the novel multiplier structure is advantageous. Such values are $m = 2, 4, 6, 10, 12, 18, 20, 28, 36, \ldots$ [4]. We observe that $m = 6$ and $m = 20$ do not qualify for the multiplier implementations of [6] or [7].

Even though the implementation of the multiplier circuit is sometimes simpler, its use in the implementation of a decoder system may not yield simpler circuitry because of the extra $n - m$ bits of the redundant representation. In any case, the extra $n - m$ bits used in the representation of the field elements offer extra error detection and/or correction capability, and the implementation of a $(2^m - 1, k)$ Reed-Solomon code with the redundant representation corresponds to a concatenation of the $(2^m - 1, k)$ Reed-Solomon code with the binary irreducible cyclic code $\langle g(X)\rangle / \langle X^n + 1\rangle$. In the next section we investigate the case $m = 10$ ($n = 11$); then we have $X^n + 1 = X^{11} + 1$, $g(X) = X + 1$ and $\langle X + 1\rangle / \langle X^{11} + 1\rangle \cong GF(2^{10})$ (even parity code). In this case, the polynomial $\sum_{i=0}^{m} X^i$ is irreducible, and both [6] and [7] apply; the novel multiplier is simpler than [6], but slightly more complex than [7], much like [6] and [7] squaring is obtained by permuting the coefficients, but only the novel multiplier has a redundant representation which inherently yields a concatenation of codes with increased error detection and/or correction capability.

3. Proposed Concatenation

In the situation depicted by figure 1, the 10-bit information symbols are mapped to elements of the finite field $\langle X + 1\rangle / \langle X^{11} + 1\rangle$ by appending a parity bit and writing the 11-bit blocks as binary polynomials. This is followed by a $(2^{10} - 1, k)$ Reed-Solomon

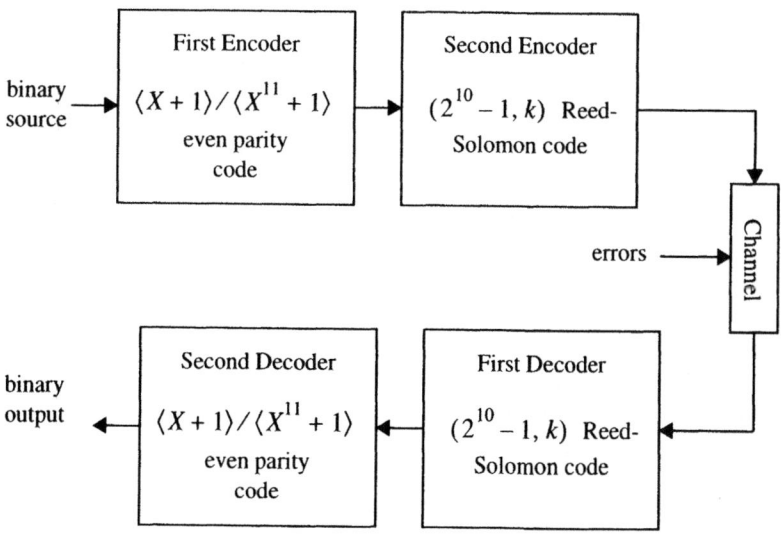

Fig. 1. Proposed Concatenated Code

encoder taking advantage of the redundant representation of the elements of the finite field GF (2^{10}) by using the arithmetic circuits of [4]. The same is done for the decoder. The parameters of the Reed-Solomon code are presented in table 1 together with the parameters of the concatenated code. Based on the fact that multiplier and inverter circuits are simpler and have a shorter longest delay path, [4] claims that a simpler and faster decoder circuitry can be obtained then by using conventional architectures; the concatenation serves both to increase the error detection and/or correction capabilities as well as simplify the hardware complexity of the Reed-Solomon encoder and decoder at the cost of a slight reduction in the transmission rate.

Table 1. Code Parameters

Code	Block Length	Rate	Burst Error Correcting Capability
Reed-Solomon Code	10230 bits	$\dfrac{k}{1023}$	$10 \ \mathrm{int}\left[\left(\dfrac{1023-k}{2}\right)-1\right]$
Concatenated Code	11253 bits	$\dfrac{k}{1023} \times \dfrac{10}{11}$	$11 \ \mathrm{int}\left[\left(\dfrac{1023-k}{2}\right)-1\right]$

4. Evaluation of Hardware Implementations

Based on the proposed code concatenation, an assessment of the claimed simplification of hardware complexity for specific algorithms used in Reed-Solomon decoders was carried out [8]. Comparisons were carried out on the implementation of 2 popular algorithms used in the decoding of Reed-Solomon codes: the Berlekamp-Massey algorithm in the time domain and the Berlekamp-Massey algorithm in the frequency domain. Each one was based on two different finite field arithmetic circuit designs: the use of a conventional parallel Massey-Omura multiplier for GF (2^{10}) as in [3] (the field elements are represented with respect to an optimal normal basis [9] and the Reed-Solomon code is not concatenated) against the use of multiplier architecture [4] operating with the redundant representation $\langle X + 1 \rangle / \langle X^{11} + 1 \rangle$. Comparisons between circuit designs were limited to the data processing elements within each design where the data bus of devices was increased from 10 bits to 11 bits. This limitation in scope was based on the fact that the design of the state controller for each algorithm does not change as a result of the use of the redundant field element representation. VLSI architectures were based on a 1.2µ CMOS standard cell library and CAD tools were used to evaluate the hardware implementations. State controller designs were simulated at a behavioural level by using the VHSIC Hardware Definition Language (VHDL). Table 2 summarizes some of the VLSI attributes of the 2 finite field multiplier implementations used in the comparison of designs.

Table 2. Design Comparisons - Finite Field multipliers

Comparison	Massey-Omura Multiplier	Drolet-Séguin Multiplier	Difference and % change
Physical Layout (mm^2)	1.52	0.73	-0.79 (-52.1%)
Typical Delay (ns)	6.5	5.3	-1.2 (-18.5%)
Gate Count	1,215	473	-742 (-61.1%)

Table 3 compares the results for the VLSI designs of Berlekamp-Massey decoders in the frequency domain and time domain.

Table 3. Results - Berlekamp-Massey Decoders

Domain	Factor	Massey-Omura Multiplier	Drolet-Séguin Multiplier	Difference
Frequency	Size-mm^2	15.17	9.71	-5.46
	Clock (ns)	24	18	-6
	Gate Count	11,585	6,490	-5,095
Time	Size-mm^2	15.90	9.66	-6.24
	Clock (ns)	28	28	0
	Gate Count	12,342	6,516	-5,826

5.   Conclusion

Results of this study show that the concatenation of a Reed-Solomon code with an irreducible code can be used to simplify the decoder's hardware as well as improve the error correction and/or detection of the code on a bursty binary communication channel. The VLSI improvements obtained are the direct result of the use of the finite field multiplier (and inverter) of [4] based on a redundant representation of field elements. If a simpler multiplier (and inverter) circuit is available, the concatenation will yield an improved error detection and/or correction but will not simplify the decoder's hardware. This would happen for example, if the multiplier architecture [7] had been used for comparison in the present study. However, the multiplier of [7] does not apply over the finite field GF(2^{20}) and the multiplier (and inverter) of [4] is then the best known to the authors. Thus, the concatenation of a Reed-Solomon code over GF(2^{20}) with the irreducible code $\langle X^5 + 1 \rangle / \langle X^{25} + 1 \rangle$ would lead to both error correction improvement and hardware simplification. The study was also limited to the context of the Berlekamp-Massey algorithm both in the frequency and in the time domain, assuming that the received symbols are converted to even parity (if necessary) prior to entering the Berlekamp-Massey processor. Such improvements on the hardware complexity of

Reed-Solomon decoders may not be obtained with other decoding algorithms or if a decoder handling erasures were to be implemented.

References

1.   C.S. Yeh, I.S. Reed, T.K. Truong, "Systolic Multipliers for Finite Fields $GF(2^m)$", *IEEE Transactions on Computers*, vol. C-33, no-4, pp 357-60, April 1984.
2.   J.L. Massey, J.K. Omura, *Computational method and Apparatus for Finite Field Arithmetic*, US Patent Application, submitted 1981.
3.   C.C. Wang, T.K. Truong, H.M. Shao, L.J. Deutsch, J.K. Omura and I.S.Reed, "VLSI Architectures for Computing Multiplications and Inverses in $GF(2^m)$", *IEEE Transactions on Computers*, vol. C-34, no-8, pp 709-17, August 1985.
4.   G. Drolet, G.E. Séguin, Manuscript, June 1995.
5.   R.E. Blahut, *Theory and Practice of Error Control Codes*, Don Mills: Addison-Wesley, 1984.
6.   T. Itoh, S. Tsujii, "Structure of Parallel Multipliers for a Class of Fields $GF(2^m)$", *Inform. Comp.*, vol. 83, pp. 21-40, 1989
7.   M.A. Hasan, M.Z. Wang, V.K. Bhargava, "A Modified Massey-Omura Parallel Multiplier for a Class of Finite Fields," *IEEE Trans. Comput.*, vol. 4, pp. 1278-1280, Oct. 1993.
8.   D.B. Gravel, *Improved VLSI Design for Decoding Concatenated Codes Comprising an Irreducible Cyclic Code and a Reed-Solomon (N,K) Code*, M.Eng. Thesis, Royal Military College Kingston, May 1995.
9.   R.C. Mullin, I.M. Onyszchuk, S.A. Vanstone, R.M. Wilson, "Optimal Normal Bases in $GF(p^n)$", *Discrete Applied Mathematics*, vol. 22, pp. 149-161, 1989.

Non-minimal Trellises for Linear Block Codes

Ryan Chi-Kong Lee and Frank R. Kschischang

Department of Electrical and Computer Engineering
University of Toronto

Abstract. The technique of atomic span modification is used to construct non-minimal trellises for linear block codes. Non-minimal trellises may have properties like regularity and vertex localization that make the corresponding trellis-based decoding algorithms better-suited for VLSI or multiprocessor implementation. A simple class of "universal" trellises with a regular structure is defined, from which soft decision decoders can be built by parallel combination of identical trellis search processors. Coset decoding is proposed as a general soft-decision decoding technique for multiprocessor implementation, and algorithms for selecting a subcode to achieve large speedup are described.

1 Introduction

Trellises were introduced by Forney [1] (see also [2]) in 1967 as an aid for explaining the Viterbi algorithm for decoding convolutional codes. It was recognized fairly early that linear block codes, like convolutional codes, have a well-defined trellis structure [3], and that the Viterbi algorithm can be used for soft-decision decoding of block codes [4, 5]. The appendix of [6] sparked a renewed interest in the topic of block code trellises, prompting a flurry of research papers in this area (see, e.g., [7–21]). With the notable exception of [21], all of these papers focus on the *minimal* trellis which has a smaller number of trellis vertices and edges than any other trellis description of the same code.

However, the minimal trellis for a block code may *not* exhibit a regularity of structure that is well-suited for a VLSI or multiprocessor implementation of a trellis-based decoding algorithm. For example, Fig. 1 illustrates the minimal trellis obtained for the (16,7,6) binary lexicode [22], with "trellis-oriented" (see Section 2) generator matrix [6, 11]

$$G = \begin{bmatrix} 1\ 1\ 1\ 1\ 1\ 1\ 0\ 0\ 0\ 0\ 0\ 0\ 0\ 0\ 0\ 0 \\ 0\ 1\ 1\ 0\ 1\ 0\ 1\ 1\ 1\ 0\ 0\ 0\ 0\ 0\ 0\ 0 \\ 0\ 0\ 1\ 1\ 1\ 0\ 0\ 1\ 0\ 1\ 1\ 0\ 0\ 0\ 0\ 0 \\ 0\ 0\ 0\ 0\ 1\ 1\ 0\ 1\ 1\ 0\ 1\ 1\ 0\ 0\ 0\ 0 \\ 0\ 0\ 0\ 0\ 0\ 1\ 1\ 1\ 0\ 0\ 0\ 1\ 0\ 1\ 1\ 0 \\ 0\ 0\ 0\ 0\ 0\ 0\ 0\ 1\ 1\ 1\ 0\ 1\ 1\ 1\ 0\ 0 \\ 0\ 0\ 0\ 0\ 0\ 0\ 0\ 0\ 0\ 0\ 1\ 1\ 1\ 1\ 1\ 1 \end{bmatrix} \tag{1}$$

This version of the lexicode is distinguished by the fact that, because the code meets the dimension/length profile (DLP) bound [15], no trellis description of a

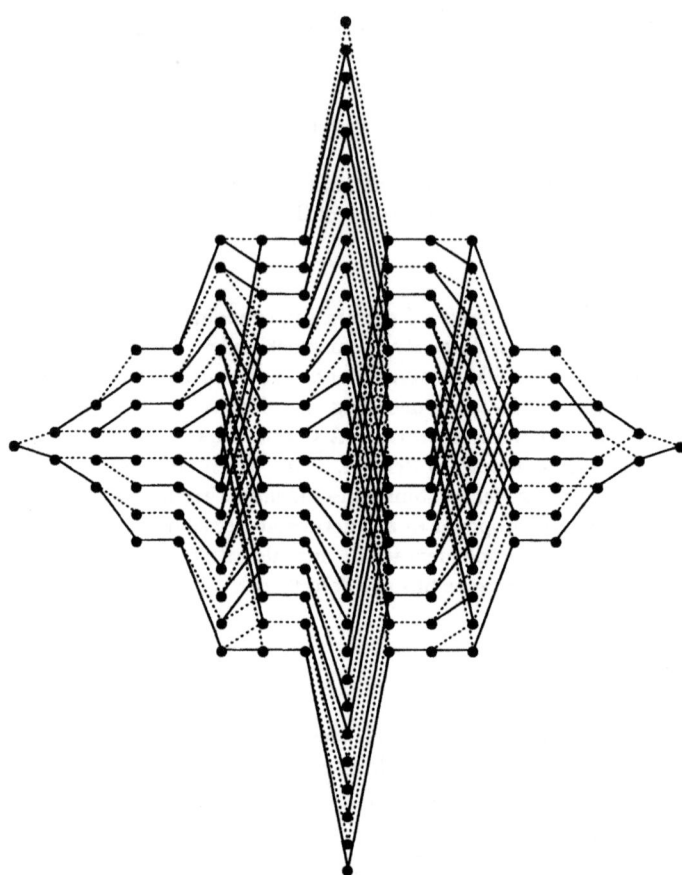

Figure 1. Minimal trellis for the (16,7,6) lexicode. Solid edges have label '0' and dotted edges have label '1'.

binary (16,7,6) code can have fewer vertices at any trellis depth than the trellis shown in Fig. 1.

However, we see that the minimal trellis in Fig. 1 is irregular, i.e., there is no regular (periodic) pattern of vertices and edges. If more than one processor is available for decoding, the time required for decoding will be less than that required for a single processor because some decoding steps can be performed in parallel. However, there is a problem in allocating the processors to the various steps involved in Viterbi decoding because of the irregularity and the fact that the computations are not localized in the trellis. The pattern of trellis edges in some sections would require that the various processors exchange accumulated path metrics, and this "global communication" will reduce decoding speed. The lack of regularity and computational localization may limit the usefulness of this minimal trellis for practical decoder implementations.

In this work we explore potential uses for *non-minimal* trellises. Whereas minimal trellises often have an irregular structure, non-minimal trellises for the same code can sometimes be designed with a regular structure. Whereas minimal trellises often lack the property of computational localization among the vertices of the trellis, non-minimal trellises for the same code can sometimes be designed that possess this property. These properties of regular structure and vertex localization become important when soft-decision block decoders are to be implemented in VLSI or on a multiprocessor. In other words, in practical implementations of decoders, minimality of the code trellis may not be as important as other desirable features like regularity and local communication.

Our aim is to design non-minimal trellises which are suitable for VLSI and multiprocessor implementations of trellis-based soft-decision decoding algorithms for linear block codes. We take two different approaches to this task. The first approach is to develop trellis structures which are "universal" in the sense that every linear block code (within a given range of parameters) can be represented by essentially the same trellis, consisting only of uniform identical trellis structures in parallel. The second approach is to develop a particular trellis for each linear block code, "optimized" for parallel decoding.

The first approach may be preferable for VLSI circuit manufacturers because only a few fundamental trellis circuits need to be implemented. On the other hand, since only one particular code is used in a typical transmission system, decoding speed is more important than whether the trellis is universal or not. As a result, the second approach is preferable when it is possible to sacrifice "universality" for decoding speed.

This paper is organized as follows. In Section 2, we review the basic theory of the trellis structure of block codes, and show that the technique of "atomic span modification" can yield a wide variety of non-minimal trellis presentations for a given code. In Section 3, we describe one approach for achieving a "universal" trellis structure. In Section 4, we describe methods for choosing a subcode of a binary linear block code so that parallel coset decoding results in maximum speed up over a single-processor Viterbi algorithm search of the minimal trellis. Finally, some concluding remarks are made in Section 5.

2 Atomic Span Modification

Algorithms for constructing the minimal trellis for linear block codes were described by Bahl, et al. [3], Massey [5] and Forney [6]. Since linear codes have minimal trellises unique up to isomorphism [10, 23], these construction procedures all yield essentially the same trellis. Kschischang and Sorokine [11] describe another equivalent trellis construction procedure based on the notion of a trellis product. In this section we describe the trellis product construction, and the notion of "atomic span modification" that yields a wide variety of non-minimal trellises.

2.1 Minimal Trellises

Let $T = (V, E, F)$, $E \subset V \times A \times V$ be a length n trellis with vertex set V, edge set E, with each edge labelled with a symbol from a finite field F. For $0 \leq d \leq n$, let V_d denote the set of trellis vertices at depth d in T, and for $1 \leq d \leq n$, let E_d denote the set of trellis edges in the dth section of T, i.e., the set of edges connecting the vertices at depth $d-1$ with the vertices at depth d.

The *product* $T = T' \times T''$ of two length n trellises $T' = (V', E', F)$ and $T'' = (V'', E'', F)$ is the trellis $T = (V, E, F)$, with

$$V = \bigcup_{d=0}^{n} V_d' \times V_d'', \quad \text{and} \quad E = \bigcup_{d=1}^{n} \omega(E_d' \times E_d'')$$

where $\omega(v_1', a', v_2', v_1'', a'', v_2'') = ((v_1', v_1''), a' + a'', (v_2', v_2''))$. If $C(X)$ is the code presented by a trellis X, then $C(T) = C(T') + C(T'')$, i.e., T represents the sum of the codes presented by T' and T'' [11].

An $[n, k]$ linear code C over the finite field F is a k-dimensional subspace of F^n. As such, it is generated by a basis $\{v_1, v_2, \ldots, v_k\}$ of k linearly independent vectors from F^n. If we denote by $\langle v_i \rangle$ the one-dimensional subspace spanned by v_i, then $C = \langle v_1 \rangle + \langle v_2 \rangle + \cdots + \langle v_k \rangle$. Let $T\langle v_i \rangle$ be a trellis presentation for $\langle v_i \rangle$. Then a trellis presentation for C is obtained as the trellis product

$$T\langle v_1 \rangle \times T\langle v_2 \rangle \times \cdots \times T\langle v_k \rangle. \tag{2}$$

Depending on the choice of basis for C, this trellis presentation may or may not be minimal.

Let $v = (v_1, v_2, \ldots, v_n)$ be a nonzero vector with components from F. As in [11], define

$$\text{start}(v) = \min\{i : v_i \neq 0\},$$

and

$$\text{end}(v) = \max\{i : v_i \neq 0\}.$$

The *span* of v is the interval $[\text{start}(v), \text{end}(v)] = \{\text{start}(v), \ldots, \text{end}(v)\}$, and the *span length* of v is the cardinality of its span, namely $\text{end}(v) - \text{start}(v) + 1$.

In [11] it is shown that the trellis product (2) yields the minimal trellis for C if and only if for all choices of i and j, $1 \leq i < j \leq k$, we have

$$\text{start}(v_i) \neq \text{start}(v_j) \tag{3}$$

and

$$\text{end}(v_i) \neq \text{end}(v_j), \tag{4}$$

and $T\langle v_i \rangle$ is the *minimal* trellis for the corresponding one-dimensional subspace. Minimal trellises for one-dimensional codes are easily found (see [11, Fig. 7]). The minimal trellis for an $(n, 1)$ linear code over $GF(q)$ with a generator having span $[a, b]$ has q paths through the trellis, with a "divergence" from the all-zeroes path in position a and "convergence" in position b.

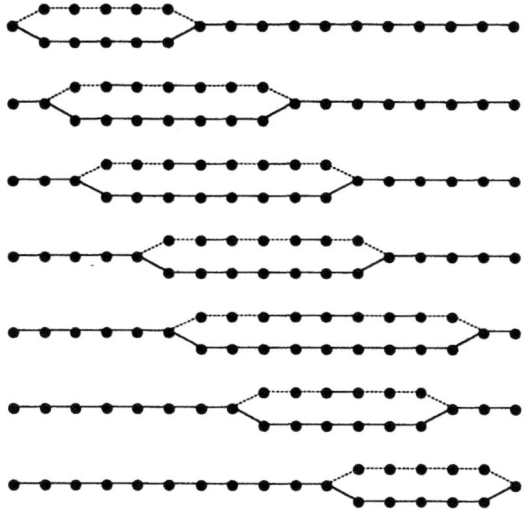

Figure 2. The minimal trellis for the (16,7,6) lexicode is obtained as the product of the seven "elementary" trellises illustrated.

A basis for C that satisfies conditions (3) and (4) is called an *atomic basis*. The elements of the basis are called *atomic generators* because each basis element has the property that it cannot be expressed as a linear combination of codewords having strictly smaller span length. (In other words, atomic generators cannot be "split" into "pieces" of small span length.) Although a code may in general have more than one atomic basis, the set of spans in every atomic basis for a given code is unique.

A generator matrix with rows that are all atomic is known as a trellis-oriented generator matrix [6]. The generator matrix (1) for the (16,7,6) lexicode is an example, because no two rows of the matrix either start in the same position or end in the same position. Fig. 2 illustrates the sub-trellises representing the one-dimensional sub-codes generated by the rows of the trellis-oriented generator matrix (1). (We refer to the trellises describing these sub-codes as "elementary trellises.") Using elementary row operations, it is always possible to convert an arbitrary generator matrix to a trellis-oriented one in polynomial time [11, 17].

Kschischang and Sorokine also showed that the set of atomic spans determine the structure of the corresponding trellis, in the sense that any two codes have the same minimal structural trellis (with edge labels suppressed) if and only if they have the same set of atomic spans. The set of atomic spans can be displayed in an upper triangular chess board, where an element located in row i and column j represents the span $[i, j]$. Conditions (3) and (4) imply that at most one element can occupy any given row or column. In other words, the set of atomic spans create a configuration of non-attacking rooks in the triangular chess board. (This is why the chess board terminology is used.) Fig. 3 illustrates the chess board corresponding to the (16,7,6) lexicode example.

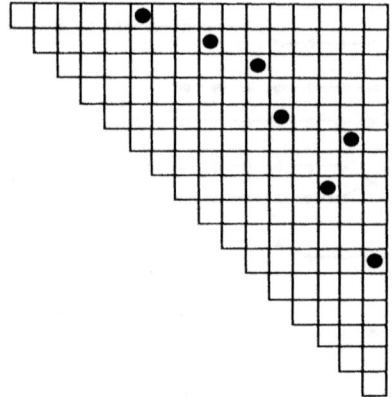

Figure 3. Chess board representation of the atomic spans in the (16,7,6) lexicode.

2.2 Non-minimal Trellises

Non-minimal trellises can be obtained by modifying the elementary trellises corresponding to the atomic generators. Although it is not possible to shorten the span of an atomic generator, it *is* possible artificially to lengthen the span. In a code of length n, a generator of span $[a, b]$ can be modified to "appear" to have span $[c, d]$ where $1 \leq c \leq a$ and $b \leq d \leq n$, simply by creating a divergence in the elementary trellis at position c and a convergence at position d. For example, the fourth row of the matrix in (1) is an atomic generator with span $[5,12]$, corresponding to the fourth elementary trellis of Fig. 2. This elementary trellis can be modified so that the generator appears to have a longer span, as illustrated in Fig. 4.

Such a span modification is equivalent to moving an element located in row a, column b to row c, column d in the chess board. In general, a "legal" span modification for an atomic generator cannot increase the row number or decrease the column number, and so must take place in the "northeast" direction, anywhere within the rectangle having lower left corner in row a and column b. For example, the span modification corresponding to Fig. 4 is illustrated on the chess board of Fig. 5.

Note that the sub-codes generated by the original generator and the "span modified" generator are the same; as a result, the same code is generated by span modification and only the structure of the product trellis changes. (In other words, only the trellises are modified, not the generators themselves.)

In the remainder of this paper, we will only consider the case of binary linear block codes. Generalization to the q-ary case is, in most cases, immediate.

One simple property of the trellis product is that a generator of span $[1, n]$ splits the product trellis into two non-communicating isomorphic sub-trellises. The structure of the sub-trellises depends on the subcode generated by the generators not having span $[1, n]$. Since $[1, n]$ is located at the far "northeast" of the chess board, any generator can be modified to have span $[1, n]$.

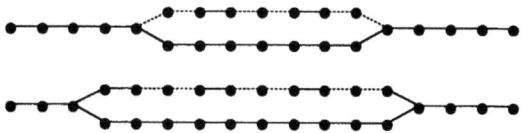

Figure 4. Minimal and non-minimal "elementary trellises" for generator 0000110110110000. The non-minimal trellis is equivalent to that of a generator with span [3,13].

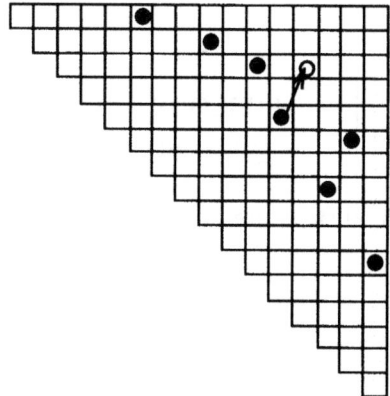

Figure 5. An atomic span modification is equivalent to moving a rook "northeast" on the chess board.

Thus, given a binary linear (n, k) code, for $0 \leq m \leq k$, we can construct a non-minimal trellis with 2^m parallel non-communicating sub-trellises by

1. modifying m generators to have span $[1, n]$;
2. designing the structure of the sub-trellis by modifying the other $(k - m)$ generators, which generate a $(k - m)$-dimensional subcode of the original code; and
3. multiplying the k modified elementary trellises to obtain the resulting non-minimal trellis.

The general structure of the resulting non-minimal trellis is shown in Fig. 6. It is important to note that the 2^m parallel sub-trellises all have, apart from their edge labels, precisely the same structure. In other words, these sub-trellises are structurally isomorphic [11].

Trellis-based decoding based on the the non-minimal trellis is performed by applying a processor to each non-communicating sub-trellis. This approach is well known, and is known as coset decoding [24], since each sub-trellis represents a coset of a subcode of the given code. Each processor performs the operations necessary to decode the given coset and determine a "survivor" for that coset. The survivors are then compared to make an overall decoding decision.

Because the code is decoded in parallel, the decoding time depends on the number of operations required for decoding one sub-trellis and on the number

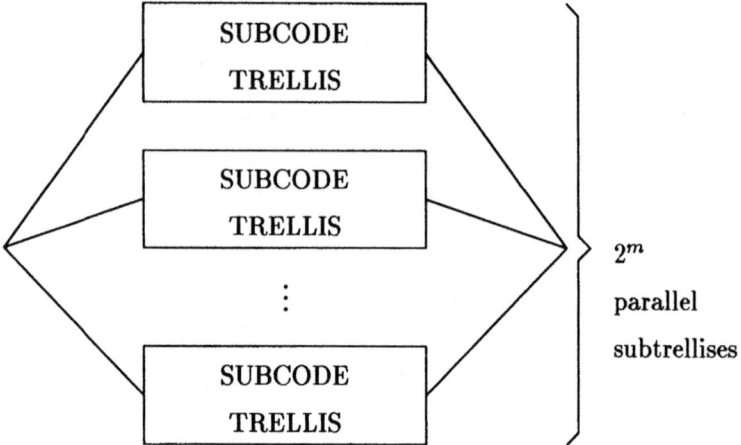

Figure 6. General structure of the resulting non-minimal trellis.

of comparisons required for choosing the best codeword among the survivors from each coset. In general, if $|E|$ denotes the number of edges in a given trellis and $|V|$ denotes the number of vertices, the number of additions performed by a Viterbi algorithm search of the trellis is $|E|$ and the number of comparisons is $|E| - |V| + 1$ [17], or a total of $2|E| - |V| + 1$ binary operations.

Let $|E_m|$ and $|V_m|$ denote the number of edges and vertices in the minimal trellis for a given code C and let $|E_s|$ and $|V_s|$ denote the number of edges and vertices in a trellis description for a subcode of C_s of C, and assume that C_s has 2^m cosets in C. We define the *speedup* σ of a decoder that searches these cosets in parallel as

$$\sigma = \frac{2|E_m| - |V_m| + 1}{2|E_s| - |V_s| + 1 + m}. \tag{5}$$

Thus σ is the ratio of the number of operations required by the minimal trellis search and the number of operations required in each sub-trellis plus m. This quantity represents the possible processing acceleration that can be achieved through the use of 2^m parallel processors, communicating with each other only after performing the computation associated with each sub-trellis. Of course, in a given application, this speedup must be weighed against the increase in hardware complexity that will typically result from parallelizing of the decoder.

In the next section, we describe methods for designing "universal" trellis structures. In Section 4, we describe methods for choosing a subcode so that this parallel trellis architecture results in maximum speedup for a given code.

3 Universal Trellises

In this section, a fundamental trellis structure is described such that every linear block code can be represented by a trellis which consists of parallel copies of the fundamental trellis. Only one VLSI circuit for decoding a fundamental trellis is required to be implemented. For all linear block codes whose trellises consist of copies of a fundamental trellis, a universal decoder can be constructed by combining VLSI circuits corresponding to the fundamental trellis. However, as we will see later, the speedup (5) is smaller than that achieved by the minimal-complexity subcode approach discussed in Section 4.

For an (n, k) linear block code, we develop a universal trellis based on the steps stated in Section 2. We move m generators to span $[1, n]$, $(k - m)/2$ generators to span $[l + 1, n]$ and $(k - m)/2$ generators to span $[1, n - l]$, where $m \leq k$ and $(n - l) \geq (l + 1)$, i.e., $2l \leq (n - 1)$. It is important to note that the values of l and m that can be chosen depend on the span lengths of the atomic generators. Also, $k - m$ must be even. However, a given value of l and m may be suitable for a wide range of codes.

The modified generators not having span $[1, n]$ generate a $(k-m)$-dimensional subcode of the original code. The resultant trellis will consist of 2^m parallel copies of a 3-section fundamental trellis where the structure of the fundamental trellis depends on the spans of the generators that generate the subcode. The first and the last sections of the fundamental trellis represent the first l bits and the last l bits of each codeword, respectively, and the center section represents the $(n - 2l)$ bits in the middle. The middle portion is a complete bipartite graph $K_{a,a}$, where $a = 2^{(k-m)/2}$. The chess board configuration and the structure of the resulting trellis are shown in Fig. 7.

The total number $E_s(n, k, l, m)$ of edges in each sub-trellis is given by

$$E_s(n, k, l, m) = 2l 2^{(k-m)/2} + (n - 2l) 2^{(k-m)}$$

for an (n, k) code trellis parameters m and l, $2l < n$. Similarly, the total number of binary comparisons required to decode each sub-trellis is $a^2 - 1$, or is given by

$$C_s(k, m) = 2^{(k-m)} - 1.$$

Thus, from (5), the speedup given by this approach is

$$\sigma = \frac{2|E_m| - |V_m| + 1}{E_s + C_s + m}$$

relative to a single-processor decoding of the minimal trellis with $|E_m|$ edges and $|V_m|$ vertices. To maximize speedup, it is desirable to maximize the value of l. However, the set of atomic spans and the value of m determine the maximum value of l than can be chosen.

For the (16,7,6) lexicode with generator matrix given in (1), different but similar trellises can be obtained by choosing different values for l and m. Note that the sum of the number of additions and the number of comparisons required for decoding the minimal trellis of the (16,7,6) lexicode by the Viterbi algorithm is $2|E| - |V| + 1 = 299$.

The resulting trellis and the corresponding chess board representation in each of the following three cases are shown in Fig. 8.

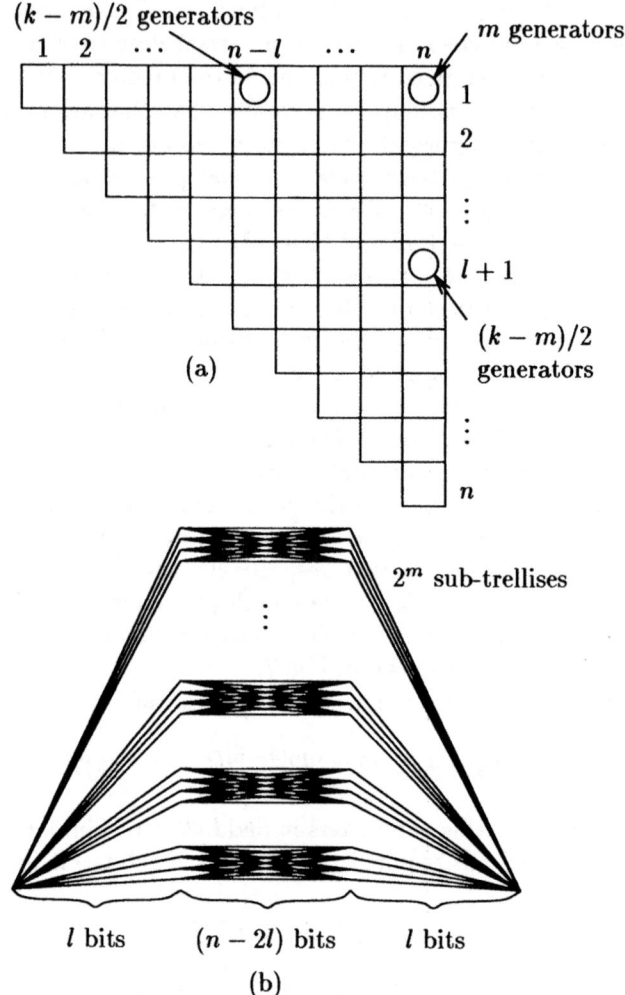

(a)

(b)

Figure 7. Structure of (a) the chess board and (b) the resulting trellis for a "universal" trellis decoding scheme.

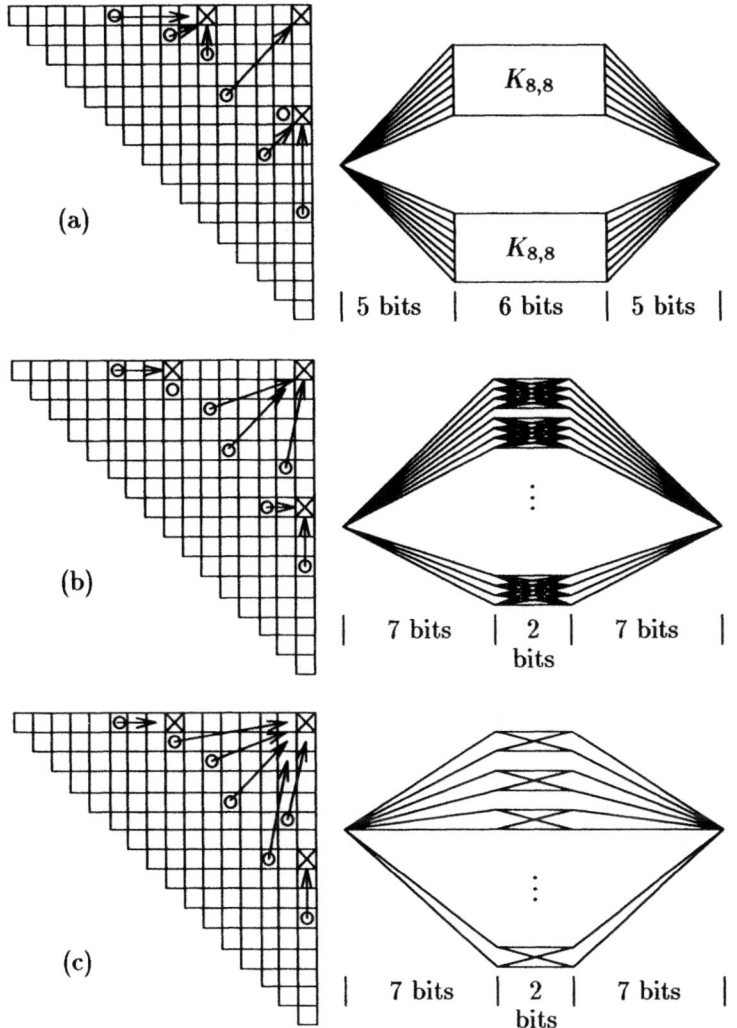

Figure 8. Trellis and chess board representation for (a) $l = 5$, $m = 1$; (b) $l = 7$, $m = 3$; (c) $l = 7$, $m = 5$.

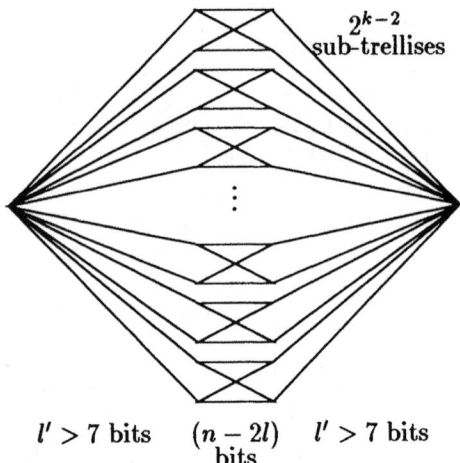

$$2^{k-2}$$
sub-trellises

$l' > 7$ bits $(n - 2l)$ $l' > 7$ bits
bits

Figure 9. Resulting trellis.

1. $l=5$, $m=1$. The resulting trellis consists of $2^m = 2$ sub-trellises. The sum of the number of additions and the number of comparisons required for decoding the resultant trellis is $E_s + C_s + m = 528$ ($\sigma = 0.57$), where it has been assumed that the decoding algorithm in use is the Viterbi algorithm and there are 2 processors available. The maximum value of l which can be chosen for $m = 1$ is 5.

2. $l=7$, $m=3$; then the resulting trellis consists of $2^m = 8$ sub-trellises. The sum of the number of additions and the number of comparisons required for decoding the resultant trellis is $E_s + C_s + m = 106$ ($\sigma = 2.8$), where it has been assumed that the decoding algorithm in use is the Viterbi algorithm and there are 8 processors available. Note that the maximum value of l which can be chosen in this case is 7.

3. Let us choose $l=7$, $m=5$, then the resultant trellis consists of $2^m = 32$ sub-trellises. The sum of the number of additions and the number of comparisons required for decoding the resultant trellis is $E_s + C_s + m = 44$ ($\sigma = 6.8$), where it has been assumed that the decoding algorithm in use is the Viterbi algorithm and there are 32 processors available. The maximum value of l which can be chosen in this case while maintaining a central "butterfly" structure is 7.

Suppose we wish to decode an (n, k) linear block code with $n > 16$ (the length of the lexicode). Similar to what we have done in the third case of the previous example, we modify one generator to span $[1, n - l']$, one generator to span $[1 + l', n]$ and $(k - 2)$ generators to span $[1, n]$ where $l' > 7$. Then the resulting trellis consists of 2^{k-2} non-communicating sub-trellises in parallel as shown in Fig. 9.

The fundamental trellis is the same as the one in the previous example except that the number of bits in corresponding sections is different. It is not necessary to design another VLSI circuit for this fundamental trellis due to this difference.

One could design a single VLSI circuit for all fundamental trellises with the same structure provided that the number of bits in each section which can be handled by the circuit is sufficiently large. For example, if a VLSI circuit is designed for the fundamental trellis in the third case of the previous example, and the number of bits which can be handled is 10 in each section, then this circuit can be used to decode the (16,7) lexicode. Since there are 7 bits, 2 bits and 7 bits in the first section, the center section and the last section of the sub-trellis, respectively, one can simply assign (10-7)=3 zeros, (10-2)=8 zeros and (10-7)=3 zeros to each path in the first section, the center section and the last section respectively. By padding with extra zeros, the number of additions required for decoding increases and, consequently, the decoding speed decreases. Nevertheless, the circuit would be capable of handling block lengths up to 30, and so in this sense is a "universal" decoder.

4 Minimal Complexity Coset Decoding

In this section, we present a method to select a subcode so that parallel coset decoding results in maximum possible speedup. We start by assuming that the trellis is decoded in sections of unit length. Later we will incorporate the results of Lafourcade and Vardy [20] on optimal sectionalization.

4.1 Exhaustive Sectionalization

Given 2^m processors for parallel decoding of an (n, k) linear block code, we can modify m generators to have span $[1, n]$. To maximize speedup, we require that the minimal trellis describing the subcode generated by the remaining $k-m$ generators have smallest possible decoding complexity. Although one could search among all possible sub-codes of dimension $k - m$ to find a subcode of minimum complexity, such an exhaustive search would be computationally infeasible in all but trivial cases. Fortunately, as confirmed by the following theorems, it is sufficient to consider only those sub-codes generated by k nonzero codewords which are atomic in the code C in order to find a subcode whose minimal trellis has minimum edge complexity.

Let $\{g_1, g_2, \ldots, g_k\}$ be an atomic basis for an (n, k) linear block code C. Let start$(C) = \{$start$(g_1), \ldots,$start$(g_k)\}$, and let end$(C) = \{$end$(g_1), \ldots,$end$(g_k)\}$ denote, respectively, the start and end sets for C. Although there may be more than one atomic basis for a given code, every such atomic basis will have the same start set and end set [11]. We say that a word u is atomic in C if it is an atomic codeword in C.

Lemma: The start of each non-zero codeword in C belongs to start(C), and the end of each non-zero codeword in C belongs to end(C).

Proof: Order the atomic basis so that start$(g_1) <$ start$(g_2) < \cdots <$ start(g_k), and let $v = \sum_{i=1}^{k} a_i g_i$ be a codeword. Let j be the smallest integer such that a_j is nonzero; then start$(v) =$ start(g_j). Reversing the code yields the corresponding result for the end set. ∎

Theorem 1: Let G' be a trellis-oriented generator matrix for an m-dimensional subcode C' of an (n, k) linear block code C. If any row (generator) in G' is not atomic in C, then there exists another m-dimensional subcode C'' of C where the minimal trellis of the subcode C'' has fewer edges than the minimal trellis for the subcode C'.

Proof: No two rows of G' start in the same position. If a row r' of G' is not atomic in C, by the previous Lemma there certainly exists an atomic generator r'' from the atomic basis for C that has the same start as r'. Replace that r' with r''. Since the rows of G'' still start in different positions, G'' generates an m dimensional subcode C''. Now, since r'' is atomic in C, it it has the shortest possible span length of all words that start in the given position. In particular, r'' has shorter span length than r'. Thus, r'' contributes fewer edges to the trellis product of the elementary trellises corresponding to the rows of G'' and therefore C'' has a trellis with fewer edges than the minimal trellis for C'.   ∎

For an (n, k) linear block code C, Theorem 1 shows that it is sufficient to consider only the sub-codes generated by a subset of the atomic generators of C when one wants to minimize the number of edges in the minimal trellis corresponding to a subcode. This result is useful since the edge complexity dominates the Viterbi decoding complexity. The "exhaustive search" thus chooses the best among the $\binom{k}{m}$ sub-codes generated by an atomic subset of size m. Since every atomic basis has the same set of atomic spans, and the complexity of the subcode depends only on the spans of its atomic generators, it does not matter which atomic basis is chosen in the first place.

When $\binom{k}{m}$ is large, it is very time-consuming to consider all sub-codes generated by the nonzero codewords which are atomic in the code. In order to reduce the time required for searching, we can perform a "linear forward search" or a "linear backward search."

The "linear forward search" operates as follows. Starting from the empty set, and proceeding iteratively to construct sub-codes of dimension 1, 2, etc., adjoin an atomic generator (not already chosen) to the current set of generators such that corresponding subcode has minimal trellis complexity. The "linear backward search" operates as follows. Starting from the set of atomic generators for C, and proceeding iteratively to construct sub-codes of dimension $k-1$, $k-2$, etc., delete an atomic generator from the current set of generators such that the corresponding subcode has minimal trellis complexity.

Using the Viterbi decoding complexity as a trellis complexity measure (rather than simply the edge complexity), we have found minimal complexity sub-codes of the (16,7) lexicode by the three searching algorithms. The generators that should be chosen and the corresponding speedup are given for sub-codes of each dimension in Table 1. It is assumed that the number of processors available is equal to the number of cosets of the subcode. We have applied these search algorithms to various other codes, and have found that, at least for the codes considered, the linear search algorithms can achieve speedups which are comparable to those achieved by the exhaustive search algorithm. However, the time required for either linear searching algorithm is, in general, much shorter.

Dimension of subcode	Exhaustive search		Linear forward search		Linear backward search	
	spans	speed-up	spans	speed-up	spans	speed-up
1	[1,6]	10.3	+[1,6]	10.3	+[1,6]	10.3
2	[1,6], [11,16]	8.5	+[11,16]	8.5	+[11,16]	8.5
3	[1,6], [8,14], [11,16]	6.4	+[8,14]	6.4	+[8,14]	6.4
4	[1,6], [2,9], [8,14], [11,16]	4.7	+[2,9]	4.7	+[2,9]	4.7
5	[1,6], [2,9], [5,12], [8,14], [11,16]	3.0	+[5,12]	3.0	+[5,12]	3.0
6	[1,6], [2,9], [3,11], [5,12], [8,14], [11,16]	1.8	+[3,11]	1.8	+[3,11]	1.8
7	[1,6], [2,9], [3,11], [5,12], [6,15], [8,14], [11,16]	1.0	+[6,15]	1.0	+[6,15]	1.0

Table 1. Minimal-complexity sub-codes of the $(16,7)$ lexicode.

Fig. 10 shows the speedup values that can be obtained for a variety of binary codes. Codes that achieve a large speedup for a large number of processors are those that have several atomic codewords with a relatively large span length (for example, the $(64,22)$ Reed-Muller code). For codes not having an atomic generator with a span "near" $[1, n]$, removing a generator does not yield a subcode with a trellis of half complexity; therefore such codes yield speedup values that are sub-linear in the number of processors.

4.2 Trellis Sectionalization

The complexity of Viterbi decoding can be reduced by trellis sectionalization, i.e., by dividing the trellis into fewer than n sections in which groups of bits are considered as the relevant code symbol in a given trellis section. Sectionalization to reduce decoding complexity has been studied for particular codes [6, 8, 25, 26]. However, the problem of optimal sectionalization was not studied until the work of Lafourcade and Vardy [20], who presented an algorithm which produces a sectionalization of the trellis of a binary code to minimize the number of binary operations needed for decoding, by taking into account various branch metric computation methods. Based on the linearity properties of the labels in trellis sections, several methods for computing branch metrics and decoding a trellis are obtained in order to further reduce the decoding complexity [20]. We have developed a computer program for the sectionalization algorithm (which is given in [20]) in conjunction with the methods on computing branch metrics and decoding a trellis, which will compute the optimal sectionalization.

Unfortunately, when sectionalization is taken into account, it is no longer sufficient to consider only a subset of an atomic basis to generate a minimum complexity subcode, as the following counterexample shows.

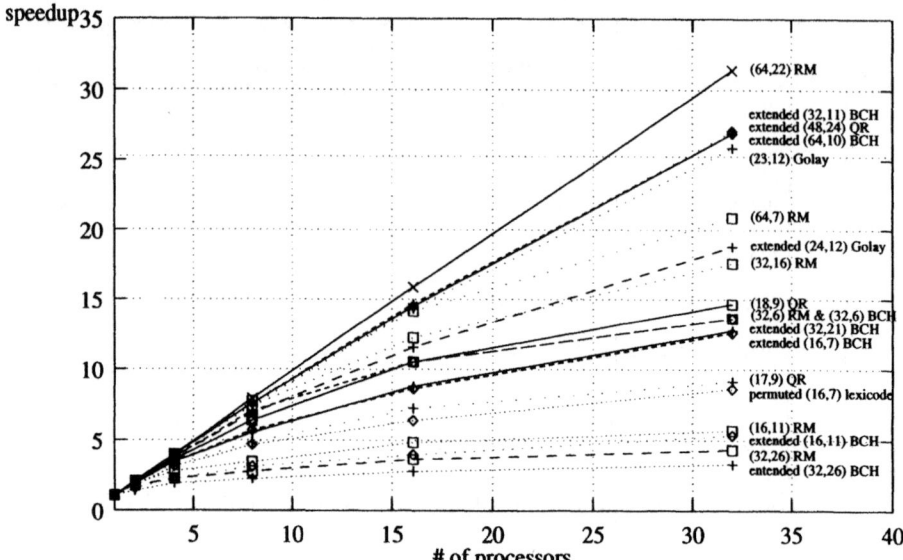

Figure 10. Speedup values versus number of processors available for parallel decoding for various binary linear block codes.

Consider the (7,4) Hamming code with trellis-oriented generator matrix

$$\begin{pmatrix} 1\,1\,0\,1\,0\,0\,0 \\ 0\,1\,1\,0\,1\,0\,0 \\ 0\,0\,1\,1\,0\,1\,0 \\ 0\,0\,0\,1\,1\,0\,1 \end{pmatrix}.$$

With sectionalization and a branch metric computation procedure that takes into account self-complementary codes [20], the one-dimensional subcode with minimum decoding complexity is generated by the codeword (1111111), which is not atomic in the code.

As a result, it is in general necessary to consider all possible sub-codes of a code in order to optimize the decoding complexity when the sectionalization technique is used. Of course, one can still consider only the sub-codes generated by atomic codewords in order to reduce the time required for searching, but such an algorithm is not guaranteed to find an optimal solution.

We have applied the forward and backward linear search algorithms together with trellis sectionalization. We find that values of speedup similar to those obtained without sectionalization can be achieved for a variety of codes. The results are omitted here, but can be found in [27].

5 Conclusions

In this paper we have shown that the simple technique of atomic span modification, combined with the trellis product construction of [11] can yield a wide variety of non-minimal trellis for linear block codes. Such trellises can be designed to have structural properties advantageous for parallel decoding or for VLSI implementation.

Using simple complexity measures based on the number of additions and comparisons required in a decoder, we have attempted to optimize the parallel trellis structure to maximize decoding speed. The trellis complexity measures that we have used in this paper are overly simplistic, though. More realistic measures might take into account survivor sequence management and VLSI wiring and layout complexity. Such an approach has been taken by Moorthy, *et al.* [21], who consider a particular (ACS) processor architecture for VLSI Viterbi decoders.

In this paper, we have also not examined codes in which a suitably sectionalized minimal trellis may itself possess parallel structure. For example, certain Reed-Muller codes have such a structure [6], which can be exploited in decoding. Another useful idea, not considered here, is to design non-minimal trellises that have useful regularity properties, but whose maximum state complexity does not exceed that of the minimal trellis [28, 29]. Indeed, we feel that the results of this paper only scratch the surface of a potentially rich research area, the results of which will lead to new practically implementable soft-decision decoder architectures.

Acknowledgments

The authors would like to thank the anonymous referees for their many helpful suggestions and the editors of this volume for their flexibility in accommodating our non-minimal delay in returning the revised paper.

References

1. G. D. Forney, Jr., "Review of random tree codes." Nasa Ames Research Center, Appendix A of Final Report on Contract NAS2-3637, NASA CR73176, Dec. 1967.

2. G. D. Forney, Jr., "Trellises old and new," in *Communications and Cryptography: Two Sides of One Tapestry* (R. E. Blahut, D. J. Costello, Jr., U. Maurer, and T. Mittelholzer, eds.), pp. 115–128, Kluwer Academic Publishers, 1994.

3. L. R. Bahl, J. Cocke, F. Jelinek, and J. Raviv, "Optimal decoding of linear codes for minimizing symbol error rate," *IEEE Trans. on Inform. Theory*, vol. 20, pp. 284–287, Mar. 1974.

4. J. K. Wolf, "Efficient maximum-likelihood decoding of linear block codes using a trellis," *IEEE Trans. on Inform. Theory*, vol. IT-24, pp. 76–80, 1978.

5. J. L. Massey, "Foundation and methods of channel encoding," in *Proc. Int. Conf. Inform. Theory and Systems*, vol. 65, (Berlin), Sept. 1978.

6. G. D. Forney, Jr., "Coset codes II: Binary lattices and related codes," *IEEE Trans. on Inform. Theory*, vol. 34, pp. 1152–1187, Sep. 1988.

7. D. J. Muder, "Minimal trellises for block codes," *IEEE Trans. on Inform. Theory*, vol. 34, pp. 1049–1053, Sept. 1988.

8. T. Kasami, T. Takata, T. Fujiwara, and S. Lin, "On the optimum bit orders with respect to the state complexity of trellis diagrams for binary linear codes," *IEEE Trans. on Inform. Theory*, vol. 39, pp. 242–245, Jan. 1993.

9. T. Kasami, T. Takata, T. Fujiwara, and S. Lin, "On complexity of trellis structure of linear block codes," *IEEE Trans. on Inform. Theory*, vol. 39, pp. 1057–1064, May 1993.

10. G. D. Forney, Jr. and M. D. Trott, "The dynamics of group codes: State spaces, trellis diagrams and canonical encoders," *IEEE Trans. on Inform. Theory*, vol. 39, pp. 1491–1513, Sept. 1993.

11. F. R. Kschischang and V. Sorokine, "On the trellis structure of block codes," *IEEE Trans. on Inform. Theory*, vol. 41, pp. 1924–1937, Nov. 1995.

12. V. Sorokine, F. R. Kschischang, and V. Durand, "Trellis-based decoding of binary linear block codes," *Lecture Notes in Computer Science*, vol. 793, pp. 270–286, 1994.

13. Y. Berger and Y. Be'ery, "Bounds on the trellis size of linear block codes," *IEEE Trans. on Inform. Theory*, vol. 39, pp. 203–209, Jan. 1993.

14. A. D. Kot and C. Leung, "On the construction and dimensionality of linear block code trellises," in *Proc. 1993 IEEE Int. Symp. on Inform. Theory*, (San Antonio, TX), p. 291, Jan. 17–22, 1993.

15. G. D. Forney, Jr., "Dimension/length profiles and trellis complexity of linear block codes," *IEEE Trans. on Inform. Theory*, vol. 40, pp. 1741–1752, Nov. 1994.

16. F. R. Kschischang and G. B. Horn, "A heuristic for ordering a linear block code to minimize trellis state complexity," in *Proc. 32nd Annual Allerton Conf. on Communication, Control, and Computing, Allerton Park, Illinois,* pp. 75–84, Sept. 1994.

17. R. J. McEliece, "On the BCJR trellis for linear block codes," *IEEE Trans. on Inform. Theory*, vol. 42, 1996. To appear.

18. A. Lafourcade and A. Vardy, "Asymptotically good codes have infinite trellis complexity," *IEEE Trans. on Inform. Theory*, vol. 41, pp. 555–559, March 1995.

19. M. Esmaeli, T. A. Gulliver, and N. P. Secord, "Trellis complexity of linear block codes via atomic codewords." Preprint, 1995.

20. A. Lafourcade and A. Vardy, "Optimal sectionalization of a trellis," *IEEE Trans. on Inform. Theory*, vol. 42, 1996.

21. H. T. Moorthy, S. Lin, and G. T. Uehara, "Good trellises for IC implementation of Viterbi decoders for linear block codes." Submitted to *IEEE Trans. on Commun.*, 1995.

22. J. H. Conway and N. J. A. Sloane, "Lexicographic codes: Error-correcting codes from Game Theory," *IEEE Trans. on Inform. Theory*, vol. IT-32, pp. 337–348, May 1986.

23. F. R. Kschischang, "The trellis structure of maximal fixed-cost codes," *IEEE Trans. on Inform. Theory*, vol. 42, 1996. To appear.

24. J. H. Conway and N. J. A. Sloane, "Soft decoding techniques for codes and lattices, including the Golay code and the Leech lattice," *IEEE Trans. on Inform. Theory*, vol. 32, pp. 41–50, 1986.

25. T. Kasami, T. Takata, T. Fujiwara, and S. Lin, "On the structural complexity of the l-section minimal trellis diagram for binary linear block codes," *IEICE Transactions*, vol. E76-A, pp. 1411–1421, 1993.

26. A. Vardy and Y. Be'ery, "More efficient soft decoding of the Golay codes," *IEEE Trans. on Inform. Theory*, vol. 37, pp. 667–672, 1991.

27. C.-K. Lee, "Nonminimal trellises for linear block codes," Master's thesis, University of Toronto, Department of Electrical and Computer Engineering, June 1996.

28. H. T. Moorthy, S. Lin, and G. T. Uehara, "Trellises with parallel structure for block codes with constraint on maximum state space dimension," in *Proc. 1995 IEEE Int. Symp. Inform. Theory*, (Whistler, B.C., Canada), p. 127, 1995.

29. M. Esmaeili, *Graphical Properties of Quasi Cyclic Codes*. PhD thesis, Ottawa-Carleton Institute of Mathematics and Statistics, 1996.

Trellis Complexity of Linear Block Codes via Atomic Codewords[1]

Morteza Esmaeili[1], T. Aaron Gulliver[2] and Norman P. Secord[3]

[1] Department of Mathematics & Statistics, Carleton University
1125 Colonel By Drive, Ottawa, Ontario, Canada K1S 5B6, mesmaeil@math.carleton.ca
[2] Department of Systems & Computer Engineering, Carleton University
1125 Colonel By Drive, Ottawa, Ontario, Canada K1S 5B6, gulliver@sce.carleton.ca
[3] Communications Research Centre, 3701 Carling Avenue
Box 11490, Station H, Ottawa, ON, Canada K2H 8S2, norman.secord@crc.doc.ca.

Abstract. The trellis complexity of a linear block code C over a field F is presented for C a subspace of the vector space $V = \prod_{i=1}^{n} V_i$ over F, where V_i $(1 \leq i \leq n)$ is a vector space over F. A generator matrix for the Reed-Muller codes is presented which is in trellis oriented form for the minimal L-section trellis diagram.

1 Introduction

The minimal trellis diagram of a linear block code has been analyzed in [1] using the group structure of the code. In [2], the state complexity of the trellis diagram is analyzed via atomic codewords. The advantage of the latter approach is that one can see all aspects of the trellis complexity of the code once the atomic classes are obtained. In this paper, we apply the concept of atomic codewords to study the branch complexity, state connectivity, and L-section trellis diagram of a linear block code. The analysis is based on the trellis multiplication operation and the principle of induction. In the next section, the basic definitions and notation are presented. In Section 3, an (n, k) linear block code C over F is considered as a subspace of $V = \prod_{i=1}^{n} V_i$, where V_i is a vector space over F. In this case, the ith component of a codeword in C is a vector in V_i, so that C may not be a subspace of F^n (this is true only if $V_i = F, 1 \leq i \leq n$). As an application, we have given the L-section trellis oriented generator matrix of the Reed-Muller codes in Section 4.

---

[1] This research was supported in part by the Natural Sciences and Engineering Research Council of Canada and the Telecommunications Research Institute of Ontario.

2 Background

Let F be a field and C an (n, k) linear block code over F. For nonnegative integers i and j, the interval $[i, j]$ is said to be the *span interval*, or simply the *span*, of codeword $c = (c_1, \cdots, c_n) \in C$ if $c_i c_j \neq 0$, and $c_l = 0$ for $l < i$ or $j < l$. In this case the codeword c is said to be *active* in the interval $[i, j - 1]$, and has span length $j - i + 1$. The time indices i and j are called the *starting* and *ending* positions of c, respectively. If $i = j$, then c is said to be *inactive* at all time indices.

Definition 1 (Atomic Codeword) [2]. A codeword $c \in C$ is said to be an atomic codeword if it cannot be expressed as a linear combination of codewords of C with span lengths strictly smaller than that of c. The set of all atomic codewords with the same span is called an atomic class.

Define an equivalence relation on C such that two codewords in C are called equivalent if they have the same span interval. An equivalence class in the partition of C obtained by this relation consists of all codewords having the same span interval. In this partition, any equivalence class containing an atomic codeword is called an atomic class. This is due to the fact that if $a \in C$ is atomic and the span interval of $c \in C$ is the same as that of a, then c is also an atomic codeword. An (n, k) linear block code has k distinct atomic classes[4] [2].

Linear block codes can be decoded using the Viterbi algorithm and a trellis [3, 4]. For this purpose, an (n, k) linear block code C can be represented by a *minimal trellis diagram G* which is a directed graph with unique initial and final vertices, v_0 and v_n, such that:

1. Each directed path connecting v_0 and v_n has length n (and is associated with one codeword of C).
2. Two paths representing two different codewords differ in at least one edge.
3. The number of vertices of G at a distance i, $1 \leq i \leq n$, from v_0 is the minimum possible among all graphical representations of C.

Using the terminology of dynamic systems [5], the vertices of G at a distance i, $0 \leq i \leq n$, from v_0 are called the *states* at time index i.

Other than the concept of atomic codewords, all that is required is the trellis multiplication operation in order to construct the minimal trellis diagram of a

---

[4] This can be considered a result of Theorem 9, which is given in Section 3.

linear block code. As a directed graph, the trellis diagram of a linear block code can be represented by the triple $G = (V, E, A)$ where V, E, and A refer to the vertex, edge, and label sets, respectively. V_i refers to the set of vertices of G at time index i and E_i denotes the set of edges of G connecting the elements of V_{i-1} to the elements of V_i. The label set A is a semigroup under the operation $'+'$. With this notation, we have the following definition of a trellis product.

Definition 2 (Trellis Product) [2]. The product $G' \times G''$ of two length n trellises $G' = (V', E', A)$ and $G'' = (V'', E'', A)$, is defined by

$$G' \times G'' = (\bigcup_{i=0}^{n} V_i' \times V_i'', \bigcup_{i=1}^{n} E_i' \times E_i'', A)$$

with

$$(e', e'') = ((v_1', a', v_2'), (v_1'', a'', v_2'')) =: ((v_1', v_1''), a' + a'', (v_2', v_2''))$$

where

$$e' \in E', e'' \in E'', v' \in V', v'' \in V'', \text{ and } a', a'' \in A.$$

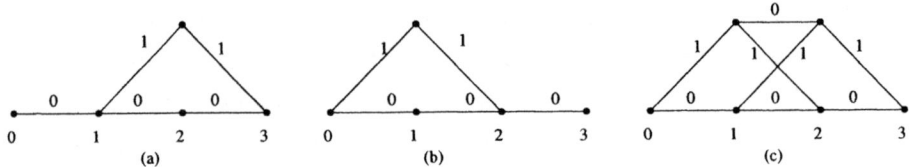

Fig. 1. (a) : G'; (b) : G''; (c) : $G' \times G''$

Definition 2 is illustrated by the trellis product of Fig. 1. The product of k trellises of the same length follows by induction on the definition. Note that the components of codeword c are the labels of a path in the minimal trellis diagram of the code. In this work the labeling is suppressed, as we are only concerned with the trellis structure.

For $0 \leq i \leq n - 1$, let

$$i^- = \{0, 1, \cdots, i\} \quad , \quad i^+ = \{i+1, i+2, \cdots, n\} \tag{1}$$

and define C_{i-} and C_{i+} as

$$\begin{aligned} C_{i-} &= \{c \in C : c_j = 0, \, i+1 \leq j \leq n\}, \\ C_{i+} &= \{c \in C : c_j = 0, \quad 1 \leq j \leq i\}. \end{aligned} \tag{2}$$

In addition, define $C_{0+} = C_{n-} = C$.

The set of all atomic classes of C is denoted by $\mathcal{A}(C)$, or simply $\mathcal{A}$. If $\hat{C}$ is a linear subcode of C, then $\mathcal{A}(\hat{C})$ denotes the set of all atomic classes of C contained in $\hat{C}$. For C_{i-} and C_{i+}, this is denoted by $A([1, i])$ and $A([i + 1, n])$, respectively. In general, $A([i, j])$ will denote the set of all atomic classes of C with span in $[i, j]$, $E([i, j])$ will denote the set of atomic classes of C with span which ends in $[i, j]$, and $I([i, j])$ will denote the set of all atomic classes of C with span which begins in $[i, j]$. Given that an atomic class with span $[a, b]$ is active in the interval $[a, b - 1]$, no element of either $A([1, i])$ or $A([i + 1, n])$ can be active at time i, and in fact atomic codeword c is active at time i if and only if

$$c \notin A([1, i]) \bigcup A([i + 1, n]).$$

Therefore we have the following theorem.

Theorem 3 (State Space Theorem) [5]. *Let C be an (n, k) linear block code over the finite field $GF(q)$. Then $S_i(C) = k - |A([1, i]) \bigcup A([i + 1, n])|$, where $q^{S_i(C)}$ is the number of states at time index i in the minimal trellis diagram of C. $S_i(C)$, or simply S_i, is defined as the dimension of the state space at time index i.*

Note that

$$\begin{aligned}\mathcal{K}(C_{i-}) &= |A([1, i])|, \\ \mathcal{K}(C_{i+}) &= |A([i + 1, n])|,\end{aligned} \tag{3}$$

where $\mathcal{K}(C)$ denotes the dimension of C. $\mathcal{K}(C_{0-})$ and $\mathcal{K}(C_{n+})$ are defined to be zero while $\mathcal{K}(C_{0+}) = \mathcal{K}(C_{n-}) = n$. In short, according to Theorem 3, the number of states at time index i in the minimal trellis diagram G of linear block code C is equal to the cardinality of the quotient group $\frac{C}{C_{i-}C_{i+}}$.

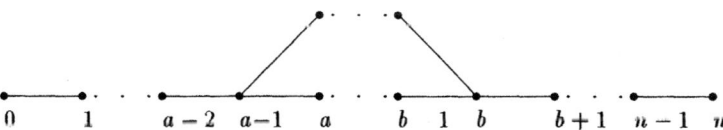

Fig. 2. The minimal trellis diagram of the $(n, 1)$ linear block code C generated by codeword c with span interval $[a, b]$.

It has been shown in [2] that if G_i, $1 \leq i \leq k$, is the minimal trellis diagram of atomic codeword $a^{(i)}$, where the $a^{(i)}$ are from k distinct atomic classes of C, then $G = \prod_{i=1}^{k} G_i$ is the minimal trellis diagram of C. Fig. 2 shows the minimal trellis diagram of the $(n,1)$ linear code C generated by codeword c with span interval $[a, b]$.

Definition 4 (Trellis Oriented Generator Matrix). Let C be an (n, k) linear block code. Any $k \times n$ matrix in which each row is a representative of one of the k distinct atomic classes of C, is called a trellis oriented generator matrix $(TOGM)$ for C.

Since all atomic codewords in one atomic class, say A, have the same minimal trellis diagram, hereafter we will use the term "minimal trellis diagram of atomic class A".

Theorem 5 [2]. *The generator matrix M of a q-ary linear block code C is a TOGM iff no two rows of M either start or end in the same position.*

Example 1. The linear (8,3) code C with TOGM,

$$M = \begin{bmatrix} 1 * * * 1\,0\,0\,0 \\ 0\,0\,1 * * * * 1 \\ 0\,0\,0\,1 * 1\,0\,0 \end{bmatrix}$$

has parameters given in Table 1.

i	0	1	2	3	4	5	6	7	8
$|A([1, i])|$	0	0	0	0	0	1	2	2	3
$|A([i+1, n])|$	3	2	2	1	0	0	0	0	0
$|A([1, i]) \bigcup A([i+1, n])|$	3	2	2	1	0	1	2	2	3
S_i	0	1	1	2	3	2	1	1	0

Table 1. Parameters of the (8,3) binary linear code with generator matrix M given in Example 1.

The entries in M denoted by $*$ have no effect on the structure of the trellis for C.

3 Trellis Complexity of Generalized Linear Block Codes

Let $\mathcal{V}_i$, $1 \leq i \leq n$, be a vector space over F and C (a linear block code) be a finitely generated subspace of the vector space $\mathcal{V} = \prod_{i=1}^{n} \mathcal{V}_i$. Note that this is a more general definition of a linear block code C than was used in the previous sections where $V_i = GF(q)$ for all i, $1 \leq i \leq n$. In [7], Forney states that any generator matrix for a linear block code C can be transformed into a trellis oriented form. In this section, we prove this statement for the general class of linear block codes just defined and determine the atomic codewords of C. As a special case, we determine the L-section trellis diagram of a q-ary linear block code when the sections are of arbitrary length.

A general definition of a row reduced generator matrix is used to define the trellis oriented generator matrix. Let M be an $m \times n$ matrix in which the entries in the jth column, $1 \leq j \leq n$, are in the vector space V_j. Let M_j denote the $m \times j$ matrix formed from the first j columns of M, $1 \leq j \leq n$.

Definition 6 (Row Reduced Matrix). The matrix M is said to be in row reduced form if the nonzero rows of M_i, $1 \leq i \leq n$, are linearly independent.

Any matrix M, with rows in $\prod_{i=1}^{n} V_i$, can easily be transformed into an equivalent row reduced matrix. If p_i, $1 \leq i \leq n$, is the rank of M_i, then by elementary row operations M can be transformed into a matrix M^1 such that the nonzero entries in the first column are linearly independent and are located in the upper portion of the column. Applying row operations on those rows of M^1 that are zero in the first column, we can transform M^1 into M^2 such that

1. M^1 and M^2 are the same in their first p_1 rows
2. for the $k - p_1$ rows of M^2 with zero entries in the first column, the nonzero entries in the second column are linearly independent and located in the upper portion of the column below the $p_1{}^{th}$ row.

The row reduced matrix $\bar{M}$ can be obtained in this manner with a finite number of row operations.

Example 2. Let $n = 4$, $F = GF(2)$, $V_1 = F^3$, $V_2 = F^5$, $V_3 = F^2$, $V_4 = F^6$, and C be a subspace of $V = \prod_{i=1}^{4} V_i$ with generator matrix

$$M = \begin{bmatrix} 111 \ 10111 \ 10 \ 000000 \\ 100 \ 01111 \ 10 \ 001111 \\ 100 \ 00000 \ 11 \ 110111 \\ 100 \ 00000 \ 00 \ 001111 \end{bmatrix}.$$

Note that C is a $(4, 4)$ linear block code, and M is a 4×4 matrix whose i^{th} column contains elements from V_i. Adding the second row to the third and fourth rows results in the matrix

$$M^1 = \begin{bmatrix} 111\ 10111\ 10\ 000000 \\ 100\ 01111\ 10\ 001111 \\ 000\ 01111\ 01\ 111000 \\ 000\ 01111\ 10\ 000000 \end{bmatrix}.$$

Finally, by adding the third row to the fourth in M^1, we obtain

$$M^2 = \begin{bmatrix} 111\ 10111\ 10\ 000000 \\ 100\ 01111\ 10\ 001111 \\ 000\ 01111\ 01\ 111000 \\ 000\ 00000\ 11\ 111000 \end{bmatrix}$$

which is, according to the definition, in row reduced form.

Definition 7 (Canonical Form of a Matrix). Let M be an $m \times n$ row reduced matrix whose entries in the jth column are in the vector space $\mathcal{V}_j$ over F. For each j, $1 \leq j \leq n$, consider the set of rows of M whose last nonzero entry is in the jth column. If these nonzero entries form a linearly independent set for all j, then M is said to be in canonical form.

A row reduced matrix M can be transformed into canonical form by elementary row operations. This can be done by adding scalar multiples of the lower rows to upper rows as required, beginning from the last column.

Example 2 (continued). Consider the last column of matrix M^2 above. Adding the last row to the third gives

$$M^3 = \begin{bmatrix} 111\ 10111\ 10\ 000000 \\ 100\ 01111\ 10\ 001111 \\ 000\ 01111\ 10\ 000000 \\ 000\ 00000\ 11\ 111000 \end{bmatrix}.$$

Clearly the remaining nonzero entries in the last column of M^3 are independent. The first and third rows of M^3 have their last nonzero entries in the third column, and these entries are dependent. Adding the third row in M^3 to the first one, we obtain the canonical matrix

$$M^4 = \begin{bmatrix} 111\ 11000\ 00\ 000000 \\ 100\ 01111\ 10\ 001111 \\ 000\ 01111\ 10\ 000000 \\ 000\ 00000\ 11\ 111000 \end{bmatrix}.$$

In this section, the definition of atomic codewords (Definition 1) is still valid and these codewords are characterized in Theorem 9. However, Theorem 5 determines the TOGM only for q-ary linear block code, and is not valid for the class of codes considered in this section. Instead we have the following lemma.

Lemma 8. *Let C be a linear (n, k) block code over F, and M a generator matrix of C in canonical form. In addition, let A_{ij} denote the set of rows of M which have the same span interval $[i, j]$. Then any nonzero linear combination of the elements of A_{ij} is an atomic codeword.*

Proof Let v be a nonzero linear combination of elements of A_{ij}. If v is not atomic, then it must be possible to express v as a linear combination of atomic codewords, i.e.

$$v = \sum_p u_p \tag{4}$$

where each u_p is atomic with span length strictly smaller than that of v.

Let u_p be an arbitrary fixed element in this expression of v. Since M is a generator matrix, u_p has a unique expression as a linear combination of the rows of M. Suppose it has the following expression

$$u_p = \sum_{v_t \in M} \lambda_t v_t \tag{5}$$

where by $v_t \in M$ we mean v_t is a row of matrix M. We show that $\lambda_t = 0$ in (5), for any $v_t \in M \cap A_{ij}$. u_p is atomic with span length smaller than that of v. Assume u_p has span interval $[s, e]$. Hence interval $[s, e]$ does not include interval $[i, j]$.

Let $e < j$. Define B_l, $j \leq l \leq n$, to be the set of rows of M with ending position l. According to the definition of canonical matrix, the l-th components of the elements of B_l are linearly independent. Since $j \leq n$ and $e < j$, in (5) all the coefficients corresponding to the elements of B_n are zero, otherwise u_p will have ending position $n > e$, which is a contradiction. Getting rid of B_n, we apply the same argument to show that in (5) all the coefficients corresponding to the elements of B_{n-1} are zero, if $j \leq n - 1$. Repeating this procedure, we see that in (5) the coefficients corresponding to the elements of B_j are zero. Since A_{ij} must be a subset of B_j, under the assumption that $e < j$, no element of A_{ij} is involved in the expression of u_p given by (5).

Suppose $j \leq e$. Since the span of u_p is smaller than that of v, it follows that $i < s$. Now define B'_l, $1 \leq l \leq i$, to be the set of rows of M with starting position *l.* An argument similar to the one just used for the case $e < j$ shows that in (5) all the coefficients corresponding to the elements of B'_l, $l \leq i$, are zero. Again A_{ij} must be a subset of B'_l and therefore no element of A_{ij} is involved in the expression of u_p given by (5).

Now by replacing each u_p in (4) with its unique expression (5) as a linear combination of the rows of M, we obtain an expression of v as a linear combination of the rows of M without making use of the elements of A_{ij}. This says that vector v has at least two distinct representations in terms of being expressed as a linear combination of the rows of M, which is a contradiction, since the rows of M are linearly independent. Hence v must be an atomic codeword.♣

For the q-ary linear block codes defined in section 2, if a is an atomic codeword and the span of codeword c is the same as that of a, then c is also an atomic codeword. However, this is not true for the general class of linear block codes considered in this section. For example, in the canonical matrix M^4, the codeword corresponding to the sum of the first and fourth rows is not an atomic codeword even though its span interval is the same as that of the atomic codeword corresponding to the second row.

The following theorem characterizes the set of atomic codewords of a general linear block code C.

Theorem 9. *With the assumptions of Lemma 8, let interval $[i, j]$ be the span-interval of codeword $c \in C$ and let*

$$c = \sum_{t=1}^{k} \lambda_t v_t \tag{6}$$

where the v_t are rows of the canonical generator matrix M. Then c is an atomic codeword iff at least one of the vectors in (6) with nonzero coefficients has span-interval $[i, j]$.

Proof From the structure of the canonical generator matrix M, the span-interval of all of the vectors (codewords) with nonzero coefficients in (6) are subintervals of $[i, j]$. Equation (6) has the following unique form

$$c = \sum_{t=1}^{k} \lambda_t v_t = \sum_{v_t \in A_{ij}} \lambda_t v_t + \sum_{v_t \notin A_{ij}} \lambda_t v_t = c' + \sum_{v_t \notin A_{ij}} \lambda_t v_t \tag{7}$$

with

$$c' = \sum_{v_t \in A_{ij}} \lambda_t v_t.$$

If $c' \neq 0$, then according to Lemma 8 c' is atomic with span-interval $[i, j]$. This along with (7) implies that c is also atomic.

On the other hand if $c' = 0$ then

$$c = \sum_{v_t \notin A_{ij}} \lambda_t v_t$$

meaning that c is not atomic. ♣

To construct a minimal trellis diagram, we must consider codes over a finite field. Hence for the remainder of this section, we assume that C is a linear block code over a finite field F. For each row of the canonical generator matrix M of C with span interval $[a, b]$, we associate a minimal trellis diagram G_{ab} as in Fig. 2 but with $q = |F|$ parallel paths connecting vertices $a - 1$ and b. One can easily check its minimality by comparing the number of states at time index i and the cardinality of the quotient group $\frac{D}{D_{i-}D_{i+}}$, where D is the one-dimensional code generated by the corresponding row of M. Now, we can introduce the minimal trellis diagram of a general linear block code.

Theorem 10. *Let C be an (n, k) linear block code over the finite field $GF(q)$, and M a canonical generator matrix for C with rows denoted by $g_1, g_2, \ldots, g_k$. Moreover, suppose that $G_i, 1 \leq i \leq k$, is a minimal trellis diagram of the one-dimensional code generated by g_i. Then, $G = \prod_{i=1}^{k} G_i$ is a minimal trellis diagram of C.*

Proof The proof is by induction on k, the dimension of C. For codes of dimension 1, the proof is trivial as previously mentioned. Suppose that the statement holds for any linear block code of dimension not greater than $k - 1$. Now, let $\bar{C}$ and $\hat{C}$ be 2 subcodes of C of dimension $k - 1$ and 1, respectively, such that $C = \bar{C}\hat{C}$ and $\hat{C}$ is generated by the codeword corresponding to the last row of M. $\bar{C}$ is generated by the first $k - 1$ rows of M. Suppose the generator of $\hat{C}$ (the last row vector of M) has span-interval $[j, l]$. Then according to the definition,

$$\hat{C}_{i-} = \begin{cases} 0 & \text{if } i \leq l - 1 \\ \hat{C} & \text{if } l \leq i \end{cases} \quad \text{and} \quad \hat{C}_{i+} = \begin{cases} \hat{C} & \text{if } i \leq j - 1 \\ 0 & \text{if } j \leq i \end{cases} \quad (8)$$

where $i \in [0, n]$. Since $C_{i-} = \bar{C}_{i-}\hat{C}_{i-}$ and $C_{i+} = \bar{C}_{i+}\hat{C}_{i+}$, it follows from (8) that

$$C_{i-} = \bar{C}_{i-}\hat{C}_{i-} = \begin{cases} \bar{C}_{i-} & \text{if } i \leq l - 1 \\ \bar{C}_{i-}\hat{C} & \text{if } l \leq i \end{cases} \quad (9)$$

and

$$C_{i+} = \bar{C}_{i+}\hat{C}_{i+} = \begin{cases} \bar{C}_{i+}\hat{C} & \text{if } i < j \\ \bar{C}_{i+} & \text{if } j \leq i. \end{cases} \quad (10)$$

For convenience let $|\frac{\bar{C}}{\bar{C}_i - \bar{C}_{i+}}|$ be denoted by s. Let $i \in [j, l-1]$, then

$$|\frac{C}{C_i - C_{i+}}| = |\frac{\bar{C}\hat{C}}{\bar{C}_i - \hat{C}_i - \bar{C}_{i+} \hat{C}_{i+}}| = |\frac{\bar{C}\hat{C}}{\bar{C}_i - \bar{C}_{i+}}| = q|\frac{\bar{C}}{\bar{C}_i - \bar{C}_{i+}}| = qs. \qquad (11)$$

By the induction hypothesis, the number of states of $\bar{G} = \prod_{t=1}^{k-1} G_t$ at time index i is s, therefore $G = \bar{G} \times G_k$ has qs states at time index i, $j \leq i \leq l-1$. Thus the number of states of G at time i is equal to the cardinality of $\frac{C}{C_i - C_{i+}}$, for $j \leq i \leq l-1$. Now, suppose $i < j$, then

$$|\frac{C}{C_i - C_{i+}}| = |\frac{\bar{C}\hat{C}}{\bar{C}_i - \hat{C}_i - \bar{C}_{i+} \hat{C}_{i+}}| = \frac{|\bar{C}||\hat{C}|}{|\bar{C}_i - ||\bar{C}_{i+}||\hat{C}|} = s. \qquad (12)$$

On the other hand for $i < j$, $G = \bar{G} \times G_k$ and $\bar{G}$ have the same number of states at time index i. Hence for $i < j$, the number of states of G at time index i is equal to $|\frac{C}{\bar{C}_i - \bar{C}_{i+}}|$. The same argument can be applied for $l \leq i$, and the proof is complete by the principle of induction.♣

Example 2 (continued). From Theorem 10, the canonical matrix M^4 is a trellis oriented generator matrix for the linear code C. The minimal trellis diagram for C is given in Fig. 3.

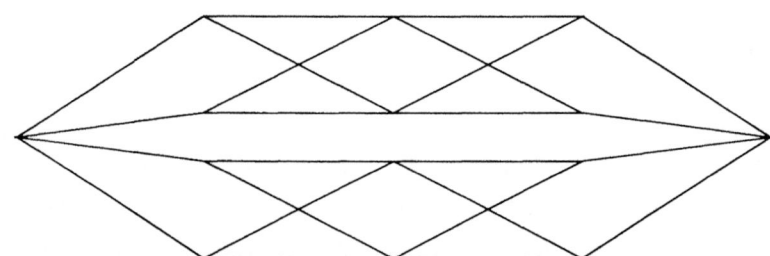

Fig. 3. The minimal trellis diagram of C with canonical generator matrix M^4.

Remark. If in Example 2, M is considered to be the generator matrix of a $(16,4)$ binary linear code C', then M^4 will not be a TOGM for the full 16-section trellis diagram of C' since the first two rows have the same starting position. However, M^4 does correspond to a 4-section minimal trellis diagram for C', with section lengths 3, 5, 2 and 6. In fact, as mentioned previously, an L-section minimal trellis diagram of a general linear block code can be obtained using the results of this section.

Let V be the set of vertices of the minimal trellis diagram, G, of a linear (n, k) block code C, and let V_i be the set of vertices at time index i, $0 \leq i \leq n$. For $1 \leq i \leq n$, let $T_i(G) \equiv G - (V_i \bigcup V_{i-1})^c$ where B^c denotes the complement of the set B [6]. In other words, $T_i(G)$ is the subgraph of G consisting of vertices at time index i and $i - 1$ with edges connecting these two sets of vertices in G. Considering G as an oriented graph, denote the number of edges entering vertex $v \in V$ by $ed(v)$, and the number of edges leaving v by $ld(v)$.

For convenience, let $e(i)$ denote the number of atomic classes with ending position i, and let $s(i)$ denote the number of atomic classes with starting position i.

Theorem 11. *Under the assumptions of Theorem 10, the following statements are true for $0 \leq i \leq n$.*

1. *In V_i, the set of vertices at time index i of the minimal trellis diagram G of C, all vertices have the same entering and leaving degrees, i.e.,*

$$ed(v) = ed(u) = q^{e(i)} \tag{13}$$

and

$$ld(v) = ld(u) = q^{s(i+1)} \tag{14}$$

for all v, $u \in V_i$, except $ld(v_n) = ed(v_0) = 0$. In addition,

$$b(T_i(G)) = q^{s_{i-1}+s(i)} \tag{15}$$

where s_{i-1} is the number of atomic classes active at time $i - 1$.

2. *The number of components [6] of $T_i(G)$, denoted by $t_i(C)$, is*

$$t_i(C) = q^{|I([1,i-1]) \bigcap E([i+1,n])|} \tag{16}$$

for $1 < i < n$. Moreover, each component is a complete bipartite graph with $q^{e(i)}$ vertices on one side and $q^{s(i)}$ vertices on the other. $t_0(C)$ is defined to be zero.

3. *G consists of $q^{|A_{1n}|}$ identical (in terms of the graph structure) parallel sub-trellises, where A_{1n} is the set of rows in M having span $[1, n]$*

Proof Let a_1, a_2, $\cdots$, a_k be atomic codewords from distinct atomic classes A_1, A_2, $\cdots$, A_k of C with minimal trellis diagrams G_1, G_2, $\cdots$, G_k, respectively. Without loss of generality we may assume that the atomic codewords a_1, a_2, $\cdots$, $a_{e(i)}$ end at position i. From the graph product operation it is clear that in $\bar{G} = G_{e(i)+1} \times G_{e(i)+2} \times \cdots \times G_{e(i)+k}$ every vertex of depth i has entering degree 1. On the other hand, by induction on $e(i)$, it follows that in $\hat{G} = G_1 \times G_2 \times \cdots \times G_{e(i)}$ every vertex at time i has entering degree $q^{e(i)}$. Now (13) follows from the fact that G, the minimal trellis diagram of C is given by $G = \bar{G} \times \hat{G}$. The same argument is applied to (14).

To derive (15), it is enough to see that $b(T_i(G))$ is the product of the number of vertices at time $i - 1$, given by $|V_{i-1}| = q^{s_{i-1}}$, and the leaving degree at that position, which is $q^{s(i)}$ according to (14).

Part 2 is an immediate result of Definition 2, (13) and (14).

Suppose in the minimal trellis diagram G' of a length n code C', for any $(e, e') \in T_1(G') \times T_n(G')$ there is a path (codeword) containing e and e' as its first and last edges, respectively. Then obviously G' has no parallel subtrellises. Consider now G'', the minimal trellis diagram of the length n linear block code C'' such that all the paths in G'' share their last edges. It follows from the graph product operation that for any pair $(e, e') \in T_1(G' \times G'') \times T_n(G' \times G'')$, there is a path in $G' \times G''$ having e and e' as its initial and final edges, respectively. Hence $G = G' \times G''$ has no parallel subtrellises.

It is clear that all the paths of the minimal trellis diagram of each element of $S = \{a_1, a_2, \cdots, a_k\} \setminus A_{1n}$ share either their last or first edges. Therefore, using the argument above and applying induction on the cardinality of S shows that the product of the minimal trellis diagrams of the elements of S has no parallel subtrellises. On the other hand, it is obvious that the product of the minimal trellis diagrams of the elements of A_{1n} consists of $q^{|A_{1n}|}$ parallel paths. This completes the proof of the third part of the theorem.♣

Example 2 (continued). The parameters for code C from Theorem 11 are given in Table 2 and are consistent with Fig. 3.

4 Reed-Muller Codes

As an application of the theory developed in Section 4, the trellis structure of the Reed-Muller codes is analyzed based on the TOGM.

i	$e(i)$	$s(i)$	s_i	$b(T_i(G))$	$ed(v)$	$ld(v)$	$a(i)$	$t_i(C)$
0	0	0	0	0	0	4	0	0
1	0	2	2	4	1	2	0	1
2	1	1	2	8	2	2	1	2
3	1	1	2	8	2	1	1	2
4	2	0	0	4	4	0	0	1

Table 2. Relevant parameters for determining the complexity of the code C with TOGM M^4.

Let $a = (a_1, a_2, \cdots, a_n)$ and $b = (b_1, b_2, \cdots, b_n)$ be two n-tuples. The Boolean product of a and b is defined by $ab := (a_1 b_1, a_2 b_2, \cdots, a_n b_n)$. The product of i n-tuples is called a Boolean product of degree i.

For nonnegative integer m, consider the 2^m-tuples $v_0, v_1, \cdots,$ and v_m such that v_0 has Hamming weight 2^m, and v_i, $1 \le i \le m$, is the concatenation of 2^{m-i} identical blocks of length 2^i, each of which is divided into 2 sub-blocks of length 2^{i-1} such that the first sub-block is the string of all zeros, and the second sub-block is the string of all ones.

Definition 12 (Reed-Muller Codes). Let $0 \le r \le m$, and A be the set consisting of v_0 and all the Boolean products of the elements of $D = \{v_1, v_2, \ldots, v_m\}$ up to degree r. The subspace of $GF(2)^{2^m}$, generated by A is defined as the rth order Reed-Muller code of length 2^m, and is denoted by $\mathcal{R}(r, m)$.

Since A is a linearly independent set of vectors, it follows from Definition 12 that $\mathcal{R}(r, m)$ has dimension

$$\sum_{i=0}^{r} \binom{m}{i},$$

and a generator matrix G_m^r whose rows are precisely the elements of A.

Example 3. $\mathcal{R}(2, 3)$ has the following generator matrix G_3^2

$$G_3^2 = \begin{bmatrix} v_0 \\ v_3 \\ v_2 \\ v_1 \\ v_2 v_3 \\ v_1 v_3 \\ v_1 v_2 \end{bmatrix} = \begin{bmatrix} 11\ 11\ 11\ 11 \\ 00\ 00\ 11\ 11 \\ 00\ 11\ 00\ 11 \\ 01\ 01\ 01\ 01 \\ 00\ 00\ 00\ 11 \\ 00\ 00\ 01\ 01 \\ 00\ 01\ 00\ 01 \end{bmatrix}. \tag{17}$$

From the $(u, u + v)$-construction of $\mathcal{R}(r, m)$ [8], it is known that G_m^r, the generator matrix of $\mathcal{R}(r, m)$, has the following form

$$G_m^r = \begin{bmatrix} G_{m-1}^r & G_{m-1}^r \\ \square & G_{m-1}^{r-1} \end{bmatrix}. \tag{18}$$

If we apply this definition to the components on the righthand side of (18), we may express G_m^r in the following way

$$G_m^r = \begin{bmatrix} G_{m-2}^r & G_{m-2}^r & G_{m-2}^r & G_{m-2}^r \\ \square & G_{m-2}^{r-1} & \square & G_{m-2}^{r-1} \\ \square & \square & G_{m-2}^{r-1} & G_{m-2}^{r-1} \\ \square & \square & \square & G_{m-2}^{r-2} \end{bmatrix}. \tag{19}$$

Let B_m^i represent the set of all Boolean products of the elements of $D = \{v_1, v_2, \cdots, v_m\}$ of degree precisely i. Then for $0 \le i \le r \le m$, G_m^r can be expressed in the form

$$G_m^r = \begin{bmatrix} B_m^r \\ B_m^{r-1} \\ \vdots \\ B_m^{i+1} \\ G_m^i \end{bmatrix} \tag{20}$$

The case $i = r - 1$ in (20) is given by

$$G_m^r = \begin{bmatrix} B_m^r \\ G_m^{r-1} \end{bmatrix} \tag{21}$$

The number of elements of B_m^r is given by

$$\binom{m}{r} = \begin{cases} \frac{m!}{r!(m-r)!} & 0 \le r \le m, \\ 0 & \text{otherwise} \end{cases}$$

Combining (19) and (20) with $i = r - 2$ results in the following form for G_m^r

$$G_m^r = \begin{bmatrix} B_{m-2}^r & B_{m-2}^r & B_{m-2}^r & B_{m-2}^r \\ B_{m-2}^{r-1} & B_{m-2}^{r-1} & B_{m-2}^{r-1} & B_{m-2}^{r-1} \\ G_{m-2}^{r-2} & G_{m-2}^{r-2} & G_{m-2}^{r-2} & G_{m-2}^{r-2} \\ \square & B_{m-2}^{r-1} & \square & B_{m-2}^{r-1} \\ \square & G_{m-2}^{r-2} & \square & G_{m-2}^{r-2} \\ \square & \square & B_{m-2}^{r-1} & B_{m-2}^{r-1} \\ \square & \square & G_{m-2}^{r-2} & G_{m-2}^{r-2} \\ \square & \square & \square & G_{m-2}^{r-2} \end{bmatrix} \tag{22}$$

Note that the arrangement of elements in G_m^r given by (22) is such that the nonzero blocks in each row are identical. For example, the elements of B_{m-2}^r are repeated 4 times in the first row of G_m^r.

If $\mathcal{R}(r, m)$ is considered as a subspace of $\prod_{i=1}^4 F^{2^{m-2}}$, then it is clear that G_m^r given by (22) is in row reduced form according to Definition 6. This matrix can easily be transformed into the following matrix M, which is, according to Definition 7 and Theorem 10, a TOGM for the 4-section minimal trellis diagram of $\mathcal{R}(r, m)$

$$
M = \begin{bmatrix}
B_{m-2}^r & B_{m-2}^r & B_{m-2}^r & B_{m-2}^r \\
B_{m-2}^{r-1} & B_{m-2}^{r-1} & \square & \square \\
\square & B_{m-2}^{r-1} & B_{m-2}^{r-1} & \square \\
\square & \square & B_{m-2}^{r-1} & B_{m-2}^{r-1} \\
G_{m-2}^{r-2} & \square & \square & \square \\
\square & G_{m-2}^{r-2} & \square & \square \\
\square & \square & G_{m-2}^{r-2} & \square \\
\square & \square & \square & G_{m-2}^{r-2}
\end{bmatrix}
\tag{23}
$$

In the same way, the following trellis oriented generator matrix M', for the minimal 8-section trellis diagram of $\mathcal{R}(r, m)$ can be obtained

$$
M' = \begin{bmatrix}
B_{m-3}^r & B_{m-3}^r & B_{m-3}^r & B_{m-3}^r & B_{m-3}^r & B_{m-3}^r & B_{m-3}^r & B_{m-3}^r \\
B_{m-3}^{r-1} & B_{m-3}^{r-1} & B_{m-3}^{r-1} & B_{m-3}^{r-1} & \square & \square & \square & \square \\
\square & \square & B_{m-3}^{r-1} & B_{m-3}^{r-1} & B_{m-3}^{r-1} & B_{m-3}^{r-1} & \square & \square \\
\square & \square & \square & \square & B_{m-3}^{r-1} & B_{m-3}^{r-1} & B_{m-3}^{r-1} & B_{m-3}^{r-1} \\
\square & B_{m-3}^{r-1} & \square & B_{m-3}^{r-1} & B_{m-3}^{r-1} & \square & B_{m-3}^{r-1} & \square \\
B_{m-3}^{r-2} & B_{m-3}^{r-2} & \square & \square & \square & \square & \square & \square \\
\square & B_{m-3}^{r-2} & B_{m-3}^{r-2} & \square & \square & \square & \square & \square \\
\square & \square & B_{m-3}^{r-2} & B_{m-3}^{r-2} & \square & \square & \square & \square \\
\square & \square & \square & B_{m-3}^{r-2} & B_{m-3}^{r-2} & \square & \square & \square \\
\square & \square & \square & \square & B_{m-3}^{r-2} & B_{m-3}^{r-2} & \square & \square \\
\square & \square & \square & \square & \square & B_{m-3}^{r-2} & B_{m-3}^{r-2} & \square \\
\square & \square & \square & \square & \square & \square & B_{m-3}^{r-2} & B_{m-3}^{r-2} \\
G_{m-3}^{r-3} & \square & \square & \square & \square & \square & \square & \square \\
\square & G_{m-3}^{r-3} & \square & \square & \square & \square & \square & \square \\
\square & \square & G_{m-3}^{r-3} & \square & \square & \square & \square & \square \\
\square & \square & \square & G_{m-3}^{r-3} & \square & \square & \square & \square \\
\square & \square & \square & \square & G_{m-3}^{r-3} & \square & \square & \square \\
\square & \square & \square & \square & \square & G_{m-3}^{r-3} & \square & \square \\
\square & \square & \square & \square & \square & \square & G_{m-3}^{r-3} & \square \\
\square & \square & \square & \square & \square & \square & \square & G_{m-3}^{r-3}
\end{bmatrix}
\tag{24}
$$

We also have matrix M'' given by (25) in which 0, 1, 2, 3, and 4 denote B_{m-4}^r, B_{m-4}^{r-1}, B_{m-4}^{r-2}, B_{m-4}^{r-3}, and G_{m-4}^{r-4}, respectively, and the dashes denote blocks of zeros. This matrix is a TOGM for the 16-section minimal trellis diagram of $\mathcal{R}(r, m)$.

$$M'' = \begin{bmatrix}
0 & 0 & 0 & 0 & 0 & 0 & 0 & 0 & 0 & 0 & 0 & 0 & 0 & 0 & 0 & 0 \\
1 & 1 & 1 & 1 & 1 & 1 & 1 & 1 & - & - & - & - & - & - & - & - \\
- & - & - & - & 1 & 1 & 1 & 1 & 1 & 1 & 1 & 1 & - & - & - & - \\
- & - & - & - & - & - & - & - & 1 & 1 & 1 & 1 & 1 & 1 & 1 & 1 \\
- & 1 & - & 1 & - & 1 & 1 & - & 1 & - & 1 & - & 1 & - & & \\
- & - & 1 & 1 & - & - & 1 & 1 & 1 & 1 & - & - & 1 & 1 & - & - \\
2 & 2 & 2 & 2 & - & - & - & - & - & - & - & - & - & - & - & - \\
- & - & 2 & 2 & 2 & 2 & - & - & - & - & - & - & - & - & - & - \\
- & - & - & - & 2 & 2 & 2 & 2 & - & - & - & - & - & - & - & - \\
- & - & - & - & - & - & 2 & 2 & 2 & 2 & - & - & - & - & - & - \\
- & - & - & - & - & - & - & - & 2 & 2 & 2 & 2 & - & - & - & - \\
- & - & - & - & - & - & - & - & - & - & 2 & 2 & 2 & 2 & - & - \\
- & - & - & - & - & - & - & - & - & - & - & - & 2 & 2 & 2 & 2 \\
- & 2 & - & 2 & 2 & - & 2 & - & - & - & - & - & - & - & - & - \\
- & - & - & - & - & 2 & - & 2 & 2 & - & 2 & - & - & - & - & - \\
- & - & - & - & - & - & - & - & - & 2 & - & 2 & 2 & - & 2 & - \\
- & - & - & 2 & 2 & 2 & 2 & - & - & 2 & 2 & 2 & 2 & - & - & - \\
3 & 3 & - & - & - & - & - & - & - & - & - & - & - & - & - & - \\
- & 3 & 3 & - & - & - & - & - & - & - & - & - & - & - & - & - \\
- & - & 3 & 3 & - & - & - & - & - & - & - & - & - & - & - & - \\
- & - & - & 3 & 3 & - & - & - & - & - & - & - & - & - & - & - \\
- & - & - & - & 3 & 3 & - & - & - & - & - & - & - & - & - & - \\
- & - & - & - & - & 3 & 3 & - & - & - & - & - & - & - & - & - \\
- & - & - & - & - & - & 3 & 3 & - & - & - & - & - & - & - & - \\
- & - & - & - & - & - & - & 3 & 3 & - & - & - & - & - & - & - \\
- & - & - & - & - & - & - & - & 3 & 3 & - & - & - & - & - & - \\
- & - & - & - & - & - & - & - & - & 3 & 3 & - & - & - & - & - \\
- & - & - & - & - & - & - & - & - & - & 3 & 3 & - & - & - & - \\
- & - & - & - & - & - & - & - & - & - & - & 3 & 3 & - & - & - \\
- & - & - & - & - & - & - & - & - & - & - & - & 3 & 3 & - & - \\
- & - & - & - & - & - & - & - & - & - & - & - & - & 3 & 3 & - \\
- & - & - & - & - & - & - & - & - & - & - & - & - & - & 3 & 3 \\
4 & - & - & - & - & - & - & - & - & - & - & - & - & - & - & - \\
- & 4 & - & - & - & - & - & - & - & - & - & - & - & - & - & - \\
- & - & 4 & - & - & - & - & - & - & - & - & - & - & - & - & - \\
- & - & - & 4 & - & - & - & - & - & - & - & - & - & - & - & - \\
- & - & - & - & 4 & - & - & - & - & - & - & - & - & - & - & - \\
- & - & - & - & - & 4 & - & - & - & - & - & - & - & - & - & - \\
- & - & - & - & - & - & 4 & - & - & - & - & - & - & - & - & - \\
- & - & - & - & - & - & - & 4 & - & - & - & - & - & - & - & - \\
- & - & - & - & - & - & - & - & 4 & - & - & - & - & - & - & - \\
- & - & - & - & - & - & - & - & - & 4 & - & - & - & - & - & - \\
- & - & - & - & - & - & - & - & - & - & 4 & - & - & - & - & - \\
- & - & - & - & - & - & - & - & - & - & - & 4 & - & - & - & - \\
- & - & - & - & - & - & - & - & - & - & - & - & 4 & - & - & - \\
- & - & - & - & - & - & - & - & - & - & - & - & - & 4 & - & - \\
- & - & - & - & - & - & - & - & - & - & - & - & - & - & 4 & - \\
- & - & - & - & - & - & - & - & - & - & - & - & - & - & - & 4
\end{bmatrix} \qquad (25)$$

Applying Theorem 11 and the trellis oriented matrices of (23) and (24), we can easily compute the complexity parameters of the minimal 4-section and 8-section trellis diagrams of the Reed-Muller code $\mathcal{R}(r, m)$. These complexity parameters are given in Tables 3 and 4, respectively.

5 Summary

The trellis oriented generator matrix and its trellis complexity have been presented for the case when C is a linear subspace of the product space $\prod_{i=1}^{n} \mathcal{V}_i$

i	1	2	3	4
$e(i)$	$\sum_{j=0}^{r-2}\binom{m-2}{j}$	$\sum_{j=0}^{r-1}\binom{m-2}{j}$	$\sum_{j=0}^{r-1}\binom{m-2}{j}$	$\sum_{j=0}^{r}\binom{m-2}{j}$
$s(i)$	$\sum_{j=0}^{r}\binom{m-2}{j}$	$\sum_{j=0}^{r-1}\binom{m-2}{j}$	$\sum_{j=0}^{r-1}\binom{m-2}{j}$	$\sum_{j=0}^{r-2}\binom{m-2}{j}$
s_i	$\binom{m-2}{r}+\binom{m-2}{r-1}$	$\binom{m-2}{r}+\binom{m-2}{r-1}$	$\binom{m-2}{r}+\binom{m-2}{r-1}$	0
$Log_2(t_i)$	0	$\binom{m-2}{r}$	$\binom{m-2}{r}$	0

Table 3. Complexity parameters of the 4-section minimal trellis diagram of $\mathcal{R}(r,m)$.

i	$e(i)$	$s(i)$	s_i	$Log_2(t_i)$
1	$\sum_{j=0}^{r-3}\binom{m-3}{j}$	$\sum_{j=0}^{r}\binom{m-3}{j}$	$\sum_{j=0}^{2}\binom{m-3}{r-j}$	0
2	$\sum_{j=0}^{r-2}\binom{m-3}{j}$	$\sum_{j=0}^{r-1}\binom{m-3}{j}$	$\binom{m-3}{r-1}+\sum_{j=0}^{2}\binom{m-3}{r-j}$	$\binom{m-3}{r}+\binom{m-3}{r-1}$
3	$\sum_{j=0}^{r-2}\binom{m-3}{j}$	$\sum_{j=0}^{r-1}\binom{m-3}{j}$	$2\binom{m-3}{r-1}+\sum_{j=0}^{2}\binom{m-3}{r-j}$	$\binom{m-3}{r}+2\binom{m-3}{r-1}$
4	$\sum_{j=0}^{r-1}\binom{m-3}{j}$	$\sum_{j=0}^{r-2}\binom{m-3}{j}$	$\binom{m-3}{r-1}+\sum_{j=0}^{2}\binom{m-3}{r-j}$	$\binom{m-3}{r}+2\binom{m-3}{r-1}$
5	$\sum_{j=0}^{r-2}C_{m-3}^{j}$	$\sum_{j=0}^{r-1}\binom{m-3}{j}$	$2\binom{m-3}{r-1}+\sum_{j=0}^{2}\binom{m-3}{r-j}$	$\binom{m-3}{r}+2\binom{m-3}{r-1}$
6	$\sum_{j=0}^{r-1}\binom{m-3}{j}$	$\sum_{j=0}^{r-2}\binom{m-3}{j}$	$\binom{m-3}{r-1}+\sum_{j=0}^{2}\binom{m-3}{r-j}$	$\binom{m-3}{r}+2\binom{m-3}{r-1}$
7	$\sum_{j=0}^{r-1}\binom{m-3}{j}$	$\sum_{j=0}^{r-2}\binom{m-3}{j}$	$\sum_{j=0}^{2}\binom{m-3}{r-j}$	$\binom{m-3}{r-1}+\binom{m-3}{r}$
8	$\sum_{j=0}^{r}\binom{m-3}{j}$	$\sum_{j=0}^{r-3}\binom{m-3}{j}$	0	0

Table 4. Complexity parameters of the 8-section minimal trellis diagram of $\mathcal{R}(r,m)$.

over a field F. The concepts of a trellis oriented generator matrix and the atomic codewords of a linear block code C have been used to determine the complexity and state connectivity of the minimal L-section trellis diagram of C. The complexity of the minimal L-section trellis diagram of Reed-Muller codes has been analyzed in a simpler and more transparent manner than that given in [1].

References

1. T. Kasami, T. Fujiwara, Y. Deskai and S. Lin, On structural complexity of the L-section minimal trellis diagram for binary linear block codes, *IEICE Trans.*, **E76-A** (1993) 1411-1421.
2. F.R. Kschischang and V. Sorokine, On the trellis structure of block codes, *IEEE Trans. Infor. Theory*, **41** (1995) 1924-1937.
3. L.R. Bahl, J. Cocke, F. Jelinek and J. Raviv, Optimal decoding of linear codes for minimizing symbol error rate, *IEEE Trans. Infor. Theory*, **20** (1974) 284-287.
4. J.K. Wolf, Maximum likelihood decoding of linear block codes using a trellis, *IEEE Trans. Inf. Theory*, **24** (1978) 76-80.
5. G.D. Forney, Jr. and M.D. Trott, The dynamics of group codes: state spaces, trellis diagrams, and canonical encoders, *IEEE Trans. Infor. Theory*, **39** (1993) 1491-1513.

6. F. Harary, *Graph Theory*, New York:Addison-Wesley, 1972.

7. G.D. Forney, Jr., Coset codes part II: binary lattices and related codes, *IEEE Trans. Infor. Theory*, **34** (1988) 1152-1187.

8. S. Roman, *Coding and Information Theory*, Graduate Texts in Mathematics, **134** New York:Springer-Verlag, 1992.

Error Correction for Channels with Substitutions, Insertions, and Deletions[1]

H. Jürgensen and S. Konstantinidis

Department of Computer Science, The University of Western Ontario
London, Ontario, Canada N6A 5B7
Electronic Mail: helmut@uwo.ca, stavros@csd.uwo.ca

Abstract. We introduce a formal method for specifying the properties of discrete information channels and give a general definition of error correction. For the case of channels with substitutions, insertions and deletions, we derive the corresponding channel algebra. Given a description of such a channel and a finite set of words, it is decidable whether the set can correct all errors introduced by the channel.

1. Introduction

With few exceptions, the classical theory of error-correcting codes treats only errors that can be modelled as symbol substitutions and only codes in which all words have the same length, that is, block codes. Less has been published on error correction for other types of errors – for instance errors that can be modelled as insertions or deletions of symbols – or on the error correction capabilities of variable-length codes, that is, codes that may contain words of different lengths.

Block codes correcting insertions and deletions of symbols have been investigated by Levenshtein [12], [11], [13], Sellers [16], and by Varshamov and Tenengol'ts [19]. Our work concerns using *non-block* codes for such error situations.

Codes for the detection of insertions or deletions of symbols have recently received renewed interest because certain types of modern physical channels – optical media, satellite communication, cellular phones, for example – have errors that, regardless of their physical causes, can be suitably modelled by symbol insertions or deletions. Error-correction codes, including variable-length codes, for special error models of this class of errors are proposed and analysed in [2], [6], and [15], for instance.

We are investigating error-correction properties of variable-length codes to be used on channels which, in addition to signal substitutions, may suffer from signal insertions and deletions. Signal insertions and deletions result in and also model loss of synchronization. We restrict our investigation to isolated errors of these kinds as opposed to burst errors; we are planning to study burst errors by appropriately modifying our current model.

---

[1] We gratefully acknowledge the support of this work by the Natural Sciences and Engineering Research Council of Canada under Grant OGP0000243.

The theory of error-correcting codes requires the following two components: a formal model of the communication channel including a precise definition of its error behaviour and an analysis of properties of codes that permit error detection or error correction for channels in the given model. In this paper, we focus on the first issue; the second issue is presented in [8]; see [10] and [9] for further details.

We develop a model for communication channels with isolated substitution, insertion, and deletion errors. This involves both, a formal syntax for describing such channels and a formal semantics for proving properties of such channels. On this basis, we obtain a channel algebra in which channel descriptions can be manipulated and equalities can be proven rigorously.

On the other hand, a general definition of error-correcting codes for arbitrary channels is given. When this definition is applied to our channel model, it provides a general framework in which to investigate and construct codes.

This paper is structured as follows: Following this introduction, Section 2 contains a brief summary of the notation and basic notions used. In Section 3, we define the formal channel descriptions and their informal semantics. In Section 4 we introduce a formal semantics for channel descriptions and we characterize the channel algebra.[2] On this basis, in Section 5, we define error correction and decodability and present some examples and results regarding suitable codes. Section 6 contains a few concluding remarks.

2. Basic Notions and Notation

In this section we introduce the notation used in the rest of the paper and review some basic notions. The symbols $\mathbb{Z}$ and $\mathbb{N}$ denote the sets of integers and of positive integers, respectively; let $\mathbb{N}_0 = \mathbb{N} \cup \{0\}$.

To be able to describe various operations on words – finite, one-sided infinite, and even two-sided infinite – precisely, we use the following notation and tools.

For $n \in \mathbb{N}_0$, let $\mathfrak{n} = \{0, 1, \ldots, n-1\}$ and $-\mathfrak{n} = \{-(n-1), \ldots, -1, 0\}$; by this definition $\mathfrak{o} = -\mathfrak{o} = \emptyset$ and $\mathfrak{1} = -\mathfrak{1} = \{0\}$. Let $\omega = \mathbb{N}_0$, $-\omega = \{-n \mid n \in \mathbb{N}_0\}$, and $\zeta = \omega \cup -\omega = \mathbb{Z}$. We define $-\zeta = \zeta$ and

$$\zeta > -\omega > \cdots > -2 > -\mathfrak{1} > \mathfrak{o} < \mathfrak{1} < \mathfrak{2} < \cdots < \omega < \zeta.$$

In this paper, a set I is said to be an *index set* if $I \in \{-\mathfrak{n}, \mathfrak{n} \mid n \in \mathbb{N}_0\} \cup \{-\omega, \omega, \zeta\}$. For index sets I and J, let

$$I + J = \begin{cases} \{i \mid i \in \mathbb{N}_0, i < |I| + |J|\}, & \text{if } I < \omega \text{ and } J < \omega, \\ \omega, & \text{if } I < \omega \text{ and } J = \omega, \\ \zeta, & \text{if } I = -\omega \text{ and } J = \omega, \\ -\omega, & \text{if } I = -\omega \text{ and } J < -\omega, \\ \{-i \mid i \in \mathbb{N}_0, i < |I| + |J|\}, & \text{if } I < -\omega \text{ and } J < -\omega, \end{cases}$$

---

[2] Due to space limitations, we provide only sketches of the proofs or omit them altogether as the proofs are *very* long. For the same reason we also develop a slightly simplified model only. We plan to present the fully general theory in a more comprehensive publication. See also [10] for details and further results.

and let $-(-I) = I$. The definition of addition could be extended to all symbols; however, we do not need it for any cases beyond these.[3]

A mapping ψ of an index set I into an index set J is an *index mapping* if it is injective and order-preserving and if the image of I is a convex subset of J, that is, $\psi(i) < j < \psi(i')$ for $i, i' \in I$ and $j \in J$ implies that $j = \psi(i'')$ for some $i'' \in I$.

We denote by $|S|$ the cardinality of a set S. For a subset of a set $S_1 \times S_2$, the first and second projections are denoted by π_1 and π_2, respectively.

An *alphabet* is a non-empty set of symbols. In the sequel, let X be an arbitrary *finite*, but fixed alphabet with $|X| > 1$. A *word* w over X is a mapping of an index set I_w into X. A word is specified by its index set I_w and the symbols $w(i) \in X$ for $i \in I_w$. When there is no risk of ambiguity, we write w simply as the string – finite, one-sided infinite, or two-sided infinite depending on I_w – $\cdots w(i-1)w(i)w(i+1)\cdots$ as usual.

A word w is *finite* if $|I_w|$ is finite, that is $I_w = \mathfrak{n}$ or $I_w = -\mathfrak{n}$ for some $n \in \mathbb{N}_0$ and $n = |I_w|$ is the *length* of w denoted by $|w|$. Two distinct finite words w and v are considered equivalent if $I_w = -I_v$ and

$$v(i) = \begin{cases} w(i+n-1), & \text{if } I_w = \mathfrak{n} \text{ for some } n \in \mathbb{N}, \\ w(i-n+1), & \text{if } I_w = -\mathfrak{n} \text{ for some } n \in \mathbb{N}, \end{cases}$$

for $i \in I_v$. Without special mention we identify finite words with their equivalence classes. As usual, let X^* be the set of finite words including the empty word λ, and let $X^+ = X^* \setminus \{\lambda\}$.

Let X^ω be the set of *right-infinite* words over X, that is, words w with $I_w = \omega$ written in the form $w(0)w(1)w(2)\cdots$. Let $X^{-\omega}$ be the set of *left-infinite* words, that is, words w with $I_w = -\omega$ written in the form $\cdots w(-2)w(-1)w(0)$. Let X^ζ be the set of *bi-infinite* words over X, that is, the set of words w with $I_w = \zeta$ written in the form $\cdots w(-2)w(-1)w(0)w(1)w(2)\cdots$. As usual in the case of bi-infinite words, we identify bi-infinite words that can be obtained from each other by a finite shift; thus, when referring to bi-infinite words, we are really referring to equivalence classes with respect to finite shifts. Let $X^\blacklozenge = X^* \cup X^\omega \cup X^{-\omega} \cup X^\zeta$. For $\eta \in \{*, -\omega, \omega, \zeta\}$, a word is an *$\eta$-word* if it is in X^η.

For any index set I, let X^I be the set of all words w with $I_w = I$, let $X^{<I} = \bigcup_{J < I} X^J$, and let $X^{\leq I} = X^I \cup X^{<I}$. Thus, for example, $X^{\leq \omega} = X^* \cup X^\omega$, $X^{<\omega} = X^*$, and $X^\blacklozenge = X^{\leq \zeta}$. Moreover, we define $X^{\leq *} = X^*$.

For $u \in X^* \cup X^{-\omega}$ and $v \in X^* \cup X^\omega$, uv is the concatenation of u and v. When writing concatenations we assume without special mention that the factors are of the required form, that is, when we write uwv with $u, v, w \in X^\blacklozenge$ then it is assumed that u and v are as above and that $w \in X^*$.

The concatenation $z = uv$ satisfies $I_z = I_u + I_v$ and the mapping z is derived from the mappings u and v in the obvious fashion.

Let $Y \subseteq X^\blacklozenge$ and let φ be a mapping of an index set I_φ into Y. If the concatenation of all the images of φ in the order given by I_φ is defined then the

---

[3] The similarity with ordinal numbers is intended. It is exploited in [10].

resulting word is denoted by $[\varphi]$. For $y \in X^{\blacklozenge}$, a *factorization of y over Y* is a mapping φ of an index set I_φ into Y such that $[\varphi] = y$.

For any index set I, Y^I is the subset of all those words in $X^{\blacklozenge}$ having a factorization φ over Y with $I_\varphi = I$. Let $Y^* = \bigcup_{n \in \mathbb{N}_0} Y^n$, $Y^+ = Y^* \setminus Y^\circ$, and $Y^{\blacklozenge} = Y^* \cup Y^{-\omega} \cup Y^\omega \cup Y^\varsigma$. Thus, for example, Y^n is the set of all words in X^* that can be decomposed into a concatenation of n words in Y and, hence, can be considered as a word of length n over the alphabet Y.

For example, if $X = \{0,1\}$, $Y = \{01, 00, 000\}$ then $y = 0100$ has a unique factorization over Y, that is, $I_\varphi = 2$ and $\varphi(0) = 01$, $\varphi(1) = 00$. On the other hand, the words $010000 \cdots \in X^\omega$ and $\cdots 0000010000 \cdots$ have many different factorizations over Y.

For general background regarding the theory of codes we refer to [1] and [17]; the survey [9] explains mechanisms for defining natural classes of codes. For the classical theory of error-correcting codes we refer to [4] or [14], for example.

3. Channel Expressions

In this section, we define the notion of channel expression, we formulate properties which every 'reasonable' physical channel should be expected to have, and we define channel descriptions and their informal semantics.

Definition 3.1 Let X be an alphabet with $|X| > 1$. A *channel over X* is a relation $\gamma \subseteq X^{\blacklozenge} \times X^{\blacklozenge}$.

If γ is a channel over X then we interpret $(y', y) \in \gamma$ to mean that, upon input y, the channel could output y'. To suggest this interpretation we write $(y' \mid y)$ instead of (y', y) in analogy with the notation used for conditional probabilities, for example. For $y \in \pi_2(\gamma)$ let $\langle y \rangle_\gamma = \{y' \in X^{\blacklozenge} \mid (y' \mid y) \in \gamma\}$. Moreover, for $Y \subseteq \pi_2(\gamma)$, let $\langle Y \rangle_\gamma = \bigcup_{y \in Y} \langle y \rangle_\gamma$. Thus $\langle Y \rangle_\gamma$ is the set of all possible outputs of γ when words in Y are used as inputs.

In this definition of a channel, we follow the approach common in the theory of error-correcting codes, that is, we model the reasonably likely transmission errors only and omit probabilities altogether.

Example 3.2 Consider the case of $X = \{0,1\}$ and the binary symmetric channel with error probability p. For a threshold $\varepsilon > 0$, let γ be the set

$$\gamma = \{(y' \mid y) \mid y, y' \in X^n, n \in \mathbb{N}, d_{\mathrm{H}}(y', y)/n \leq p + \varepsilon\}$$

where d_{H} is the Hamming distance. This is, essentially, the approach taken by the classical theory of error-correcting uniform codes for the binary symmetric channels. The word y' of length n can be the output of the channel for a word y, if also y has length n and, moreover, if the number of errors in y' does not exceed a certain threshold. The theory of error-correcting block codes further simplifies this idea by assuming that no more than m errors are likely to occur in blocks of L consecutive symbols for some $m \in \mathbb{N}_0$ and $L \in \mathbb{N}$ with $m < L$.

Consider a channel γ over X. Let $(y' \mid y) \in \gamma$ and $Y \subseteq X^+$ and let φ be a factorization of y over Y. A factorization φ' of y' over $\langle Y \rangle_\gamma$ is γ-admissible for φ if $I_\varphi = I_{\varphi'}$ and, for every non-empty index set I and every index mapping ψ of I into I_φ one has

$$[\varphi'\psi] \in \langle[\varphi\psi]\rangle_\gamma.$$

Intuitively, γ-admissibility of φ' for φ means the following: For any consecutive factors of y with respect to φ, $y_i y_{i+1} \cdots y_{i+k}$ for example, the corresponding factors of y' with respect to φ', $y'_i y'_{i+1} \cdots y'_{i+k}$ say, are obtained as output from γ, that is,

$$y'_i y'_{i+1} \cdots y'_{i+k} \in \langle y_i y_{i+1} \cdots y_{i+k} \rangle_\gamma.$$

For physical reasons it seems natural to require that a channel γ satisfy at least the following conditions:

$(\mathcal{P}_0)$ *A channel preserves finiteness and the type of infiniteness,* that is, if one has $(y' \mid y) \in \gamma$ and $y \in X^\eta$ for $\eta \in \{*, \omega, -\omega, \zeta\}$ then $y' \in X^\eta$.

$(\mathcal{P}_1)$ *Input factorizations have corresponding factorizations of the output,* that is, if $(y' \mid y) \in \gamma$ and φ is a factorization of y over $Y \subseteq \pi_2(\gamma) \cap X^+$ with $I_\varphi \neq \emptyset$ then there is a factorization of y' over $\langle Y \rangle_\gamma$ which is γ-admissible for φ.

In most physical channels, noise affects portions of messages independently of the context in which these portions occur. Moreover, usually, if no message is transmitted – namely, the input to the channel is λ – then no message is received – namely, the output is λ. For this reason we also consider the following condition on channels:

$(\mathcal{P}_2)$ *Error-freeness does not depend on the context,* that is, $(\lambda \mid \lambda) \in \gamma$ and, if $(y' \mid y) \in \gamma$ then $(xy'z \mid xyz) \in \gamma$ for all $x, z \in X^\blacklozenge$ with $xyz \in \pi_2(\gamma)$.

In this paper we only consider channels the error behaviour of which can be modelled as a combination of three basic operations: substitution, insertion, and deletion of symbols, denoted by σ, ι, and δ, respectively. We use the symbol ε to denote the error-free channel type. We define a syntax involving these symbols for channel type expressions defining such channels. We then introduce the formal semantics for such expressions. It will follow that channels described by such expressions satisfy the conditions $\mathcal{P}_0$–$\mathcal{P}_2$.

Definition 3.3 An *SID channel type expression* is a symbol $\tau \in \{\sigma, \iota, \delta, \varepsilon\}$ or an expression of the form $(\tau_1 \odot \tau_2)$ or $(\tau_1 \oplus \tau_2)$ where τ_1 and τ_2 are SID channel type expressions. Let $\mathfrak{T}_0$ be the set of SID channel type expressions not involving the symbol $\oplus$ and let $\mathfrak{T}$ be the set of all SID channel type expressions.

Definition 3.4 An *SID channel expression* is an expression, either of the form $\tau_1(m, L)$ or of the form $(\tau_1(m, L) \oplus \gamma_2)$ where τ_1 is an SID channel type expression, $L \in \mathbb{N}$, $m \in \mathbb{N}_0$, $m < L$, and γ_2 is an SID channel expression. Let $\mathfrak{C}_1$ be the set of all SID channel expressions of the first kind and let $\mathfrak{C}$ be the set of all SID channel expressions.

Informally, every SID channel expression γ defines an *SID channel* $\mathfrak{c}(\gamma)$ as follows: Suppose that $\gamma = \tau(m, L)$ where $\tau \in \mathfrak{T}$. If $\tau = \varepsilon$ then $\mathfrak{c}(\gamma) = \{(y \mid y) \mid y \in X^{\blacklozenge}\}$. If $\tau = \sigma$ then $\mathfrak{c}(\gamma)$ is the set of all pairs $(y' \mid y)$ with $y', y \in X^{\blacklozenge}$ such that y' and y can be aligned in such a way that among any L consecutive symbols of y at most m are different from the corresponding symbols in y'. If $\tau = \iota$ then $\mathfrak{c}(\gamma)$ is the set of all pairs $(y' \mid y)$ with $y', y \in X^{\blacklozenge}$ such that y' and y can be aligned in such a way that between any L consecutive symbols of y at most m symbols need to be inserted to obtain the corresponding symbols in y'. If $\tau = \delta$ then $\mathfrak{c}(\gamma)$ is the set of all pairs $(y' \mid y)$ with $y', y \in X^{\blacklozenge}$ such that y' and y can be aligned in such a way that among any L consecutive symbols of y at most m symbols need to be deleted to obtain the corresponding symbols in y'.

Now if $\tau = (\tau_1 \odot \tau_2)$ for $\tau_1, \tau_2 \in \mathfrak{T}$, then $\mathfrak{c}(\gamma)$ is the set of all pairs $(y' \mid y)$ with $y', y \in X^{\blacklozenge}$ such that y' and y can be aligned in such a way that among any L consecutive symbols of y at most a total of m symbols need to be changed according to τ_1 or τ_2 to get the corresponding symbols of y'. If $\tau = (\tau_1 \oplus \tau_2)$ for $\tau_1, \tau_2 \in \mathfrak{T}$, then $\mathfrak{c}(\gamma)$ is the set of all pairs $(y' \mid y)$ with $y', y \in X^{\blacklozenge}$ such that y' and y can be aligned in such a way that, among any L consecutive symbols of y, at most m symbols need to be changed according to τ_1 and, independently, according to τ_2 to get the corresponding symbols of y'.

Finally, if $\gamma = (\tau_1(m, L) \oplus \gamma_2)$ then $\mathfrak{c}(\gamma)$ is the set of all pairs $(y' \mid y)$ with $y', y \in X^{\blacklozenge}$ such that y' and y can be aligned in such a way that among any L consecutive symbols of y at most m symbols need to be changed according to τ_1 in addition to independent changes according to γ_2 to get the corresponding symbols of y'.

Different SID channel expressions may describe the same channel. For instance, intuitively, the expressions $\delta(1, 5) \oplus \delta(1, 5)$, $(\delta \oplus \delta)(1, 5)$, and $\delta(1, 5)$ all denote the same channel. Similarly, the SID channel type expressions $\iota \odot \sigma$ and $\sigma \odot \iota$ should decsribe the same SID channel type. These statements are supported by the informal definitions of SID channel and SID channel type provided above. However, these informal definitions are quite insufficient as a basis for any rigorous proofs of such properties. They leave far too many ambiguous notions up to intuition. They do serve, however, the purpose of illustrating the meaning of the formal definitions and they should help the reader in understanding the results.

In essence, from the informal definition one can conjecture that SID channels satisfy conditions $\mathcal{P}_0$–$\mathcal{P}_2$ and that representing SID channel types by SID channel type expressions will permit abstract computations with channel types. We develop these ideas in the next section.

4. Channels

In this section we outline a formal semantics for channel type expressions and channel expressions and we characterize the channel type algebra. A completely rigorous presentation is beyond the space allowed for this paper. We resort to some intuitive bridging, but try to keep this at a minimum.[4]

---

[4] Intuitive reasoning about infinite sequences is very error-prone after all.

In the sequel, to simplify the presentation, we only consider SID channel expressions in $\mathfrak{C}_1$. For the more general case of $\mathfrak{C}$, see [10].

To describe the effect of information transmission through a physical channel we introduce the notion of channel functions. In essence, a channel is modelled by a set of channel functions such that, if y is an input and f is a function associated with the channel in question, then $f(y)$ is a possible output of the channel. In this setting, every channel will have many channel functions according to all of its possible behaviours.

For example, if $y = 010$ then the positions for possible symbol substitutions and deletions can be thought of as marked by the symbols themselves. We mark the positions for possible insertions by λ. Thus, we write y as $\lambda 0\lambda 1\lambda 0\lambda$. Now we can express the effect of insertions, deletions, and substitutions as a mapping of this string over the alphabet $X \cup \{\lambda\}$ with λ considered as a symbol into the set of strings over $X \cup \{\lambda\}$ where, this time, λ is considered as the empty word. For example, with i_1 meaning mapping λ to 1, e meaning the identity, and d meaning mapping any symbol to λ, the string $i_1 eeeede$ denotes the function that maps the first symbol to 1, leaves the next four symbols alone, deletes one symbol, and keeps the last one. When applied to $\lambda 0\lambda 1\lambda 0\lambda$, this yields $10\lambda 1\lambda\lambda = 101$.

We use the following *basic error function symbols:*

d to denote the mapping $X \to \{\lambda\}$;

i_x for $x \in X^+$ to denote the mapping of λ onto x;

s to denote a mapping of X into X, such that $s(x) \neq x$ for all $x \in X$;

e to denote the identity function on $X \cup \{\lambda\}$.

Let $G_\delta = \{d\}$, $G_\iota = \{i_x \mid x \in X^+\}$, $G_\sigma = \{s \mid s : X \to X, s(x) \neq x \text{ for all } x\}$, $G_\varepsilon = \{e\}$, and $G = G_\delta \cup G_\iota \cup G_\sigma \cup G_\varepsilon$. For $n \in \mathbb{N}_0$, an n-*error function symbol* is a word h of length $2n + 1$ over G such that

$$h(i) \in \begin{cases} G_\varepsilon \cup G_\iota, & \text{if } i \text{ is even,} \\ G_\varepsilon \cup G_\delta \cup G_\sigma, & \text{if } i \text{ is odd,} \end{cases}$$

for $i \in 2n + 1$. For $\eta \in \{\omega, -\omega, \zeta\}$, an η-*error function symbol* is an η-word h over G satisfying the condition given above for all $i \in \eta$.

For any index set I, an I-error function symbol defines an I-*error function* of X^I into $X^{\blacklozenge}$ as indicated in the example above.[5]

We now define a predicate Q which is true if and only if a finite error function symbol h permits at most m errors of a given type. For an SID channel type expression τ and a finite error function symbol h, let G_τ be the smallest subset of G such that, $d \in G_\tau$ if δ occurs in τ, $G_\sigma \subseteq G_\tau$ if σ occurs in τ, and $G_\iota \subseteq G_\tau$ if ι occurs in τ. Now let $\mathcal{N}(\tau, h)$ be the following sum:

- the number of occurrences of symbols in $G_\tau \cap (G_\delta \cup G_\sigma)$ in h
- plus, for each $i_u \in G_\tau$, the number of occurrences of i_u in h times the length of u.

While defined for all τ and h, $\mathcal{N}(\tau, h)$ is going to be used only when $\tau \in \mathfrak{T}_0$ and $h \in (G_\tau \cup G_\varepsilon)^{\blacklozenge}$. In this case, it is the number of errors introduced by a channel

---

[5] We omit the straighforward, but tedious formal definition. For ζ some special care needs to be taken.

described by τ when the channel's actual input-output behaviour is modelled by h.

For $m \in \mathbb{N}_0$ we require $\mathcal{Q}(\tau, h, m)$ to satisfy the following conditions:

(1) For $\tau \in \mathfrak{T}_0$, $\mathcal{Q}(\tau, h, m)$ is true if and only if $\mathcal{N}(\tau, h) \leq m$.

(2) For $\tau = \tau_1 \oplus \tau_2$, $\mathcal{Q}(\tau, h, m)$ is true if and only if both, $\mathcal{Q}(\tau_1, h, m)$ and $\mathcal{Q}(\tau_2, h, m)$ are true.

(3) $\mathcal{Q}(\tau_1 \odot (\tau_2 \oplus \tau_3), h, m)$ is true if and only if $\mathcal{Q}((\tau_1 \odot \tau_2) \oplus (\tau_1 \odot \tau_3), h, m)$ is true, and analogously for $\mathcal{Q}((\tau_2 \oplus \tau_3) \odot \tau_1, h, m)$.

(4) If $\mathcal{Q}(\tau_1, h, m) = \mathcal{Q}(\tau_2, h, m)$ then $\mathcal{Q}(\tau \odot \tau_1, h, m) = \mathcal{Q}(\tau \odot \tau_2, h, m)$ and $\mathcal{Q}(\tau_1 \odot \tau, h, m) = \mathcal{Q}(\tau_2 \odot \tau, h, m)$.

Condition (1) expresses a total bound on the number of errors as implied by the intuitive meaning of $\odot$. Similarly, condition (2) expresses that the bound applies to error types separately, when $\oplus$ is involved. Conditions (3) and (4) are consistency conditions. In principle, there could be more than one predicate satisfying conditions (1)–(4). Below we demonstrate that this is not the case; therefore, the predicate $\mathcal{Q}$ introduced above is well-defined.

We consider the set $\mathfrak{T}$ as a free algebra generated by the set $\{\varepsilon, \sigma, \iota, \delta\}$ with the two binary operation symbols $\oplus$ and $\odot$. Let $\cong$ be the smallest congruence on $\mathfrak{T}$ such that, in the factor algebra $\mathfrak{T}/\cong$, the operations are associative, commutative, $\odot$ is idempotent on $\mathfrak{T}_0$, $\oplus$ is idempotent on $\mathfrak{T}$, and the following equations are satisfied

(E1) $\quad \tau \odot \varepsilon = \tau$ (E2) $\quad \tau \oplus \varepsilon = \tau$

(E3) $\quad (\tau_1 \odot \tau_2) \oplus \tau_2 = \tau_1 \odot \tau_2$ (E4) $\quad (\tau_1 \oplus \tau_2) \odot \tau_3 = (\tau_1 \odot \tau_3) \oplus (\tau_2 \odot \tau_3)$

for all $\tau, \tau_1, \tau_2, \tau_3 \in \mathfrak{T}$.

The postulated properties of the congruence $\cong$ are chosen with the following intuition in mind. Channel type expressions are intended for specifying the kinds of errors occurring in various communication channels. Thus, it seems plausible to conjecture that, in the specification of a channel, the order in which the errors are listed does not matter nor in which way independently restricted and dependently restricted errors are grouped together. Hence, one would expect that the congruent types $\tau_1 \circ \tau_2 \cong \tau_2 \circ \tau_1$, $\circ \in \{\oplus, \odot\}$, specify the same error situation and similarly for $\tau_1 \circ (\tau_2 \circ \tau_3) \cong (\tau_1 \circ \tau_2) \circ \tau_3$; that is, one would expect $\oplus$ and $\odot$ to be commutative and associative. In a similar fashion, intuitive reasons can also be given for the other properties of $\cong$. For example, the SID channel expression $((\delta \odot \iota) \oplus \iota)(3, 7)$ is intended to denote a channel with the following properties: In every 7 consecutive symbols, a total of up to 3 deletions and insertions is permitted and, independently, up to three insertions are permitted. This channel would already be completely specified by the expression $(\delta \odot \iota)(3, 7)$, and this implies (E3). These and similar arguments suggest that congruent SID channel type expressions describe the same error situations. In fact, the soundness of $\cong$ will be verified after we have derived the definitions of SID channel and of equivalence between SID channels. In the notation we now omit parentheses and assume that $\odot$ binds more strongly that $\oplus$.

Theorem 4.1 *Every SID channel type expression is effectively congruent to one of the following:*

(1)	ε	*(2)*	σ		
(3)	ι	*(4)*	δ		
(5)	$\sigma \oplus \iota$	*(6)*	$\sigma \oplus \delta$	*(7)*	$\iota \oplus \delta$
(8)	$\sigma \oplus \iota \oplus \delta$				
(9)	$\sigma \odot \iota$	*(10)*	$\sigma \odot \iota \oplus \delta$	*(11)*	$\sigma \odot \iota \oplus \sigma \odot \delta$
(12)	$\sigma \odot \delta$	*(13)*	$\sigma \odot \delta \oplus \iota$		
(14)	$\iota \odot \delta$	*(15)*	$\iota \odot \delta \oplus \sigma$	*(16)*	$\iota \odot \delta \oplus \iota \odot \sigma$
(17)	$\iota \odot \delta \oplus \sigma \odot \delta$	*(18)*	$\iota \odot \sigma \oplus \iota \odot \delta \oplus \sigma \odot \delta$	*(19)*	$\sigma \odot \iota \odot \delta$

Moreover, no two SID channel type expressions in this list are congruent.

Proof idea: The fact that every $\tau \in \mathfrak{T}$ is congruent to one of the listed above type expressions follows easily if we note that when (E4) is used together with the associativity and commutativity of $\oplus$ and $\odot$, the type expression τ can be transformed to a $\oplus$-sum of type expressions in $\mathfrak{T}_0$.

On the other hand, the fact that any two of the types (1)–(19) are not congruent relies on the following observations which can be shown by a careful examination of the properties of $\cong$. Let $\tau \cong \tau'$.

(i) For every $\varrho \in \{\sigma, \iota, \delta\}$, if ϱ occurs in τ then ϱ occurs in τ'.

(ii) For every $\varrho_1, \varrho_2 \in \{\sigma, \iota, \delta\}$ with $\varrho_1 \neq \varrho_2$, if $\varrho_1 \odot \varrho_2$ occurs in τ then there are $\tau_1, \tau_2 \in \mathfrak{T}$ such that $\tau_1 \odot \tau_2$ occurs in τ', ϱ_{i_1} occurs in τ_1, and ϱ_{i_2} occurs in τ_2, where $\{\varrho_{i_1}, \varrho_{i_2}\} = \{\varrho_1, \varrho_2\}$.

(iii) If $\{\varrho_1, \varrho_2, \varrho_3\} = \{\sigma, \iota, \delta\}$ and $\varrho_1 \odot \varrho_2 \odot \varrho_3$ occurs in τ then there are $\tau_1, \tau_2, \tau_3, \mu \in \mathfrak{T}$ such that $\tau_1 \odot \mu$ or $\mu \odot \tau_1$ occurs in τ', ϱ_{i_1} occurs in τ_1, $\tau_2 \odot \tau_3$ occurs in μ, ϱ_{i_2} occurs in τ_2, and ϱ_{i_3} occurs in τ_3, where $\{\varrho_{i_1}, \varrho_{i_2}, \varrho_{i_3}\} = \{\sigma, \iota, \delta\}$.

The first statement implies that, when $\tau \cong \tau'$, τ and τ' contain exactly the same type expressions from $\{\varepsilon, \sigma, \iota, \delta\}$. The third statement ensures that (19) cannot be congruent to any of (1)–(18). Finally, the second statement implies that, for any τ and τ' in (1)–(19), if $\tau \cong \tau'$ and $\sigma \odot \iota$ occurs in τ, for instance, then $\sigma \odot \iota$ must occur in τ' as well. Thus, when the three statements are considered together the claim of the theorem follows. $\square$

The SID channel type expressions listed in Theorem 4.1 are said to be the SID *normal forms* of channel type expressions. In particular, in normal form, an SID channel type can be represented in the form of a $\oplus$-sum of terms not involving $\oplus$.

Theorem 4.2 *The predicate Q is well-defined.*

Proof idea: We define a predicate $\mathcal{R}$ as follows:

(i) For all $\tau \in \mathfrak{T}_0$ in normal form and for all h and m, $\mathcal{R}(\tau, h, m)$ is true if and only if $\mathcal{N}(\tau, h) \leq m$.

(ii) For all $\tau \in \mathfrak{T} \setminus \mathfrak{T}_0$ in normal form and for all h and m, $\mathcal{R}(\tau, h, m)$ is true if and only if $\mathcal{R}(\tau_i, h, m)$ is true for every $i \in \mathfrak{n}$, where $\tau = \tau_0 \oplus \cdots \oplus \tau_{n-1}$ and $2 \leq n \leq 3$.

(iii) For all $\tau \in \mathfrak{T}$ not in normal form and for all h and m, $\mathcal{R}(\tau, h, m)$ is true if and only if $\mathcal{R}(\tau', h, m)$ is true, where τ' is the unique normal form of τ.

The claim of the theorem follows as a result of the next two statements the proofs of which are omitted.

- The predicate $\mathcal{R}$ satisfies the conditions (1)–(4) of a predicate $\mathcal{Q}$.
- If $\mathcal{R}'$ is another predicate satisfying conditions (1)–(4) then $\mathcal{R}' = \mathcal{R}$.

$\square$

Definition 4.3 The *SID η-channel* $\mathfrak{c}_\eta(\tau(m, L))$ is the set of pairs $(y' \mid y)$ satisfying the following conditions:

(1) $y \in X^{\leq \eta}$ and $y' = h(y)$ for some I_y-error function symbol h;

(2) h is a word over the alphabet $(G_\tau \cup G_\varepsilon)$;

(3) for every factor g of length $2L + 1$ of h starting with e or with i_u for some $u \in X^+$ one has $\mathcal{Q}(\tau, g, m)$.

By condition (1) of Definition 4.3, y' is obtained from y by an error function; by condition (2), this error function involves only errors occurring in τ and, by condition (3), there are at most m such errors among each consecutive L symbols of the message. Nothing can be assumed about the length of y', that is, at this point we can only state that $y' \in X^\blacklozenge$.

Lemma 4.4 *For $\tau \in \mathfrak{T}$, $\eta \in \{*, -\omega, \omega, \zeta\}$, $L \in \mathbb{N}$, and $m \in \mathbb{N}_0$ with $m < L$, let $\gamma_\eta = \mathfrak{c}_\eta(\tau(m, L))$. For $y \in \pi_2(\gamma_\eta)$ one has $\langle y \rangle_{\gamma_\eta} = \langle y \rangle_{\gamma_\zeta}$ and, therefore, $\pi_1(\gamma_\eta) = \langle \pi_2(\gamma_\eta) \rangle_{\gamma_\zeta}$.*

In view of this observation, it is sufficient to consider the case of $\eta = \zeta$.

Definition 4.5 An *SID channel* is a channel γ such that $\gamma = \mathfrak{c}_\zeta(\gamma')$ for some channel expression[6] $\gamma' \in \mathfrak{C}_1$.

Intuitively, we consider two SID channel type expressions as equivalent if they describe the same SID channels.

Definition 4.6 Two SID channel type expressions τ_1 and τ_2 are said to be *equivalent*, $\tau_1 \equiv \tau_2$, if, for all $L \in \mathbb{N}$ and all $m \in \mathbb{N}_0$ with $m < L$, the SID channels $\mathfrak{c}_\zeta(\tau_1(m, L))$ and $\mathfrak{c}_\zeta(\tau_2(m, L))$ are equal.

Our next result shows that the intuition of the congruence $\cong$ is correct, namely the relations $\cong$ and $\equiv$ are equal.

Theorem 4.7 *One has $\tau_1 \equiv \tau_2$ if and only if $\tau_1 \cong \tau_2$.*

By Theorem 4.1, the congruence of SID channel type expressions is decidable: Given two SID channel type expressions τ_1 and τ_2, one computes their normal forms using the equations which define the congruence $\cong$. One has $\tau_1 \cong \tau_2$ if and only if their normal forms are equal. The decidability of $\cong$ implies, by Theorem 4.7, that also $\equiv$ is decidable.

Example 4.8 Consider the SID channel type expression[7]

---

[6] This definition and the subsequent results about SID channels can be extended to $\mathfrak{C}$; see [10].

[7] As before, we omit outermost parentheses.

$$\tau = \Big(\big((\sigma \oplus \iota) \odot \iota\big) \odot \delta\Big) \oplus (\sigma \odot \delta).$$

First, one has

$$\big((\sigma \oplus \iota) \odot \iota\big) \odot \delta \cong (\sigma \oplus \iota) \odot (\iota \odot \delta) \cong \big(\sigma \odot (\iota \odot \delta)\big) \oplus \big(\iota \odot (\iota \odot \delta)\big)$$

by the associativity of $\odot$ and by (E4),

$$\cong \big(\sigma \odot (\iota \odot \delta)\big) \oplus \big((\iota \odot \iota) \odot \delta\big) \cong \big(\sigma \odot (\iota \odot \delta)\big) \oplus (\iota \odot \delta)$$

by the associativity of $\odot$ and by the idempotence of $\odot$ on $\mathfrak{T}_0$,

$$\cong \big(\sigma \odot (\iota \odot \delta)\big) \cong \big((\sigma \odot \iota) \odot \delta\big)$$

by (E3) and by the associativity of $\odot$,

$$\cong \big((\iota \odot \sigma) \odot \delta\big) \cong \big(\iota \odot (\sigma \odot \delta)\big)$$

by the commutativity of $\odot$ and by the associativity of $\odot$. Thus,

$$\tau \cong \big(\iota \odot (\sigma \odot \delta)\big) \oplus (\sigma \odot \delta) \cong \big(\iota \odot (\sigma \odot \delta)\big) \cong \big((\iota \odot \sigma) \odot \delta\big)$$

by (E3) and by the associativity of $\odot$

$$\cong \big((\sigma \odot \iota) \odot \delta\big)$$

by the commutativity of $\odot$. Thus τ has the normal form (19).

Theorem 4.9 *Every SID channel satisfies* $\mathcal{P}_0$, $\mathcal{P}_1$, *and* $\mathcal{P}_2$.

We have, in this section, given a precise meaning to the notion of a channel involving substitutions, insertions, and deletions. This formalism permits us to prove rigorously – rather than argue intuitively – properties of channels. Theorem 4.1 states that there are really only 19 different SID channel types. By Theorem 4.9, all SID channels conform to our basic requirements for a mathematical model of physical channels. We have stated this result here only for channel expressions in $\mathfrak{C}_1$. It can, however, be extended to channel expressions in $\mathfrak{C}$ with some effort.

5. Error Correction and Decoding

In information transmission, the sender *encodes* the message in order for the receiver to be able to regain the message uniquely (with probability close to 1). With this idea in mind, we formulate the notions of decodability and error correction with respect to a given channel γ.

In the context of this paper, a *code* is any set[8] $K \subseteq X^+$. For

$$\eta \in \{*, \omega, -\omega, \zeta, \leq-\omega, \leq\omega, \leq\zeta\}$$

an η-*message* over K is any word in K^η. The code K is *uniquely η-decodable* if every η-message over K has a unique factorization[9] over K. For a channel γ, a γ-*received η-message* is any word in $\langle K^\eta \rangle_\gamma$.

---

[8] In adopting this terminology we go back to the early days of coding theory when unique decodability was not implied by the term of *code*.

[9] Usually, uniquely *-decodable codes are just called codes in the literature. In the case of $\eta = \zeta$, one has to observe some subtle details which we omit here as they are not needed. See [10].

Clearly, if K is a uniquely ω-decodable code or a uniquely $(-\omega)$-decodable code then it is also uniquely $*$-decodable; if it is a uniquely ζ-decodable code it is also uniquely ω-decodable and uniquely $(-\omega)$-decodable. The converse implications are not true in general.[10]

A set $K \subseteq X^+$ is a *uniform code*[11] if $K \subseteq X^n$ for some $n \in \mathbb{N}$. Every uniform code is uniquely ω-decodable; however, there are uniform codes which are not uniquely ζ-decodable. For example, using $X = \{0,1\}$, $K = \{01,10\}$, the ζ-word $\cdots 01010 \cdots \in K^\zeta$ has two distinct factorizations over K, that is $\cdots (01)(01)(0 \cdots$ and $\cdots 0)(10)(10) \cdots$.

A set $K \subseteq X^+$ is a *prefix code* if no word in K has another word in K as a prefix, that is, $K \cap KX^+ = \emptyset$. Every prefix code is uniquely ω-decodable. The set K is an *infix code* if no word in K has another word in K as a factor, that is, $(K \cap X^+ KX^*) \cup (K \cap X^* KX^+) = \emptyset$. Thus, every infix code is a prefix code and every uniform code is an infix code. The set K is a *solid code* if it is an infix code such that, for all $u, v \in K$, there is no $w \in X^+$ such that $u \in wX^+$ and $v \in X^+ w$. Every solid code is uniquely ζ-decodable [10, 9]. For a survey of and further information on the construction methods for such classes of codes see [7, 9] and the references provided there.

The decoding problem for a given channel γ and a given code K is as follows. Given any γ-received η-message $y' \in \langle K^\eta \rangle_\gamma$, find $w \in K^\eta$ such that $(y' \mid y) \in \gamma$. For y to be unique, one must have $\langle u \rangle_\gamma \cap \langle w \rangle_\gamma = \emptyset$ for any two distinct words $u, w \in K^\eta$. Let $\mathcal{K}^\eta_\gamma$ be the set of uniquely η-decodable codes K such that, for any $u, w \in K^\eta$, $\langle u \rangle_\gamma \cap \langle w \rangle_\gamma \neq \emptyset$ implies that $u = w$.

Definition 5.1 A code K is (γ, η)-*correcting* if $K \in \mathcal{K}^\eta_\gamma$.

In [8] we describe general methods for constructing (γ, η)-correcting codes for channels $\gamma = \delta(1, L)$. For example, the code

$$K = \{0022, 02122, 001122\}$$

over the alphabet $X = \{0,1,2\}$ is (γ, ζ)-correcting for such a channel with $L = 6$. Note that K is a solid code which is not uniform.

Example 5.2 Let $X = \{0,1\}$ and $K = \{0011, 010111\}$. K is a solid, but not a uniform code. Let $\gamma = \delta(1,6)$. Then

$$0011(10111)^\omega = 00111(01111)^\omega \in \langle 0011(010111)^\omega \rangle_\gamma \cap \langle (010111)^\omega \rangle_\gamma.$$

Therefore, K is not (γ, ω)-correcting. On the other hand, one can show that K is $(\gamma, *)$-correcting.

Theorem 5.3 [8] *Let γ be an SID channel. For a given finite code it is decidable whether it is $(\gamma, *)$-correcting.*

---

[10] For the difference between the various cases of unique decodability see [5], [18], [3], and [9].

[11] Called *block code* in the literature on error-correcting codes.

This result is an analogue to the famous Sardinas-Patterson theorem for noiseless channels; its proof is significantly more difficult, however, and uses completely different ideas. In the rest of the section we give an informal description of the notion of *overlap* which provides the basic tools for proving Theorem 5.3 and, more generally, for studying the behaviour of codes with SID channels. Consider an SID channel $\gamma = \tau(m, L)$ and two distinct $*$-messages $w = w_0 \cdots w_{n-1}$ and $u = u_0 \cdots u_{m-1}$ over a code K such that $w_0, \ldots, w_{n-1}, u_0, \ldots, u_{m-1} \in K$. If there is a $w' \in X^*$ which can be received through γ from both w and u, then there are error function symbols $h, g \in (G_\varepsilon \cup G_\tau)^+$ such that $w' = h(w) = g(u)$. Since h is a word over G and γ satisfies $\mathcal{P}_1$, one can decompose h such that $h = h_0 \cdots h_{n-1}$ and $h(w) = h_0(w_0) \cdots h_{n-1}(w_{n-1})$. On the other hand, for SID channels, it can be shown that when $g(u) = h_0(w_0) \cdots h_{n-1}(w_{n-1})$, g and u can be decomposed such that the following statements hold true:

(i) $u = z_0 \cdots z_{n-1}$, $z_0 \in K^* P(K)$, $z_1, \ldots, z_{n-2} \in S(K) K^* P(K) \cup F(K)$, and $z_{n-1} \in K^* S(K)$, where $P(K)$ is the set of all proper prefixes of the words in K, $S(K)$ is the set of all proper suffixes of the words in K, and $F(K)$ is the set of all proper infixes of the words in K.

(ii) $g = g_0 \cdots g_{n-1}$ and $g(z_0), \ldots, g(z_{n-1})$ are well-defined such that $h(w) = g_0(z_0) \cdots g_{n-1}(z_{n-1})$.

An *elementary* (γ, K)-*overlap* is a quadruple $[h, g, u, z]$ such that $u \in K$, $z \in K^* P(K) \cup S(K) K^* P(K) \cup F(K) \cup S(K) K^*$, and $h(u) = g(z)$. For a given SID channel γ and finite code K, the set of all elementary (γ, K)-overlaps is computable. Moreover, under certain conditions,[12] two (γ, K)-overlaps $[h, g, u, z]$ and $[h', g', u', z']$ can be multiplied to produce a new overlap $[hh', gg', uu', zz']$. Overlaps and the partial multiplication between overlaps provide the tools for investigating the behaviour of codes with SID channels. As a result, a necessary and sufficient condition for a code to be $(\gamma, *)$-correcting can be obtained which subsequently leads to the decidability of error correction for SID channels.

6. Concluding Remarks

In defining SID channel expressions and their semantics we have established a formal method for specifying the idealized non-probabilistic behaviour of discrete channels. We believe that this approach can serve as a guiding example for modelling other types of error situations in channels.

With such a formal specification, the statements about channel properties are mathematically unambiguous and, assuming the adequacy of the respective model for the given physical situation, one can *prove* properties of such channels and of codes used on such channels.

Concerning the first point, we establish a formal system in which SID channel specifications can be manipulated without recourse to their informal intuitive interpretations. The algebra of channel types provides an algorithm for proving the equivalence of SID channel type expressions, for example. Moreover, using

---

[12] Observe that when h and h' are error function symbols satisfying $\mathcal{Q}(\tau, h, m) \wedge \mathcal{Q}(\tau, h', m)$, it is not necessarily the case that $\mathcal{Q}(\tau, hh', m)$ is true.

our formal model only, one can prove that SID channels as defined have certain basic properties which one would expect every physical channel to satisfy. This strengthens our claim that our mathematical SID channels are true models of physical SID channels. It also points to some general techniques for modelling other kinds of channels.

Given our definition of a channel, one can define the notion of error correction with respect to this channel without reference to probabilities. In the case of substitution errors, this is the approach taken by the theory of error-correcting codes from its very beginning. For the case of other error models, the relevant notions have, to our knowledge, never been formulated as rigorously as to render rigorous proofs of general properties of error-correcting codes possible. For insertion and deletion errors, all proofs in the literature rely on an intuitive understanding of these error models. Our Theorem 5.3, establishing the decidabilty of error correction, would be impossible to prove without a formal error model and a formal definition, as would be the decidability of unique decodability in the case of noiseless channels without a rigorous definition of unique decodability.

The techniques developed in this paper are used in [8] and [10] to derive sufficient and necessary properties of a code to be (γ, η)-correcting. Undoubtedly, much more work is required to arrive at general characterizations. In view of the physical characteristics of modern types of communication channels and in view of the fact that new communication media and methods are likely to require new error models, our approach might serve as a prototype and a general framework of how to model such physical channels and their error patterns.

References

1. J. Berstel, D. Perrin: *Theory of Codes*. Academic Press, Orlando, 1985.
2. P. A. H. Bours: Construction of fixed-length insertion/deletion correcting runlength-limited codes. *IEEE Trans. Inform. Theory*, **IT-40** (1994), 1841–1856.
3. J. Devolder, M. Latteux, I. Litovsky, L. Staiger: Codes and infinite words. *Acta Cybernet.*, **11** (1994), 241–256.
4. J. Duske, H. Jürgensen: *Codierungstheorie*. BI Wissenschaftsverlag, Mannheim, 1977.
5. S. W. Golomb, B. Gordon: Codes with bounded synchronization delay. *Inform. and Control*, **8** (1965), 355–372.
6. H. D. L. Hollman: A relation between Levenshtein-type distances and insertion-and-deletion correcting capabilities of codes. *IEEE Trans. Inform. Theory*, **IT-39** (1993), 1424–1427.
7. H. Jürgensen, S. Konstantinidis: The hierarchy of codes. In Z. Ésik (editor): *Fundamentals of Computation Theory, 9th International Conference, FCT'93. Lecture Notes in Computer Science* **710**, 50–68, Berlin, 1993. Springer-Verlag.
8. H. Jürgensen, S. Konstantinidis: Variable-length codes for error correction. In Z. Fülöp, F. Gécseg (editors): *Automata, Languages and Programming,*

22nd International Colloquium, ICALP95, Proceedings. Lecture Notes in Computer Science **944**, 581–592, Berlin, 1995. Springer-Verlag.

9. H. Jürgensen, S. Konstantinidis: Codes. In G. Rozenberg, A. Salomaa (editors): *Handbook of Formal Language Theory.* Springer-Verlag, Berlin, to appear.

10. S. Konstantinidis: Error correction and decodability. PhD thesis, Department of Computer Science, The University of Western Ontario, 1996.

11. V. I. Levenshtein: Binary codes capable of correcting deletions, insertions, and reversals. *Dokl. Akad. Nauk. SSSR,* **163** (1965), 845–848, in Russian. English translation: *Soviet Physics Dokl.,* **10** (1966), 707–710.

12. V. I. Levenshtein: Binary codes capable of correcting spurious insertions and deletions of ones. *Problemy Peredachi Informatsii,* **1**(1) (1965), 12–25, in Russian. English translation: *Problems Inform. Transmission,* **1**(1) (1966), 8–17.

13. V. I. Levenshtein: Asymptotically optimum binary code with correction for losses of one or two adjacent bits. *Problemy Kibernet.,* **19** (1967), 293–298, in Russian. English translation: *Systems Theory Research,* **19** (1970), 298–304.

14. W. W. Peterson, E. J. Weldon, Jr.: *Error-Correcting Codes.* MIT Press, Cambridge, second ed., 1972.

15. R. M. Roth, P. H. Siegel: Lee-metric BCH codes and their application to constrained and partial-response channels. *IEEE Trans. Inform. Theory,* **40** (1994), 1083–1096.

16. F. F. Sellers, Jr.: Bit loss and gain correction code. *IRE Trans. Inform. Theory,* **IT-8** (1962), 35–38.

17. H. J. Shyr: *Free Monoids and Languages.* Hon Min Book Company, Taichung, second ed., 1991.

18. L. Staiger: On infinitary finite length codes. *RAIRO Inform. Théor. Appl.,* **20** (1986), 483–494.

19. R. R. Varshamov, G. M. Tenengol'ts: Codes capable of correcting single asymmetric errors. *Avtomat. i Telemekh.,* **26** (1965), 288–292, in Russian.

Reduced Complexity Soft-Output Maximum Likelihood Sequence Estimation of 4-ary CPM Signals Transmitted over Rayleigh Flat-Fading Channels

Daniel Boudreau and Yanick Viens

Communications Research Centre
Industry Canada, Ottawa
Ontario, K2H 8S2
Tel.: (613) 990-6278
Fax: (613) 990-0316
email: dan.boudreau@crc.doc.ca

Abstract. Maximum Likelihood Sequence Estimation (MLSE) of Continuous Phase Modulated (CPM) signals can be realized in a few different manners. For the special case of transmission over a Rayleigh flat-fading channel, an attractive receiver is in the form of the Viterbi algorithm with the trellis updates being computed using a bank of FIR filters and square operations. The FIR filters are functions of the CPM pulse shape, of the channel model and of the transmitted sequence. The channel modeling being performed by using a finite order linear predictor, the receiver has been called a *linear predictive receiver*. For signaling with a symbol set larger than two symbols, this type of receiver can become too complex to even contemplate a practical implementation. The problems of complexity reduction and Viterbi soft-output decoding for quaternary CPM signals are examined. State-reduction techniques coupled with a q-ary Soft-Output Viterbi Algorithm (QSOVA) are presented. The performance of the algorithms is evaluated by simulations, showing the trade-off between performance and complexity reduction in some important practical cases.

1 Introduction

Continuous phase modulated signals are of current interest for terrestrial mobile data communication, because relatively good spectral efficiency can be achieved, and the resulting modulated signal has a constant envelope [1], [2]. A large proportion of the communication channels over which this type of signal is transmitted can be modeled, at least in a first approximation, as Rayleigh flat-fading channels, for bandwidths up to 100 kHz [3]. This is true for many land-mobile FDMA (narrowband) systems, where the coherence bandwidth of the channel is larger than the transmitted signal bandwidth. Over these types of systems, coherent demodulation is difficult, especially in high fading rates. Suboptimum demodulation techniques, such as noncoherent or differentially coherent methods, are usually adopted. When the fading rate is high, these techniques usually exhibit an irreducible error rate that cannot be improved upon at any signal-to-noise ratio [3].

It was shown in [4] that, for the special case of Rayleigh flat-fading and CPM signaling, maximum likelihood sequence estimation (MLSE) could be implemented as a *linear predictive receiver*. This type of receiver does not require any channel phase information and is in the form of the Viterbi algorithm, with the trellis updates being computed using a bank of finite-impulse response (FIR) filters and square operations. The realization of the filter bank requires estimates of the fading rate and of the channel signal-to-noise ratio. This receiver was shown to be very effective in fast fading conditions, in the sense that its performance exhibits either a very low irreducible error rate in the case of an approximately bandlimited fading power spectral density, or an arbitrarily small irreducible error rate in the strictly bandlimited case [4].

The superiority of the linear predictive receiver over simple detection schemes is obtained at the expense of computational complexity and memory requirements. In particular, if the combined channel fading and noise can be estimated by an Lth order autoregressive process, then the linear predictive receiver consists of a bank of $rq^{\lceil \alpha+\beta \rceil}$ filters [1] and a Viterbi algorithm with $q^{\lceil \alpha+\beta \rceil - 1}$ states. The parameter r represents the number of samples per symbol used in the receiver, q is the size of the symbol alphabet, α is the length, in symbols, of the CPM frequency pulse, and $\beta = L/r$.

For a *quaternary* CPM scheme ($q = 4$), assuming that: a) two samples per symbol are necessary to avoid too much aliasing, b) the CPM frequency pulse can be approximated by only three symbol periods ($\alpha = 3$), and c) a 5th-order predictor ($L = 5$ and $\beta = 2.5$) is sufficient, the receiver requires 8192 filters and a trellis of 1024 states, with 4096 branch updates.

In addition, this detector is most likely to be used in conjunction with a form of channel decoding [5]. For optimum performance, it should deliver unquantized symbol estimates (soft-outputs) to the channel decoder, which will increase the complexity further.

This article examines the problems of complexity reduction and soft-output computations for the MLSE of quaternary CPM signals. Two methods for reducing the computing requirements are evaluated and compared. In the first method, the number of states is reduced by approximating the linear prediction of the fading plus noise process with a smaller length impulse response. A second method is based on the M-algorithm, which retains, at a given trellis level, only M paths for extension to the next level [6]. This M-algorithm is coupled with a modified version of the q-ary Soft-Output Viterbi Algorithm (QSOVA) of Li *et al.* [7].

The performance of the algorithms is evaluated by simulating a system in which convolutional coding is performed before modulation, and conventional Viterbi decoding follows soft-output MLSE at the receiver. The 4-ary CPM modulation scheme chosen for evaluation is that of the Project 25 standard for public safety communications [1]. For that modulation, the pulse shape satisfies

---

[1] The notation $\lceil a \rceil$ represents the smallest integer greater than or equal to a.

Nyquist third criterion [8]. This CPM signal can then be demodulated using differential phase detection. The simulation results show the trade-off between performance and complexity attained with MLSE, compared to the use of differential detection in the receiver.

The paper is organized as follows. The linear predictive receiver and its basic performance are reviewed in Section 2, and the performance of some reduced-state implementations is given in Section 3. The soft-output linear predictive receiver is derived in Section 4 and its performance is compared with that of the hard-output version. A discussion of the results and a conclusion are provided in Section 5.

2 The Linear Predictive Receiver for MLSE of CPM Signals

In this section, the linear predictive receiver is reviewed in some detail. This presentation follows closely that of [4], with some of the intermediate steps omitted. Consider the communications system model shown in Fig. 1. It is assumed that the receiver down-converts the signal to complex baseband, passes it through an ideal antialiasing filter, and then samples it. The filter bandwidth BW is wide enough that for practical purposes all of the signal energy is passed, including the signal energy that is spread by the fading process. In this case all of the modeling and analysis can be performed using the discrete-time complex baseband model.

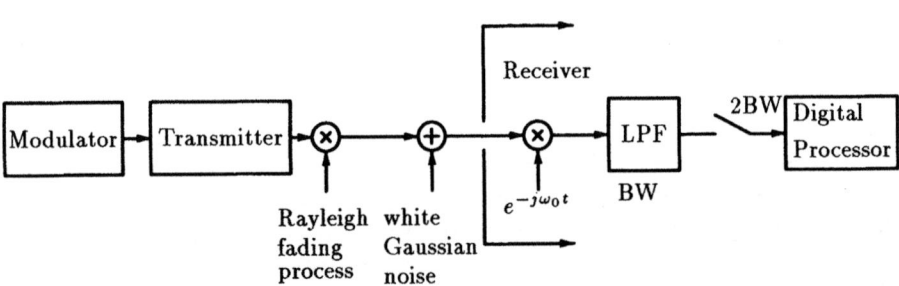

Fig. 1. An illustration of the communications systems model.

Define the set of K possible N-dimensional transmitted signal vectors as

$$\{\mathbf{s}(1), \mathbf{s}(2), \cdots, \mathbf{s}(K)\} \tag{1}$$

and the received signal vector as

$$\mathbf{y}^T = [y_N, \cdots, y_1], \tag{2}$$

where N is the number of samples observed per received vector. Define also the transformed received sample

$$\tilde{y}_k(m) = s_k^*(m)y_k, \tag{3}$$

where $s_k^*(m)$ is the complex conjugate of the kth sample of the mth transmitted signal vector $s(m) = [s_N(m), \cdots, s_1(m)]$. The generation of $\tilde{y}_k(m)$ given that $s(m)$ is transmitted is therefore modeled as in Fig. 2 (a). If the vector $\tilde{y}(m)$ represents the N-dimensional complex vector of elements $\tilde{y}_k(m)$, $k = 1, \ldots, N$, and the superscript H refers to complex conjugate transpose, the transformed vector can be written as

$$\tilde{y}(m) = P^H(m)y, \tag{4}$$

where $P(m)$ is a diagonal matrix with ii^{th} element being the ith element of $s(m)$.

If it is assumed that the CPM signal has a constant envelope of magnitude one, equation (4) represents a set of linear transformations, each with a Jacobian of magnitude one. Therefore,

$$p_{\tilde{y}|s}[\tilde{y}(m)|s(m)] = p_{y|s}[y|s(m)], \tag{5}$$

where $p[\cdot]$ is a probability density function. Note that $p_{\tilde{y}|s}[\tilde{y}(m)|s(m)]$ is introduced in order to facilitate an efficient approach to MLSE-based detection, and that from the fundamental point of view, it is still $p_{y|s}[y|s(m)]$ that is being maximized over m. Since $p_{y|s}[y|s(m)]$ is a Gaussian density function with mean zero, $p_{\tilde{y}|s}[\tilde{y}(m)|s(m)]$ is also a Gaussian density function with mean zero and covariance

$$\begin{aligned} E[\tilde{y}(m)\tilde{y}^H(m)|s(m)] &= E[P^H(m)yy^H P(m)] \\ &= E[P^H(m)(P(m)\xi + n)(P(m)\xi + n)^H P(m)] \\ &= E[\xi\xi^H] + N_o I \\ &= \mathcal{R}, \end{aligned} \tag{6}$$

where ξ represents the vector of fading process samples and $P^H(m)P(m) = I$ (because $|s_k|^2 = 1$). The vector of additive white Gaussian noise samples is represented by n and N_o is the variance of these samples. The $N \times N$ matrix $\mathcal{R}$ is a Toeplitz covariance matrix with its elements being the autocorrelation coefficients corresponding to the composite fading plus noise power spectral density (PSD) function $F(f)$. The composite PSD is given by

$$F(f) = |H(f)|^2 + N_o, \tag{7}$$

where $|H(f)|^2$ is the PSD of the Rayleigh fading process. By applying (6), the block diagram of Fig. 2 (a) can be shown to be equivalent to that of Fig. 2 (b). The appeal of Fig. 2 (b) is that the conditional probability density function $p_{\tilde{y}|s}$ can be easily determined from the model.

The conditional probability density of the vector $\tilde{y}(m)$, given $s(m)$, is

$$p_{\tilde{y}|s}[\tilde{y}(m)|s(m)] = [\pi^N |\mathcal{R}|]^{-1} e^{-\tilde{y}(m)^H \mathcal{R}^{-1} \tilde{y}(m)}, \tag{8}$$

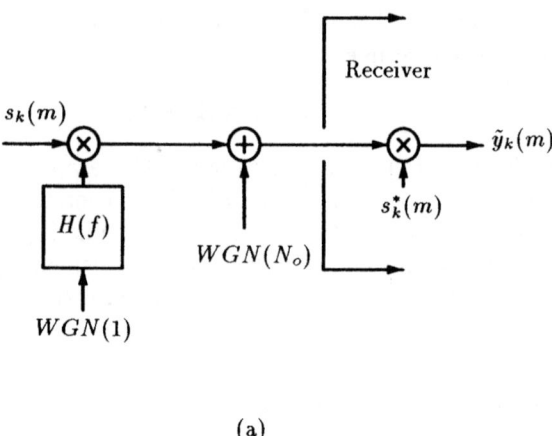

(a)

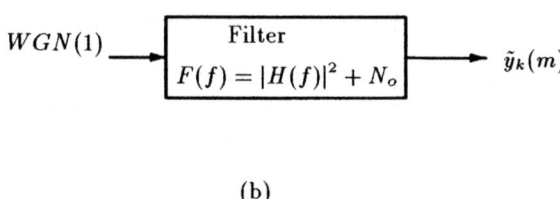

(b)

Fig. 2. (a) Generation of $\tilde{y}_k(m)$ given that $s(m)$ is transmitted. (b) An equivalent model for the generation of $\tilde{y}_k(m)$. In this figure, $WGN(x)$ represents white Gaussian noise with a power spectral density of x. The function $H(f)$ is such that $|H(f)|^2$ represents the fading process power spectral density.

where $|\mathcal{R}|$ is the determinant of $\mathcal{R}$. The MLSE detector must determine the K conditional densities of (8), or K equivalent indices.

Choosing the hypothesis with the maximum likelihood is equivalent to choosing the one that minimizes the quadratic likelihood index

$$J(m) = \tilde{y}(m)^H \mathcal{R}^{-1} \tilde{y}(m). \tag{9}$$

As indicated in [4], the matrix $\mathcal{R}^{-1}$ may be factored as

$$\mathcal{R}^{-1} = r_0^{-1} \mathcal{A}^H D \mathcal{A}, \tag{10}$$

where $\mathcal{A}$ is the $N \times N$ forward prediction matrix for the composite channel PSD $F(f)$ of (7), D is a diagonal matrix with the ii^{th} element corresponding to the inverse of the normalized expected squared prediction error of order $N - i$ and r_0

is the zero-lag channel autocorrelation coefficient. The matrix $\mathcal{A}$ is of the form

$$
\mathcal{A} = \begin{bmatrix}
1 & -a_1^{N-1} & \cdots & -a_{N-2}^{N-1} & -a_{N-1}^{N-1} \\
0 & 1 & -a_1^{N-2} & \cdots & -a_{N-2}^{N-2} \\
\cdots & \cdots & \cdots & \cdots & \cdots \\
\cdots & \cdots & 0 & 1 & -a_1^1 \\
\cdots & \cdots & \cdots & 0 & 1
\end{bmatrix},
\tag{11}
$$

where a_k^i is the kth coefficient of the ith-order linear predictor.

As shown in [4], the likelihood index, at instant k and for the mth hypothesis, can be written as

$$
\Lambda_k^m(\mathbf{y}) = \Lambda_{k-1}^m(\mathbf{y}) + |\tilde{y}_k(m) - \tilde{y}_{k|k-1}(m)|^2,
\tag{12}
$$

where

$$
\tilde{y}_{k|k-1}(m) = \sum_{j=1}^{k-1} a_j^{k-1} \tilde{y}_{k-j}(m).
\tag{13}
$$

The quantity $\tilde{y}_{k|k-1}(m)$ is then the predicted value of $\tilde{y}_k$, using the samples up to instant $k-1$.

If one assumes that the composite channel $F(f)$ corresponds to an Lth-order all-pole filter, with L much smaller than the data window N, then the matrix $\mathcal{A}$ of (11) has $N - L + 1$ rows of the form

$$
\mathcal{A}_\ell = [0 \cdots 0 \; 1 \; -a_1^{L-1} \; -a_2^{L-1} \cdots - a_{L-1}^{L-1} \; 0 \cdots 0]
\tag{14}
$$

and $L - 1$ rows with predictors of order lower than $L - 1$. The effect of these last rows can be neglected for continuous transmission and the likelihood index can be simplified as follows.

First, define the increment in (12) as

$$
B_k(m) = |\tilde{y}_k(m) - \tilde{y}_{k|k-1}(m)|^2.
\tag{15}
$$

The assumption that the composite spectrum $F(f)$ corresponds to an Lth-order all-pole filter implies that the channel has a finite memory and

$$
\begin{aligned}
B_k(m) &= \left| \tilde{y}_k(m) - \sum_{j=1}^{L} a_j^L \tilde{y}_{k-j}(m) \right|^2 \\
&= \left| \sum_{j=0}^{L} \tilde{a}_j^L s_{k-j}^*(m) y_{k-j} \right|^2,
\end{aligned}
\tag{16}
$$

for $k > L$. In (16), $\tilde{a}_0^L = 1$ and $\tilde{a}_j^L = -a_j^L$ for $1 \le j \le L$.

It can be shown that, if the modulator uses a finite length frequency pulse, the increment of (16) will only depend on the finite number of information symbols which affect the samples $s_{k-L}(m), s_{k-L+1}(m), \cdots, s_k(m)$. Under these conditions the likelihood index $\Lambda_k^m(\mathbf{y})$ can be efficiently calculated using the dynamic programming approach (Viterbi algorithm).

Assume that the samples of the CPM signal can be represented as

$$s_k(m) = \exp\left[-j2\pi h \int_0^{kT_s} \sum_n b_n(m)p(t - nT)dt\right] \tag{17}$$

where $\{b_n(m)\}$ is the information sequence corresponding to hypothesis m, T_s is the sampling period, T is the symbol period, $p(t)$ is the frequency pulse shape and h is the modulation index [9]. It is assumed that the $b_n(m)$ are selected from a set of q distinct real numbers and that the signal is sampled r times per symbol, i.e. $rT_s = T$. The pulse shape is assumed to be time-limited, i.e.

$$p(t) = 0; \quad \text{for} \quad t < 0 \quad \text{and for} \quad t > \alpha T. \tag{18}$$

Then the increment of (16) can be written as

$$B_k(m) = \left|\sum_{j=0}^{L} \tilde{a}_j^L z_{k-j}^*(m)y_{k-j}\right|^2, \tag{19}$$

where $z_\ell(m)$, for $\ell = k - L, k - L + 1, \cdots, k$, represent the locally generated transmitted samples for hypothesis m, *over the most recent set of symbols spanning the combined memory of the modulator and the channel*. It can be shown that the length of this combined memory span, in symbols, is [4]

$$Sp = \lceil \alpha + \beta \rceil, \tag{20}$$

where $\beta = LT_s/T$ is the predictor length in symbols.

The Viterbi algorithm updates the likelihood index r steps at a time. Over a single symbol interval

$$\Lambda_{k+r}^m(\mathbf{y}) = \Lambda_k^m(\mathbf{y}) + \sum_{j=0}^{r-1} B_{k+j}(m), \tag{21}$$

where the summation is the branch update. This update can be written as

$$\sum_{j=0}^{r-1} B_{k+j}(m) = \sum_{j=0}^{r-1}\left|\sum_{i=0}^{L} h_i(u, j)y_{k+j-i}\right|^2, \tag{22}$$

where $h_i(u, j) = a_i^L z_{N+j-i}^*(u)$ and u is any hypothesis for which the first $\lceil \alpha + \beta \rceil$ symbols are the same as the most recent $\lceil \alpha + \beta \rceil$ symbols of m, and $N = (\lceil \alpha + \beta \rceil - 1)r$.

For a symbol set of q elements, the branch update of (22) can then be implemented with a bank of $rq^{\lceil \alpha + \beta \rceil}$ time-invariant FIR filters, with impulse responses given by the $h(u, j)$, followed by squaring and summing operations as shown in Fig. 3. This algorithm has $q^{\lceil \alpha + \beta \rceil - 1}$ states and q branch updates per state. At instant k, the Viterbi algorithm makes a decision at instant $k - \ell$, in favor of the symbol pertaining to the surviving path with the smallest likelihood index. The value of ℓ is usually chosen to be larger than five times the memory given in (20) [10].

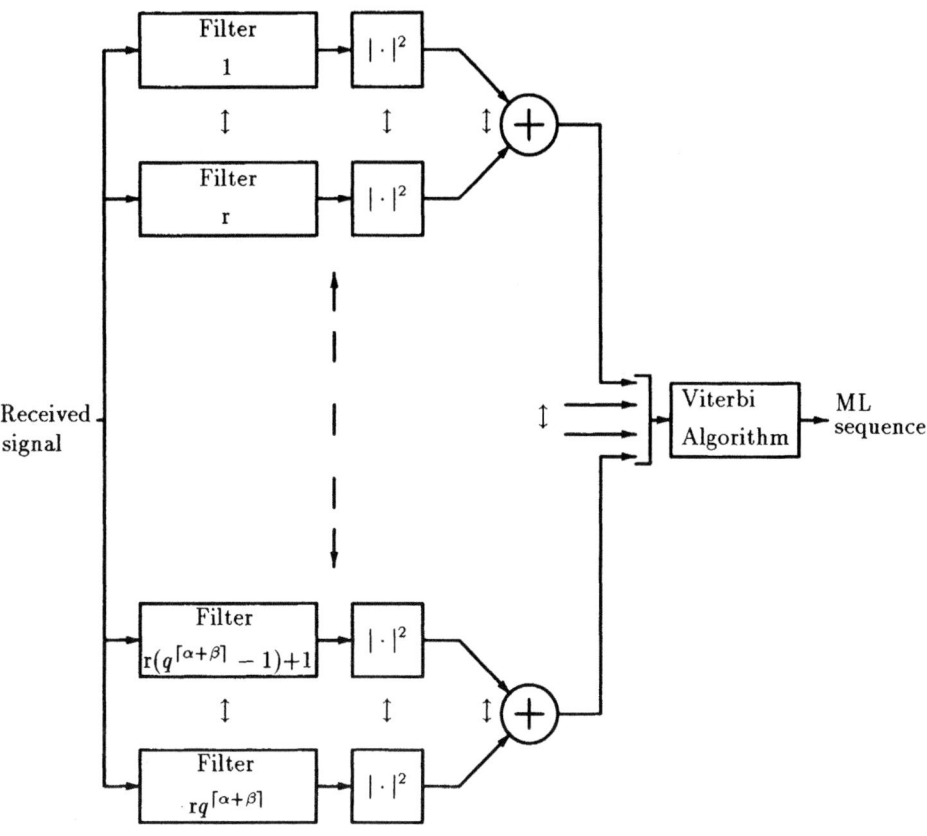

Fig. 3. Implementation of the linear predictive receiver using a bank of FIR filters.

2.1 Basic Performance of the Linear Predictive Receiver

In order to test this algorithm with a CPM modulation scheme of practical value, the pulse shape defined in the early phase of the Project 25 is considered. This pulse $p(t)$ is chosen to satisfy Nyquist's third criterion, which means that its Fourier transform $P(\omega)$ satisfies [8]

$$P(\omega) = I(\omega) \left(\frac{\sin \omega T/2}{\omega T/2} \right)^{-1}, \tag{23}$$

where $I(\omega)$ satisfies Nyquist's *first* criterion, i.e.

$$\sum_{k=-\infty}^{\infty} I\left(\omega - \frac{2\pi k}{T}\right) = 1. \tag{24}$$

The Project 25 pulse is such that the magnitude of $I(\omega)$ is that of a 20% raised cosine filter.

In the simulations, this pulse shape is sampled at two samples per symbol ($r = 2$) and is truncated to three symbols ($\alpha = 3$), in both the transmitter and the receiver. The transmitted signal therefore suffers from fairly severe aliasing and intersymbol interference distortion, compared to what could be expected from a practical implementation. The composite PSD of the Rayleigh fading channel is such that $|H(f)|$ in (7) is equal to the transfer function of a 10% raised cosine filter. In all the simulations, a Rayleigh flat-fading channel with a rate equal to 30% of the symbol rate is assumed. This composite PSD is given in Fig. 4, at Eb/No = 10 dB, along with the transfer functions of the all-pole filter, for different orders, used to approximate the channel in the MLSE algorithm. This channel is chosen because it represents a very extreme one, for which noncoherent detection schemes, such as differential detection, are almost useless.

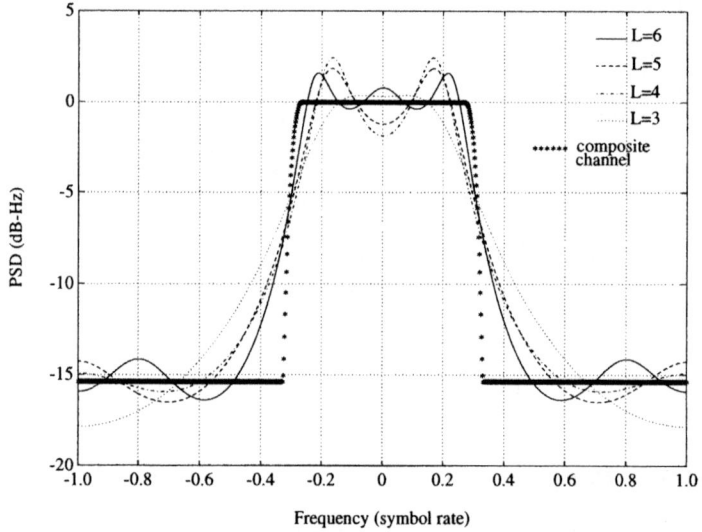

Fig. 4. The nominal composite power spectral density function, for Eb/No=10 dB, and the 3rd, 4th, 5th and 6th order approximation of it. The fading rate is 0.3R, with R representing the symbol rate.

Using channel estimators with L=5 and 6, in the linear predictive receiver, results in a trellis with 1024 states and a filter bank with 8192 filters. The two other forms of estimators (L=3,4) imply 256 states and 2048 filters. The basic performance of the receiver, using these different orders of prediction, is shown in Fig. 5. These performance curves are therefore called *basic*, and serve as benchmarks when the algorithm is simplified. Note that no state-reduction is performed in these simulations, i.e. the full linear predictive receiver implementation is used. Note also that, as for all the results presented in this paper, the channel prediction filter is fixed and optimized for all the different values of Eb/No.

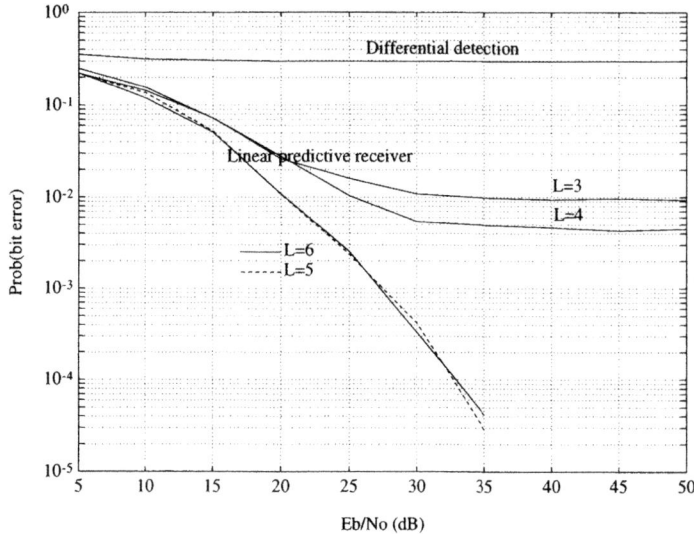

Fig. 5. The basic simulated performance, for channel estimator orders of L=3,4,5,6, on a Rayleigh flat-fading channel with a fading rate of 0.3R.

It is interesting to note that the performance for the 5th and 6th order estimators are essentially identical, and do not show any sign of irreducible error rate above 10^{-5}. This is to be compared with the performance of differential phase detection, for which the bit error rate is always larger than 20%, for any signal-to-noise ratio. This good performance of the linear predictive receiver was predicted in [4], since the transfer functions of the channel estimates model fairly well the composite channel. The irreducible error rate, when using lower orders of channel estimation reflects the fact that, with only three or four coefficients, a linear forward prediction error filter is limited in its ability to follow the decreasing composite channel noise floor. This is illustrated in Fig. 6 for a 4th order estimation, when Eb/No ranges from 25 dB to 50 dB. The estimator response stays approximately the same, while the actual noise floor drops by 5 dB at each step.

In the next section, the basic linear predictive receiver is modified in order to reduce the number of filters and trellis states required, while keeping the performance at an adequate level.

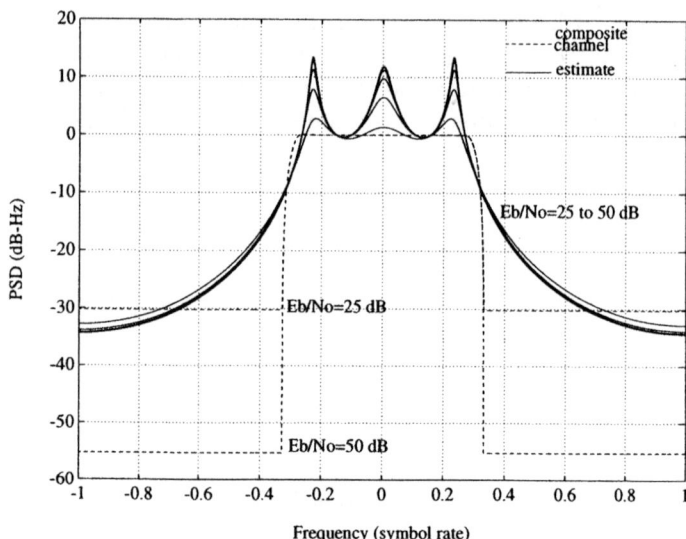

Fig. 6. The nominal composite power spectral density function, for Eb/No=25 to 50 dB, and the 4th order approximation of it.

3   Reduced-State MLSE with the M-Algorithm

The M-algorithm is a limited trellis search algorithm that essentially retains, at a given symbol interval, only M paths for extension into the succeeding intervals [6]. At a given time, the algorithm extends the M paths previously retained to a maximum of qM states at the next interval. The accumulated metrics of these newly extended paths are then searched, and the M ones with the smallest values are kept. As with the usual Viterbi algorithm, a symbol decision is taken ℓ symbol intervals in the past, by finding the best path among the M ones retained (this algorithm is therefore sometimes called the (M,ℓ) algorithm). The reduction in the number of state updates also means that a smaller number of filters needs to be computed in the filter bank. This is indicated in Fig. 7. Note that the M-algorithm is used here because it constitutes a simple method for reducing the computational burden of the linear predictive receiver, while maintaining good performance. It is not the only option, nor is it optimal in any way.

The simulated performance of the linear predictive receiver combined with the M-algorithm (M-linear predictive receiver), for a fading rate of 30% the symbol rate, is illustrated in Figs. 8 and 9. The first of these figures represents the error rate for a channel estimator with L=5 (1024 original states). The

	Order of filter bank	# of states
full	$rq^{\lceil \alpha+\beta \rceil}$	$rq^{\lceil \alpha+\beta \rceil -1}$
M-algorithm	rqM	M

Fig. 7. Complexity of the Linear Predictive Receiver.

effect of retaining 8 to 64 states is compared to the full-trellis implementation (M=1024). The case for L=4 (256 original states) is shown in Fig. 9.

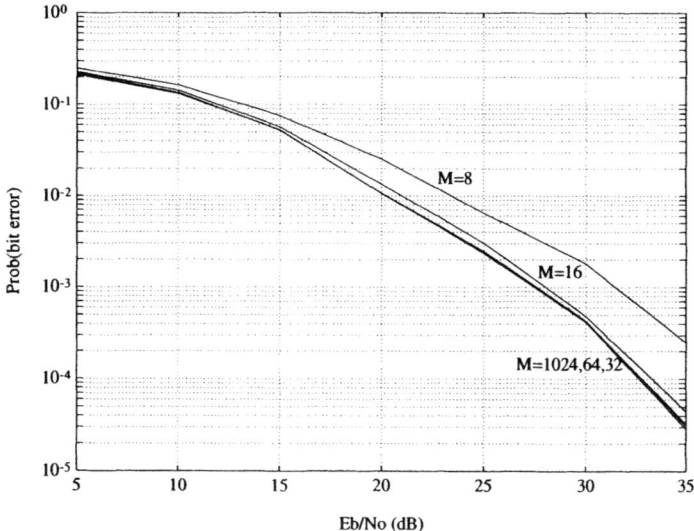

Fig. 8. The performance of the M-linear predictive receiver, for a channel estimator with L=5 and M=8, 16, 32, 64 and 1024 states retained out of 1024. The fading rate is 0.3R.

These figures show that the average loss incurred by dropping a large subset of the states can be made smaller than one dB, when M is at least equal to 16. Even for M=8, the degradation can be considered reasonable. The complexity and performance at key points is summarized in Figs. 10 and 11.

For M=16, it is observed that the number of filters to be considered is reduced by more than 98% for L=5, and by more than 93% for L=4. This improvement in the computational complexity is somewhat reduced by the sorting that must be accomplished by the M-algorithm, in order to find the M best paths out of qM. This sorting can be done in less than $(qM)^{1.5}$ operations [11], i.e. in less than 512 operations for $q=4$ and M=16. Note that this sorting effort could be reduced through the use of other techniques, as discussed in [12].

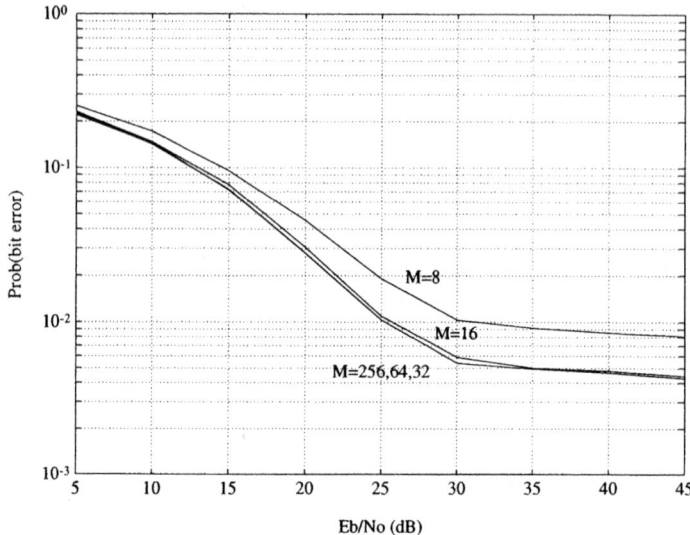

Fig. 9. The performance of the M-linear predictive receiver, for a channel estimator with L=4 and M=8, 16, 32, 64 and 256 states retained out of 256. The fading rate is 0.3R.

	# of states	Order of filter bank	Sorting Upper bound (ops)	degradation at BER $= 10^{-4}$
full	1024	8192	0	–
$M = 64$	64	512	4096	0
$M = 32$	32	256	1448	0
$M = 16$	16	128	512	~ 0.5 dB
$M = 8$	8	64	181	~ 5 dB

Fig. 10. Complexity of the M-Linear Predictive Receiver: 1024-state trellis

For a given value of M, the combination of the M-algorithm and the linear predictive receiver has a comparable complexity, when L=4 or L=5. Since the L=5 implementation does not produce a high irreducible error rate, it appears that it should be the logical choice with M=8 or 16.

If one is interested in an even better performance, then channel coding should

	# of states	Order of filter bank	sorting, upper bound (ops)	degradation at BER = 10^{-2}	irreducible BER
full	256	2048	0	–	$\sim 4 \times 10^{-3}$
$M = 64$	64	512	4096	0	$\sim 4 \times 10^{-3}$
$M = 32$	32	256	1448	0	$\sim 4 \times 10^{-3}$
$M = 16$	16	128	512	~ 0.5 dB	$\sim 4 \times 10^{-3}$
$M = 8$	8	64	181	~ 5 dB	$\sim 7.5 \times 10^{-3}$

Fig. 11. Complexity of the M-Linear Predictive Receiver: 256-state trellis

be used. In this case, an encoder follows the source of bits, and interleaving is performed on the symbols. In the receiver, linear predictive detection is followed by the corresponding deinterleaver and decoder. As it has been presented, the linear predictive receiver of Section 2 produces hard outputs, i.e. decisions on whether ones or zeros are the most likely outputs. In a scenario where channel coding is used, it is well known that the best performance is obtained when the symbol detector provides the channel decoder with unquantized decisions, or with decisions bearing some information about their reliabilities [13].

At a given time the optimum soft-decision information is the set of *a posteriori* probabilities for each symbol. The symbol-by-symbol maximum *a posteriori* probability (MAP) algorithm uses the symbol error probability as the optimization criterion and can deliver such optimum soft-decisions. The major drawback of the MAP algorithm is its high complexity [14]. This has prompted the development of sub-optimum soft-output detection schemes with good performance. In particular, the Viterbi algorithm, which inherently produces hard decisions, has been modified to also compute estimates of the decisions reliability. This algorithm was called the Soft-Output Viterbi Algorithm (SOVA) [15] and applies to the detection of binary symbols. For q-ary symbols, different forms of the algorithm have been proposed, to deal mainly with the equalization of time dispersive channels [16], [7]. In the next section, a method to obtain soft-outputs from the linear predictive receiver is outlined.

4 Soft-Output MLSE of q-ary CPM Signals

In order to modify as little as possible the operation of the linear predictive receiver, the soft-outputs are required to be produced at a fixed decision delay of ℓ symbols. At symbol interval n, the optimum set of soft-outputs would therefore be

$$\{\Pr[b_{n-\ell} = U_1 | \mathbf{y}_k], \Pr[b_{n-\ell} = U_2 | \mathbf{y}_k], \cdots, \Pr[b_{n-\ell} = U_q | \mathbf{y}_k]\}, \qquad (25)$$

where b_n is the symbol at symbol instant n, drawn from the set $\{U_1, U_2, \cdots, U_q\}$, and $\mathbf{y}_k$ represents the vector of received samples at the corresponding sampling time k.

A suboptimum set of soft-outputs, similar to the one proposed in [16], is

$$\left\{ \max_m \{p[\tilde{\mathbf{y}}_k | s_n(m), b_{n-\ell} = U_1]\}, \cdots, \max_m \{p[\tilde{\mathbf{y}}_k | s_n(m), b_{n-\ell} = U_q]\} \right\}, \quad (26)$$

where $p[\cdot]$ is the probability density function of (5), conditioned on a fixed $(n - \ell)$th symbol. This set gives the maximum likelihood probabilities of the received signal, given each possible value for the delayed symbols.

From Section 2, a set related to that of (26) is

$$\left\{ \min_{\{m|b_{n-\ell}=U_1\}} \{\Lambda_n^m(\mathbf{y})\}, \cdots, \min_{\{m|b_{n-\ell}=U_q\}} \{\Lambda_n^m(\mathbf{y})\} \right\}. \quad (27)$$

This represents the set of minimum likelihood indices, as defined in (21), conditioned on the occurrence of a specific delayed symbol. The sets of (26) and (27) are not optimum, but the simulations will show that the latter allows significant error rate improvements compared to the use of hard decisions.

4.1 The q-ary Soft-Output Viterbi Algorithm

In order to describe the QSOVA, the state x_n is defined as

$$x_n \overset{\Delta}{=} (b_{n-S}, b_{n-S+1}, \cdots, b_{n-1}) \quad (28)$$

with

$$S = \lceil \alpha + \beta \rceil - 1. \quad (29)$$

Then the likelihood index of (21) is redefined as a function of this state. The index update is

$$\tilde{\Lambda}(x_{n+1}) = \tilde{\Lambda}(x_n) + \mathcal{M}(x_n, x_{n+1}), \quad (30)$$

where $n = \lceil k/r \rceil$. The quantity $\mathcal{M}(x_n, x_{n+1})$ in (30) is defined as the branch metric between states x_n and x_{n+1}. Comparing with (21), the branch metric is

$$\mathcal{M}(x_n, x_{n+1}) \overset{\Delta}{=} \sum_{j=0}^{r-1} B_{nr+j}(m) \quad (31)$$

with m being the index of a sequence hypothesis going through states x_n and x_{n+1}.

For a given state x_n, the conditional likelihood indices of (27) are then defined as

$$\tilde{\Lambda}(x_n | b_{n-\ell}) \overset{\Delta}{=} \min_{\{m|x_n, b_{n-\ell}\}} \{\tilde{\Lambda}(x_n)\}. \quad (32)$$

With this notation, the ordinary Viterbi algorithm recursion is

$$\tilde{\Lambda}_{VA}(x_{n+1}) = \min_{\{m|x_n\}} [\tilde{\Lambda}(x_n) + \mathcal{M}(x_n, x_{n+1})] \quad (33)$$

and the QSOVA update is

$$\tilde{\Lambda}(x_{n+1}|b_{n-\ell}) = \min_{\{m|x_n, b_{n-\ell}\}} [\tilde{\Lambda}(x_n|b_{n-\ell}) + \mathcal{M}(x_n, x_{n+1})]. \tag{34}$$

The set of soft-outputs is then

$$\left\{ \min_{\{x_n\}} \{\tilde{\Lambda}(x_n|b_{n-\ell} = U_1)\}, \cdots, \min_{\{x_n\}} \{\tilde{\Lambda}(x_n|b_{n-\ell} = U_q)\} \right\}. \tag{35}$$

In order to have this set available for symbol instant $n - \ell$, the conditional likelihood indices for instants $n - \ell$ to $n - S$ must be kept in memory. This requires keeping track of a matrix for every trellis state x_n. This matrix is of the form

$$M(x_n) = \begin{bmatrix} \tilde{\Lambda}(x_n|b_{n-S} = U_1) & \cdots & \tilde{\Lambda}(x_n|b_{n-S} = U_q) \\ \tilde{\Lambda}(x_n|b_{n-S-1} = U_1) & \cdots & \tilde{\Lambda}(x_n|b_{n-S-1} = U_q) \\ \cdots & \cdots & \cdots \\ \cdots & \cdots & \cdots \\ \cdots & \cdots & \cdots \\ \tilde{\Lambda}(x_n|b_{n-\ell} = U_1) & \cdots & \tilde{\Lambda}(x_n|b_{n-\ell} = U_q) \end{bmatrix}. \tag{36}$$

For $S < \delta \leq \ell$, the conditional likelihood indices are updated as in (34), i.e.

$$\tilde{\Lambda}(x_{n+1}|b_{n-\delta}) = \min_{\{m|x_n, b_{n-\ell}\}} [\tilde{\Lambda}(x_n|b_{n-\delta}) + \mathcal{M}(x_n, x_{n+1})], \tag{37}$$

while the first row of matrix $M(x_n)$ is updated according to

$$\tilde{\Lambda}(x_{n+1}|b_{n-S}) = \tilde{\Lambda}(x_n) + \mathcal{M}(x_n, x_{n+1}), \tag{38}$$

where, because of (28), x_n is uniquely determined by b_{n-S} and x_{n+1}.

As described above, the QSOVA is a fairly simple extension to the ordinary VA. At symbol instant $n + 1$, the branch updates are computed as usual. Then the last $\ell - S$ rows of each state matrix is updated according to (37), and the soft outputs at time $n - \ell$ are obtained by using the rule of (35) over the last row of all the matrices. The state matrices are then shifted down one row, discarding the last one. The first row of each matrix is obtained with (38). The algorithm requires an initialization period in which the state matrices are filled with significant values. Also, as in the ordinary VA, the QSOVA must have its likelihood indices periodically normalized, in order to avoid computational overflows.

4.2 Reduced-State QSOVA with the M-Algorithm

The M-algorithm is applied to the QSOVA in essentially the same manner as it is with the ordinary VA. The M states retained for extension are chosen according to the VA update of (33). Note that the survivor path of each trellis state is imbedded in the corresponding matrix of (36) and does not need to be computed again. Compared to the combined Viterbi algorithm and M-algorithm, the combination with the QSOVA requires $\ell - S + 2$ times more updates and q times more storage capacity, as indicated in Fig. 12.

	Updates	Storage
VA	Mq	$M(\ell - S + 1)$
QSOVA	$Mq(\ell - S + 2)$	$Mq(\ell - S + 1)$

Fig. 12. Complexity of the VA and the QSOVA, when combined with the M-algorithm.

To evaluate the benefits of the MLSE soft-outputs, when channel coding is used, the maximum free distance rate $1/2$ convolutional code with constraint length 3 was simulated [10]. Although it has a free distance of only 5, this simple code was chosen to minimize the simulation times. It is decoded using a Viterbi decoder. Note that this is a second Viterbi algorithm, where no state reduction is applied. In order to present this Viterbi decoder with symbols that are affected by approximately uncorrelated channel effects, a 25×10 block interleaver is applied on the transmitted symbols. The MLSE outputs are processed with the corresponding deinterleaver. In a voice communications scheme, it is interesting to know the delay imposed by this combination of coding and interleaving, when coupled with a linear predictive receiver. Consider the Project 25 system [1]. Its voice bit rate is 4.4 kbits/s, and its symbol rate is 4.8 ksymbols/s. The delay ℓ imposed by the linear predictive receiver is five times the combined memory span of equation (20), which is fixed at a maximum of 6 symbols in all the simulations. The delay due to the interleaver/deinterleaver is 500 symbols, while the delay of the Viterbi decoder is, as usual, equal to five times the code constraint length, or 15 bits. The overall transceiver delay budget is therefore divided in 6.25 ms for the linear predictive receiver, 104.17 ms for the interleaver/deinterleaver and 3.41 ms for the decoder. This gives a total of 113.83 ms, which must be added to the total algorithmic voice coder delay of 78.75 ms [1]. This gives an overall link delay of little less than 200 ms, which should be adequate for normal voice transmission [17].

In the case of soft-output MLSE, the deinterleaved soft-outputs are used directly as branch metrics in the Viterbi decoder. When bit hard decisions are produced by the linear predictive receiver, symbols are recreated from the pairs of output bits, and the Viterbi decoder branch metrics are computed as the Euclidean distances between these symbols and the branch symbols.

The performance of the hard and soft-output schemes can be compared in Figs. 13 and 14, for 1024 and 256 initial states respectively. The soft-output algorithm performs consistently better than the hard-output method. The advantage, at an error rate of 10^{-4} is about 3.5 dB for L=5 and 5 dB for L=4. Although the set of soft-outputs used in the simulation is obviously suboptimum, the improvement obtained when using it is comparable with that obtained with a linear predictive receiver delivering true MAP outputs to the channel decoder [14] (for a binary CPM scheme).

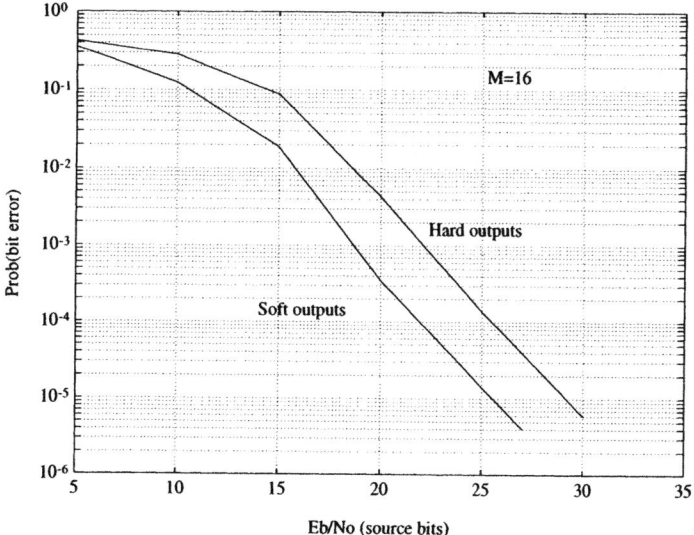

Fig. 13. The performance of the M-linear predictive receiver with channel coding and M=16, when hard and soft-outputs are passed to the channel decoder. The fading rate is 0.3R and L=5 (1024 initial states). The channel coder is the maximum free distance, rate $\frac{1}{2}$ convolutional code, with constraint length 3. A 25 × 10 block interleaver is used, along with the corresponding deinterleaver.

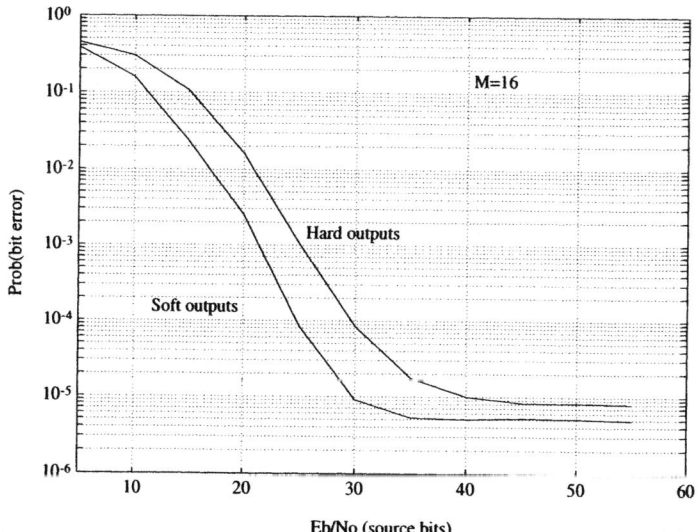

Fig. 14. The performance of the M-linear predictive receiver with channel coding and M=16, when hard and soft-outputs are passed to the channel decoder. The fading rate is 0.3R and L=4 (256 initial states). The channel coder is the maximum free distance, rate $\frac{1}{2}$ convolutional code, with constraint length 3. A 25 × 10 block interleaver is used, along with the corresponding deinterleaver.

5 Discussion and Conclusion

The simulation results provided in this paper have shown that the linear predictive receiver can be made attractive, from the implementation point of view, by using state reduction techniques. In particular, the fairly simple M-algorithm can be used to retain and extend only a small number of states, without significantly degrading the overall bit error rate.

The addition of channel coding, in an effort to improve further the performance, requires the MLSE algorithm to produce soft-outputs. One approach, the suboptimum QSOVA, can improve the coded performance by several dBs relative to the coded performance of hard output MLSE. The price to pay is an increase in computational complexity. In this case, the M-algorithm is a must, since it also significantly reduces the storage requirements of the algorithm.

For mobile terrestrial applications, the proposed soft-output version is probably not worth the additional receiver complexity and cost. A hard-output M-linear predictive receiver, with the proper error correcting code, is likely to be the best choice. In a power-limited environment, such as on a mobile satellite channel, the linear predictive receiver using the QSOVA would be of great interest.

References

1. Telecommunications Industry Association, *Project 25 System and Standards Definition*, TSB-102.
2. G. Benelli, A. Garzelli and F. Salvi, "Simplified Viterbi processors for the GSM Pan-European cellular communication system," *IEEE Trans. on Vehicular Technology* , vol. 43, no. 4, pp. 870–878, November 1994.
3. W. Y. C. Lee, *Mobile Communications Engineering*, New York: McGraw-Hill, 1982.
4. J. H. Lodge and M. L. Moher, "Maximum likelihood sequence estimation of CPM signals transmitted over Rayleigh flat-fading channels," *IEEE Trans. on Communications*, vol. 38, no. 6, pp. 787–794, June 1990.
5. D. Boudreau and J. H. Lodge, "An advanced implementation structure for a narrowband mobile radio," *17th Biennial Symposium on Communications*, Queens University, pp. 337-340, May 1994.
6. N. Seshadri and J. B. Anderson, "Decoding of severely filtered modulation codes using the (M, L) algorithm," *IEEE Journal on Selected Areas in Communications*, vol. 7, no. 6, pp. 1006–1016, August 1989.
7. Y. Li, B. Vucetic and Y. Sato, "Optimum soft-output detection for channels with intersymbol interference," *Submitted to the IEEE Trans. on Communications*, 1994.
8. S. Pasupathy, "Nyquist's third criterion," *Proceedings of the IEEE*, vol. 62, pp. 860–861, June 1974.
9. T. Aulin and C.-E. Sundberg, "Continuous phase modulation–Parts I and II," *IEEE Trans. on Commun.*, vol. COM-29, pp. 196–225, March 1981.
10. J. G. Proakis, *Digital Communications, 2nd Ed.*, McGraw-Hill, 1989.
11. W. H. Press, B. P. Flannery, S. A. Teukolsky and W. T. Betterling, *Numerical Recipes; The Art of Scientific Computing*, Cambridge University Press, 1986.

12. J. B. Anderson and S. Mohan, "Sequential coding algorithms: a survey and cost analysis," *IEEE Trans. on Communications*, vol. COM-32, no. 2, pp. 169–176, February 1984.

13. G. D. Forney, *Concatenated Codes*, The M.I.T. Press, Cambridge, Massachusetts, 1966.

14. M. J. Gertsman and J. H. Lodge, "Symbol-by-symbol MAP demodulation of CPM signals on Rayleigh flat fading channels," *Proc. of The 7th Conference on Wireless Communications*, Calgary, pp. 219–238, July 1995.

15. J. Hagenauer and P. Hoeher, "A Viterbi algorithm with soft-decision outputs and its applications," *Proc. of Globecom'89*, Dallas, pp. 1680–1686, November 1989.

16. P. Hoeher, "TCM on frequency-selective fading channels: a comparison of soft-output probabilistic equalizers," *Proc. of Globecom'90*, San Diego, pp. 376–381, December 1990.

17. A. M. Kondoz, *Digital Speech, Coding for Low Bit Rate Communications Systems*, John Wiley & Sons, Chichester, 1994.

A Novel Receiver Structure for MPSK in the Presence of Rapidly Changing Phase*

Carl R. Nassar[1] and M. Reza Soleymani[2]

[1] Department of Electrical Engineering, McGill University, Montreal, Quebec,
Canada, H3A 2A7
[2] SPAR Aerospace Limited, Satellite and Communications Systems Division,
Ste-Anne-de-Bellevue, Quebec, Canada, H9X 3R2.

Abstract. In this paper, we introduce a novel receiver for the detection of data in the presence of rapidly changing nuisance parameters and additive noise. This receiver uses a parallel structure, and its output approximates a Maximum Likelihood detection of the data. We apply this receiver to the special case of data detection of MPSK in the presence of rapidly changing unknown phase; here, the rate of phase change is such that phase change over two symbols is negligible, but it is not necessarily negligible over a longer interval. The unknown-phase receiver we propose outperforms DPSK by 1.5 dB. We also compare the performance of this receiver with that of Multiple Symbol Differential Detection (MSDD). Except under very slow phase change conditions, this receiver outperforms MSDD. Furthermore, our unknown-phase receiver is easy to implement, requiring a relatively low complexity.

1 Introduction

In many communication environments, channels can be characterized by two effects: the addition of noise and the introduction of unknown parameters. The added noise is usually well modelled statistically as Additive White Gaussian Noise (AWGN). The unknown parameters introduced by the channel correspond to channel effects such as phase offset and frequency offset. These unknown parameters are often rapidly changing (for instance, the unknown parameters introduced in many mobile communication channels change quickly).

Communication systems which operate over such channels can detect data in one of two ways. First, they can estimate the unknown parameters, remove them from the information-bearing signal, and then detect the data using a Maximum Likelihood (ML) scheme. Alternatively, the system can detect the data in the presence of both the unknown parameters and the additive noise.

An important, practical example of a communication system experiencing unknown parameters is described as follows. A transmitter sends out differentially encoded MPSK (M-ary Phase Shift Keying) symbols. These are sent over a

* Research supported under National Science and Engineering Research Council (NSERC) grant OGP/N011 and NSERC scholarship 106418.

channel, which introduces both an Additive White Gaussian Noise and a rapidly changing phase offset. This scenario is frequently encountered, especially in mobile radio communication environments.

A number of receiver schemes have been proposed for the rapidly-changing-phase environment. The traditional reception method is Differential Phase Shift Keying (DPSK) [1]. Here, the only required assumption is that phase be constant over two symbol intervals. The major drawback of DPSK is a performance degradation of up to 3 dB when compared to coherent reception on an AWGN channel.

Recently, several researchers have proposed a receiver for the rapidly-changing-phase environment which builds on the ideas of DPSK ([2],[3],[4],[5]). Their receiving scheme is generally referred to as Multiple Symbol Differential Detection (MSDD). In MSDD, the phase is assumed constant over n $(n > 2)$ symbol intervals. MSDD offers substantial performance improvements over DPSK with n as small as 3. MSDD can be implemented at an affordable complexity by using the implementation proposed in [6].

Still, in cases where the phase varies so rapidly that it can be assumed constant only over 2 symbols, DPSK remains the receiver of choice in the rapidly-changing-phase environment; despite its substantial performance degradation.

In this work, we introduce a novel receiver for the detection of data in the presence of rapidly changing unknown parameters. This receiver uses a parallel structure, and generates an output approximating an ML detection of the data.

We apply this receiver to the special case of data detection of MPSK in the presence of rapidly changing unknown phase. We show that the unknown-phase receiver we propose is able to achieve substantial performance gains over DPSK, even in the case where phase can only be assumed constant over two symbol intervals. Specifically, at error probabilities of 10^{-3} and 10^{-4}, and considering an 8-PSK constellation, this receiver offers a 1.5 dB improvement over DPSK. It achieves this gain over the entire range of channel phase conditions for which DPSK is applicable.

We also compare the performance of our unknown-phase receiver with that of Multiple Symbol Differential Detection (MSDD). Except under very slow phase change conditions, our receiver outperforms MSDD.

Our unknown-phase receiver demonstrates a relatively low complexity. Our receiver's complexity is only about eight times that of a typical PSK symbol-by-symbol decoder.

2 The Communication Environment

As a first step toward designing our receiver, we describe the communication environment mathematically. We do this by detailing the input to the receiver.

In the case of a channel that introduces both a noise and unknown parameters, the receiver input is

$$\underline{r} = (r_1, r_2, ..., r_L), \tag{1}$$

where the r_i's are samples received at a rate of one sample per symbol interval, and each r_i is described by

$$r_i = v(a_i, \underline{c_i}) + \eta_i. \tag{2}$$

Here, η_i is a random value representing the noise. We assume that this noise is AWGN. Also, the vector $\underline{c_i}$ is a random vector representing the channel nuisance parameters. It is assumed that $\underline{c_i}$ contains J elements, i.e., $\underline{c_i} = (c_{i,1}, c_{i,2}, ..., c_{i,J})$. The a_i represents the data symbol transmitted at sample time i; the entire sequence of transmitted data symbols is $\underline{a} = (a_1, a_2, ..., a_L)$. We assume that a_i is independent of its history and its future (i.e., the modulation is memoryless). Finally, $v(\cdot, \cdot)$ is a known function.

In the case of channels introducing a rapidly changing unknown phase, the input to the receiver is given by

$$\underline{r} = (r_1, r_2, ..., r_L) \tag{3}$$

where

$$r_i = a_i e^{j\theta_i} + \eta_i. \tag{4}$$

This corresponds to a special case of (1) and (2) with $\underline{c_i}$ replaced by θ_i and $v(a_i, \underline{c_i})$ set to $a_i e^{j\theta_i}$.

We are particularly interested in the unknown phase case where the data symbols a_i in (4) correspond to differentially encoded MPSK symbols, and the phase θ_i in (4) can only be assumed constant over two symbol intervals. In this case, the statistical characterization of phase sequence $\underline{\theta} = (\theta_1, ..., \theta_L)$ is generated as follows. Phase change $(\theta_i - \theta_{i-1})$ is negligible over 2 symbols; that is, mathematically, $|\theta_i - \theta_{i-1}| < \beta$, where β is small. A good choice for β is the largest value at which DPSK still performs well; we have found that this β is $\frac{2\pi}{8 \cdot M}$. We represent this information statistically by assuming that phase change is uniformly distributed in the range $|\theta_i - \theta_{i-1}| < \beta$, i.e.,

$$p(\theta_i | \theta_{i-1}) = \begin{cases} \frac{1}{2\beta}, & |\theta_i - \theta_{i-1}| < \beta \\ 0, & \text{otherwise.} \end{cases} \tag{5}$$

This characterizes the phase sequence $\underline{\theta} = (\theta_1, ..., \theta_L)$ in (4) when phase is constant over only two symbol intervals.

In the work which follows, we first design a receiver for the general unknown-parameter case described by Equations (1) and (2). We then apply this receiver to the special case of unknown phase described by Equations (3), (4), and (5). We establish the performance of the receiver, in the special case of unknown phase, by simulation, using a random walk (Gaussian) phase model to generate our results.

3 Receiver Design for Data Detection in the Presence of Unknown Parameters

3.1 Key Equation

We want to build a receiver which, given the received sequence $\underline{r}$ of (1) and (2), puts out an estimate of $\underline{a}$ which results in the minimum probability of error. A receiver which does this is known as a *Maximum A Posteriori* (MAP) receiver. It is well known that the output of such a receiver is

$$\hat{\underline{a}} = \arg\max_{\underline{a}} p(\underline{a}|\underline{r}). \tag{6}$$

Assuming that each sequence $\underline{a}$ is equally likely, we can rewrite this as the Maximum Likelihood estimate

$$\hat{\underline{a}} = \arg\max_{\underline{a}} p(\underline{r}|\underline{a}). \tag{7}$$

Incorporating the unknown parameter sequence $\underline{c} = (\underline{c}_1, ..., \underline{c}_L)$ leads to

$$\hat{\underline{a}} = \arg\max_{\underline{a}} \int p(\underline{r}|\underline{a}, \underline{c}) \cdot p(\underline{c}) \cdot d\underline{c}. \tag{8}$$

We examine each term in the integrand closely. First, we known that $p(\underline{r}|\underline{a}, \underline{c})$ is given by $p(\underline{r}|\underline{a}, \underline{c}) = \prod_{i=1}^{L} p(r_i|a_i, \underline{c}_i) = \prod_{i=1}^{L} p_{\eta_i}(r_i - v(a_i, \underline{c}_i))$. Each distribution $p(r_i|a_i, \underline{c}_i) = p_{\eta_i}(r_i - v(a_i, \underline{c}_i))$ is Gaussian, and, characteristically, this term has a peak (usually sharp) at the $\underline{c}_i$ which maximizes $p(r_i|a_i, \underline{c}_i)$. Consequently, $p(\underline{r}|\underline{a}, \underline{c}) = \prod_{i=1}^{L} p_{\eta_i}(r_i - v(a_i, \underline{c}_i))$ has a sharp peak at the $\underline{c}$ which maximizes $p(\underline{r}|\underline{a}, \underline{c})$. Additionally, the second term in the integrand, $p(\underline{c})$, either has no significant peak, or, if it has a peak, the $\underline{c}$ at the peak will coincide, with high probability, with the $\underline{c}$ at which $p(\underline{r}|\underline{a}, \underline{c})$ peaks sharply. As a result, the intergrand $p(\underline{r}|\underline{a}, \underline{c}) \cdot p(\underline{c})$ provides a dominant contribution to the integral of (8) at the value of $\underline{c}$ which maximizes $p(\underline{r}|\underline{a}, \underline{c})p(\underline{c})$. Hence, the decision rule of equation (8) can be well approximated by

$$\hat{\underline{a}} \cong \arg\max_{\underline{a}} \max_{\underline{c}} p(\underline{r}|\underline{a}, \underline{c}) \cdot p(\underline{c}). \tag{9}$$

This approximation, and the arguments leading to it, are based on ideas presented by Helstrom [7].

We introduce two equalities into Equation (9). These are $p(\underline{r}|\underline{a}, \underline{c}) = \prod_{i=1}^{L} p(r_i|a_i, \underline{c}_i)$ (generated based on the modelling of noise as white) and $p(\underline{c}) = \prod_{i=1}^{L} p(\underline{c}_i|\underline{c}_{i-1}, ..., \underline{c}_0)$ (established by repeated application of the formula $p(A, B) = p(A|B)p(B)$). This leads to

$$\hat{\underline{a}} \cong \arg\max_{\underline{a}} \max_{\underline{c}} \prod_{i=1}^{L} p(r_i|a_i, \underline{c}_i) \cdot p(\underline{c}_i|\underline{c}_{i-1}, ..., \underline{c}_0) \tag{10}$$

or, equivalently,

$$\hat{\underline{a}} \cong \arg \max_{\underline{a}} \max_{\underline{c}} \sum_{i=1}^{L} [\ln p(r_i|a_i, \underline{c}_i) + \ln p(\underline{c}_i|\underline{c}_{i-1}, ..., \underline{c}_0)]. \tag{11}$$

Interchanging the order of the joint maximization, $\hat{\underline{a}}$ corresponds to the sequence resulting from

$$\max_{\underline{c}} \max_{\underline{a}} \sum_{i=1}^{L} [\ln p(r_i|a_i, \underline{c}_i) + \ln p(\underline{c}_i|\underline{c}_{i-1}, ..., \underline{c}_0)]. \tag{12}$$

With a_i independent of its history and its future, the $\hat{\underline{a}}$ in (12) corresponds to the value resulting from

$$\max_{\underline{c}} \sum_{i=1}^{L} [[\max_{a_i} \ln p(r_i|a_i, \underline{c}_i)] + \ln p(\underline{c}_i|\underline{c}_{i-1}, ..., \underline{c}_0)]. \tag{13}$$

Additionally, in any communication environment, the correlation between channel nuisance parameter values diminishes over time. Hence, $p(\underline{c}_i|\underline{c}_{i-1}, ..., \underline{c}_0)$ is well approximated by $p(\underline{c}_i|\underline{c}_{i-1}, ..., \underline{c}_{-I})$. In the case of rapidly changing nuisance parameter values, I is small. Using this in the above equation, we have $\hat{\underline{a}}$ achieved from

$$\max_{\underline{c}} \sum_{i=1}^{L} [[\max_{a_i} \ln p(r_i|a_i, \underline{c}_i)] + \ln p(\underline{c}_i|\underline{c}_{i-1}, ..., \underline{c}_{-I})]. \tag{14}$$

3.2 Implementation

To achieve a realizable implementation, we introduce an approximation which we apply to the receiver. Specifically, we approximate the continuous channel parameter space, i.e., the space spanned by $\underline{c}_i$, with a discrete channel parameter space consisting of the m values $\tilde{C} = \{\underline{c}^1, ..., \underline{c}^m\}$.

We choose the discretization of the channel parameter space carefully. We require that the discretization satisfy the following criterion: on average, a symbol-by-symbol decoding with the $\tilde{\underline{c}}_i \in \tilde{C}$ closest to $\underline{c}_i$ will display a performance very close to that of a symbol-by-symbol decoding with the exact $\underline{c}_i$; that is, performing $argmax_{a_i} p(r_i|a_i, \tilde{\underline{c}}_i)$ at the $\tilde{\underline{c}}_i$ closest to $\underline{c}_i$ will result in a probability of error very close to that achieved by the decoding $argmax_{a_i} p(r_i|a_i, \underline{c}_i)$, on average.

We now apply this discretization to the equation describing the receiver output, equation (14). This equation becomes: select the $\hat{\underline{a}}$ generated from

$$\max_{\underline{\tilde{c}} = (\tilde{\underline{c}}_1, ..., \tilde{\underline{c}}_m) \in \tilde{C}^L} \sum_{i=1}^{L} [[\max_{a_i} \ln p(r_i|a_i, \tilde{\underline{c}}_i)] + \ln P(\tilde{\underline{c}}_i|\tilde{\underline{c}}_{i-1}, ..., \tilde{\underline{c}}_{-I})]. \tag{15}$$

Here, $P(\tilde{\underline{c}}_i|\tilde{\underline{c}}_{i-1}, ..., \tilde{\underline{c}}_{-I})$ corresponds to the probability function generated using the probability density $p(\underline{c}_i|\underline{c}_{i-1}, ..., \underline{c}_{-I})$ and the discretization of $\underline{c}_i$ to $\tilde{\underline{c}}_i$.

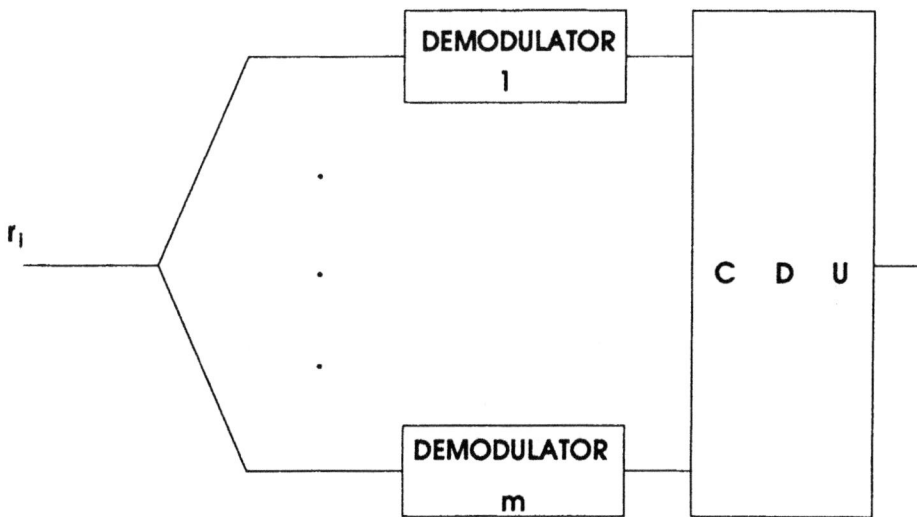

Fig. 1. The general receiver structure

A receiver can be designed to detect data according to (15). The implementation of this receiver is shown in Figure 1. Here, the kth demodulator in the set of demodulators assumes that $\underline{c}_i$ has the value $\underline{c}^k$. The kth demodulator, using r_i, and this claim of $\underline{c}_i = \underline{c}^k$, generates, at each time i, a decision regarding a_i, the transmitted symbol. The demodulator's decision corresponds to the ML decision; it is described by

$$a_i^k = \arg \max_{a_i} p(r_i | a_i, \underline{c}^k). \tag{16}$$

The kth demodulator also generates the corresponding likelihood value

$$l_i^k = \max_{a_i} p(r_i | a_i, \underline{c}^k). \tag{17}$$

The Computation and Decision Unit (CDU) selects the system output from among the demodulator decisions. At one sample time, the CDU may decide to output a decision received from demodulator A, and at the next time it may select a decision from demodulator B. Overall, the CDU's output sequence is a mixing of the decisions from the m demodulators. Let $\underline{\hat{a}}$ represent the sequence of demodulator decisions output by the CDU. This sequence is chosen according to

$$\underline{\hat{a}} = (\hat{a}_1, \hat{a}_2, ..., \hat{a}_L) \text{ where } \hat{a}_i = a_i^{k_i}, \ k_i \in \{1, 2, ..., m\}; \tag{18}$$

here, the k_i are generated using

$$\underline{k} = (k_1, k_2, ..., k_L) = \arg \max_{\underline{k}} \sum_{i=1}^{L} [\ln(l_i^{k_i}) + \ln P(c^{k_i} | c^{k_{i-1}}, ..., c^{k_{i-I}})] \tag{19}$$

This receiver is simply an implementation of equation (15).

4 Application to Rapidly Changing Phase

We now generate a receiver for the special case of data detection in the presence of rapidly changing unknown phase. This special case is described by equations (3),(4) and (5).

4.1 Key Equation for the Rapidly Changing Phase Case

We generate a key equation defining the receiver output, $\hat{a}$, by applying the same equations and approximations used to generate the key equation in the general case of Section 3. The only difference is that, here, we replace c_i with θ_i, and $v(a_i, c_i)$ with $a_i e^{j\theta_i}$. This leads to an equation for the receiver output $\hat{a}$ analogous to equation (14). This equation states that the receiver output $\hat{a}$ should be achieved from

$$\max_{\underline{\theta}} \sum_{i=1}^{L} [[\max_{a_i} \ln p(r_i|a_i, \theta_i)] + \ln p(\theta_i|\theta_{i-1}, ..., \theta_{i-I})]. \tag{20}$$

In this special case, we can further simplify this equation. This is because all the available information regarding phase is given by the $p(\theta_i|\theta_{i-1})$ of equation (5). We can substitute this into the above equation for $p(\theta_i|\theta_{i-1}, ..., \theta_{i-I})$. In this case, for $|\theta_i - \theta_{i-1}| > \beta$, the $p(\theta_i|...)$ term becomes 0. This leads to: select the $\hat{a}$ from

$$\max_{\underline{\theta}} \sum_{i=1}^{L} [\max_{a_i} \ln p(r_i|a_i, \theta_i)] \text{ subject to } |\theta_i - \theta_{i-1}| < \beta. \tag{21}$$

4.2 Implementation for the Rapidly Changing Phase Case

We apply an approximation to the receiver, analogous to the one used in Section 3, to achieve a realizable implementation. Specifically, we approximate the continuous phase space by a discrete one, choosing the discrete space carefully. We begin to determine the discrete space by first closely examining the continuous phase space. The continuous phase space is $\Phi = [0, 2\pi)$. However, since the transmitted symbols are differentially encoded, the output of our receiver can be mapped into the correct $[\frac{2\pi \cdot i}{M}, \frac{2\pi \cdot (i+1)}{M})$ sector of space by applying a differential decoder. Hence, it suffices, for the purposes of our receiver, to represent the continuous phase space by $\Theta = [0, \frac{2\pi}{M})$. We are now ready to establish the discrete phases. We assume that the continuous phase distribution for θ_i is uniform; then, borrowing from quantization theory, the best choice for m discrete phases to represent a uniform θ_i corresponds to m evenly spaced phases. Hence, the best discrete phase space is $\tilde{\Theta} = \{\theta^1, ..., \theta^m\} = \{\frac{2\pi}{M} \frac{2j-1}{2m}, j = 1, ..., m\}$. We choose a value for m to insure that the criteria for discretization, outlined in the previous section, is satisfied. We have found that $m = 8$ is sufficient in all cases considered.

Using the discrete phase space to approximate the continuous one, the equation for the receiver output $\hat{\underline{a}}$ becomes: choose $\hat{\underline{a}}$ from

$$\max_{\underline{\tilde{\theta}}=(\tilde{\theta}_1,...,\tilde{\theta}_m)\in\Theta^L} \sum_{i=1}^{L}[\max_{a_i}\ln p(r_i|a_i,\tilde{\theta}_i)] \text{ subject to } (\tilde{\theta}_i - \tilde{\theta}_{i-1}) = \delta \qquad (22)$$

where $\delta\epsilon[-\Delta\tilde{\theta}, 0, \Delta\tilde{\theta}]$, $\Delta\tilde{\theta} = \frac{2\pi}{8 \cdot M}$. Here, we have discretized the continuous constraint on phase, resulting in an equality in terms of the ternary valued variable δ.

A receiver can be built which outputs the $\hat{\underline{a}}$ from equation (22). This receiver can again be implemented as shown in Figure 1. Here, the kth demodulator generates, every T seconds, the ML decision

$$a_i^k = \arg\max_{a_i} p(r_i|a_i, \theta^k) \qquad (23)$$

and the corresponding likelihood value

$$l_i^k = \max_{a_i} p(r_i|a_i, \theta^k). \qquad (24)$$

The Computation and Decision Unit (CDU) selects its output from among the many demodulator decisions. It makes a selection according to

$$\hat{\underline{a}} = (\hat{a}_1, ..., \hat{a}_L) \text{ where } \hat{a}_i = a_i^{k_i}, \ k_i \in \{1, 2, ..., m\}; \qquad (25)$$

here, the k_i's are evaluated according to

$$\underline{k} = (k_1, ..., k_L) = \arg\max_{\underline{k}} \sum_{i=1}^{L} \ln(l_i^{k_i}) \text{ subject to } [(k_i - k_{i-1}) \bmod m] \in \{-1, 0, 1\}. \qquad (26)$$

This receiver is simply an implementation of equation (22).

We can express the CDU's operation in a way which suggests an easy construction. First, we explore the constraint $[(k_i - k_{i-1}) \bmod m] \in \{-1, 0, 1\}$. This states that if the decision of demodulator k (the demodulator assuming θ^k) is selected by the CDU at time $i - 1$ (i.e., $k_{i-1} = k$), then at time i the CDU will select as its output the decision generated by either demodulator $k - 1$ (assuming θ^{k-1}), demodulator k (assuming θ^k), or demodulator $k + 1$ (assuming θ^{k+1}). That is, the CDU chooses which demodulators' decisions to output under the restriction that the only allowed discrete phase transitions permitted are as indicated in Figure 2.

Equations (25) and (26) state that the CDU outputs the sequence of demodulator decisions corresponding to the path through the trellis of Figure 2 which maximizes $\sum_{i=1}^{L} \ln(l_i^{k_i})$. Here, $l_i^{k_i}$ are the likelihood values generated by the $m = 8$ demodulators. This can be implemented by using a rather simple Viterbi Algorithm (VA).

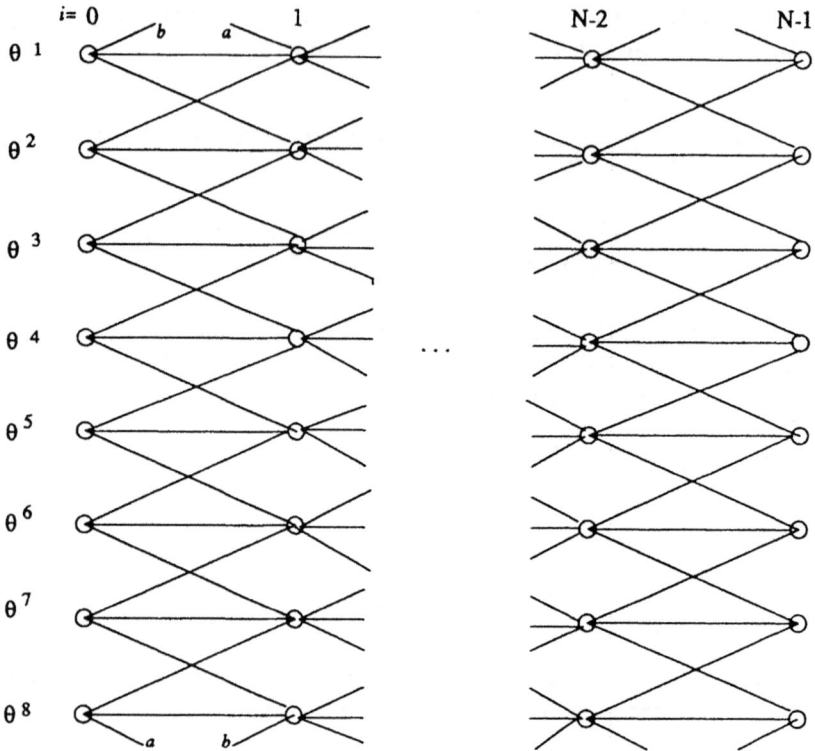

Fig. 2. The trellis indicating allowed phase transitions.

4.3 Computational Complexity

In this section, we describe the computations performed by the receiver implementation of Section 4.2. We will show that this implementation achieves a very low computational complexity.

The computations carried out by the implementation of Section 4.2 can be broken down into two parts: the computation at the $m = 8$ demodulators, and the computation at the CDU.

The computation at the demodulators consists of the computation of a_i^k and the computation of l_i^k. The a_i^k computation corresponds to, simply, a symbol-by-symbol decoding of a PSK signal with known phase offset θ^k. Additionally, it is easily shown that the computation of l_i^k can be replaced by the evaluation of $Re\{r_i^* a_i^k e^{j\theta^k}\}$. The $a_i^k e^{j\theta^k}$ component is discrete, taking on one of $m \cdot M$ values, and can be generated using a look-up table. Hence, the computation of $Re\{r_i^* a_i^k e^{j\theta^k}\}$ requires only 2 multiplications and 1 addition.

The CDU finds the best path through the trellis made up of 8 nodes, one node for each discrete phase θ^k. Three branches enter into each one of these nodes. When using a VA to determine this best path, one requires at each time i, 8 comparisons of 3 values each.

Putting it all together, the overall computation required by the implementation at each time i is: 8 MPSK symbol-by-symbol decisions, 8*2 multiplies, 8 additions, and 8 comparisons of 3 values each. That is, at each time i, the computational requirement is in the order of 8 times that of a symbol-by-symbol PSK decoder.

4.4 Phase movement from one $\frac{2\pi}{M}$ sector to another

In this section, we point out a potential trouble area for the receiver we implemented in Section 4.2. We then explain how this potential for trouble can be avoided by introducing a minor change in the receiver.

In Section 4.2, we implemented our receiver by approximating the continuous phase space by a discrete one. We carried out this approximation assuming a differential decoder was present at the receiver output to undo phase ambiguities of $\frac{2\pi}{M}$. However, at times when θ_i drifts from one $\frac{2\pi}{M}$ sector to another, a differential decoder is unable to undo the $\frac{2\pi}{M}$ phase ambiguity. At these times, the assumption made in implementing our receiver is not valid, and our system will generate errors.

Fortunately, by making a minor change to our receiver, we can undo the error effects that occur when θ_i moves from one $\frac{2\pi}{M}$ sector to another. Assuming that the best path through the trellis tracks phase correctly, then when phase θ_i moves from one $\frac{2\pi}{M}$ sector to another, the phases of the trellis move either from θ^8 to θ^1, or from θ^1 to θ^8. Hence, the movements $\theta^1 \rightarrow \theta^8$ and $\theta^8 \rightarrow \theta^1$ indicate that θ_i has changed $\frac{2\pi}{M}$ sectors. So, we can flag the times when a $\frac{2\pi}{M}$ movement has occurred. Knowing this, we are able to update the sequence $\hat{\underline{a}}$ sent to the differential decoder so that, as far as the differential decoder is concerned, all channel phases θ_i remain in the same $\frac{2\pi}{M}$ sector of space.

4.5 Performance

In this section, we detail the performance of our receiver, implemented as indicated in Section 4.2. This performance is established using plots showing probability of symbol error, $P(\epsilon)$, versus $\frac{E_b}{N_o}$. In this section, we display the performance curves for an 8-PSK constellation. We demonstrate our receiver performance under the following channel phase conditions. The phase is simulated using $\theta_i = \theta_{i-1} + w_i$ where w_i are samples from an AWGN process. This models phase as a random walk on a circle. This rather general phase model has been employed successfully in the analysis of heterodyne optical communications, as well as communication over slowly fading channels and telephone lines [8].

In our simulations, we choose the variance of w_i, σ_w^2, such that the phase θ_i can be modelled as constant over two symbol intervals, but not necessarily constant over a longer interval. Specifically, we display performance curves at the following σ_w^2. We first consider $\sigma_w^2 = 0.0$; in this case, the channel phases θ_i are constant. We also consider $\sigma_w^2 = 0.0009, 0.0025, 0.0049$, and 0.0081 radians^2; or, equivalently a standard deviation of $\sigma_w = 0.03, 0.05, 0.07$, and 0.09 radians. In degrees, $\sigma_w = 1.71, 2.88, 4.01$, and 5.15 respectively. In these cases, the phase movement from θ_{i-1} to θ_i is within $1.71, 2.88, 4.01$, and 5.15 degrees, respectively, 60% of the time; and this movement is within $6.84, 11.53, 16.04$, and 20.6 degrees, respectively, 98% of the time.

We implement our CDU using a VA which runs over 100 symbols at a time. The last symbol of one set of 100 is used as the first symbol of the next set of 100. This implementation keeps the memory requirement of the VA fairly low. Additionally, at this length, the end effects of the VA are negligible. An alternative, and more common, way to implement the VA is to use a lookback of depth 12 to 15. This has an even lower memory requirement.

Our simulated results are displayed in Figures 3 to 10. Figure 3 shows performance results under the best of circumstances. Here, the phases θ_i remain unchanged from symbol to symbol. This curve indicates that at a $P(\epsilon)$ of 10^{-3} and 10^{-4}, we gain 1.41 and 1.43dB respectively over DPSK. Similarly, in Figures 4, 5, and 6, where the phase movement has variance $0.0009, 0.0025$, and 0.0049, we observe performance gain over DPSK: at $P(\epsilon)$ of 10^{-3}, we gain 1.40dB, 1.53dB, and 1.46dB respectively, and at $P(\epsilon)$ of 10^{-4}, we gain 1.72dB, 1.50dB, and 1.32dB.

Next, we examine Figure 7, where phase movement displays a variance of $\sigma_w^2 = 0.0081$. In this case, our scheme performs poorly, at times performing worse than DPSK. However, this is a case where the phase change is so rapid that even DPSK is not applicable.

Overall, over the range of values of σ_w^2 where DPSK is applicable, our scheme is able to outperform DPSK by about 1.5dB at the $P(\epsilon)$'s of 10^{-3} and 10^{-4}. This improvement over DPSK can be explained as follows. The receiver we implement considers a longer phase history than DPSK. That is, DPSK assumes phase constant over two samples r_i and r_{i-1}, but ignores the implied phase correlation that would then exist between r_{i-2} and r_i. Our receiver considers this correlation via the trellis of Figure 2, and thus gains over DPSK.

Figures 8, 9, and 10 compare the performance of our proposed scheme with that of MSDD using $n = 3, 5$ and 10. Figure 8 compares these performances for the case when phase is constant. It is seen that, in this case, the performance of our scheme is better than MSDD with $n = 3$, matches that of the MSDD with $n = 5$, but is inferior to MSDD using $n = 10$.

Figure 9 compares the performance of the two schemes under low phase variation ($\sigma_w^2 = 0.0009$). In this case, our scheme once again outperforms MSDD with $n = 3$, and is again as good as MSDD with $n = 5$. MSDD with $n = 10$ no longer outperforms our scheme. The performance of MSDD with $n = 10$ deteriorates quite rapidly as phase variation increases, due to the fact that the assumption of constant phase over $n = 10$ symbols does not hold any longer.

Figure 10 compares our scheme with MSDD for a rapidly changing phase ($\sigma_w^2 = 0.0049$). In this case, the performance of our scheme is always better than MSDD. This is because the constant phase assumptions of MSDD prove unrealistic, even at the lower values of n.

These results demonstrate that our scheme is at least as good as MSDD under slow phase change conditions, and outperforms MSDD under rapid phase changes. The gains of our scheme over MSDD can be explained as follows. MSDD assumes that phase is constant over n symbols, where typical values of n are values such as 3, 5, or 10. Hence, as phase changes become more and more rapid, MSDD's performance degrades quickly. Our scheme considers a block of L symbols, where L is very large (e.g. $L = 100$), and, unlike MSDD, we do not assume a constant phase over the block, but rather a correlated one. This leads to two effects. For a truly constant phase (not a practical situation in many environments), MSDD with $n = 10$ will outperform our scheme because its assumption of constant phase is better than our assumption of correlated phase. However, once there is phase movement, even a small one, we achieve a better performance, because our assumption of correlated phase is superior, and, additionally, we consider a longer history of phase.

5   Conclusions

In this work, we introduce a novel receiver structure for the detection of data in the presence of unknown channel parameters. We demonstrate the applicability of this receiver by applying it to the special case of data detection of MPSK in the presence of rapidly changing unknown phase. This leads to a receiver which serves as a robust alternative to DPSK. Our scheme offers substantial performance gains. The price we pay for this performance gain is an increase in complexity. The complexity of our decoder is in the order of 8 times that of a typical PSK receiver. In any environment in which a DPSK receiver is currently employed, this receiver can be replaced by the receiver introduced here, achieving substantial performance gains at a practical complexity.

References

1. B. Sklar, *Digital Communications: Fundamentals and Applications.* Englewood Cliffs, NJ: Prentice-Hall, 1988.
2. D. Divsalar and M.K. Simon, "Multiple-symbol differential detection of MPSK," *IEEE Trans. Commun.*, Vol. 38, pp.300-308, Mar.1990.
3. S.G. Wilson, J. Freebersyser, and C. Marshall,"Multi-symbol detection of M-DPSK," presented at GLOBECOM'89, Dallas, Texas, Nov. 27-30, 1989, pp.1692-1697.
4. D. Makrakis and K. Feher,"Optimal noncoherent detection of PSK signals,"*Electronics Letters*, Vol.26, pp. 398-400, Mar.15, 1990.
5. H. Leib and S. Pasupathy,"Optimal noncoherent block demodulation of differential phase
6. K.M. Mackenthun, "A fast algorithm for multiple-symbol differential detection of MPSK," *IEEE Trans. Commun.*, Vol. 42, pp. 1471-1474, April 1994.
7. C. W. Helstrom, *Statistical Theory of Signal Detection.* Oxford, New York: Pergamon Press, 1968.
8. P.Y. Kam, K.Y. Seek, T.T. Tjhung, and P. Sinha, "Error probability of 2DPSK with phase noise," *IEEE Trans. Commun.*, Vol. 42, pp.2366-2369, July 1994.

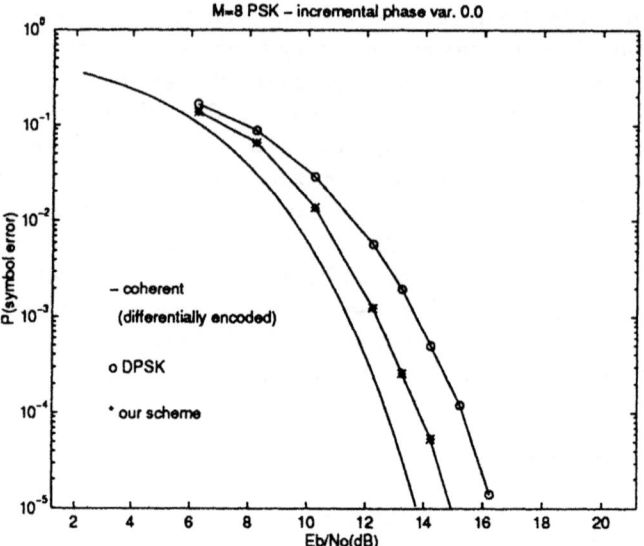

Fig. 3. $P(\epsilon)$ vs. $\frac{E_b}{N_o}$ curves for 8-PSK, $\sigma_w^2 = 0.0$.

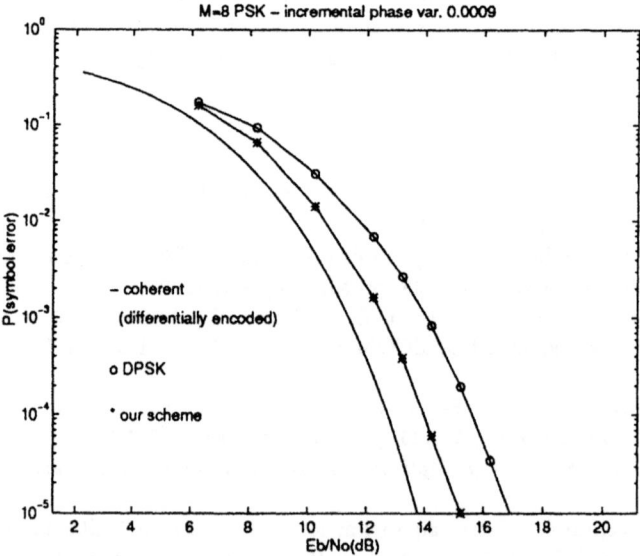

Fig. 4. $P(\epsilon)$ vs. $\frac{E_b}{N_o}$ curves for 8-PSK, $\sigma_w^2 = 0.0009$.

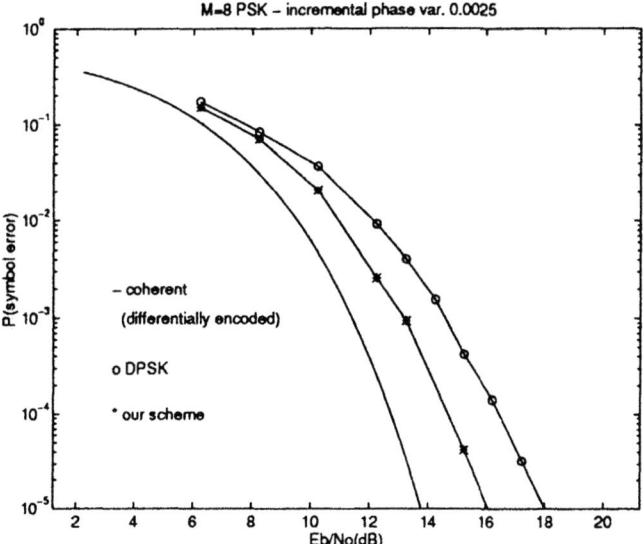

Fig. 5. $P(\epsilon)$ vs. $\frac{E_b}{N_o}$ curves for 8-PSK, $\sigma_w^2 = 0.0025$.

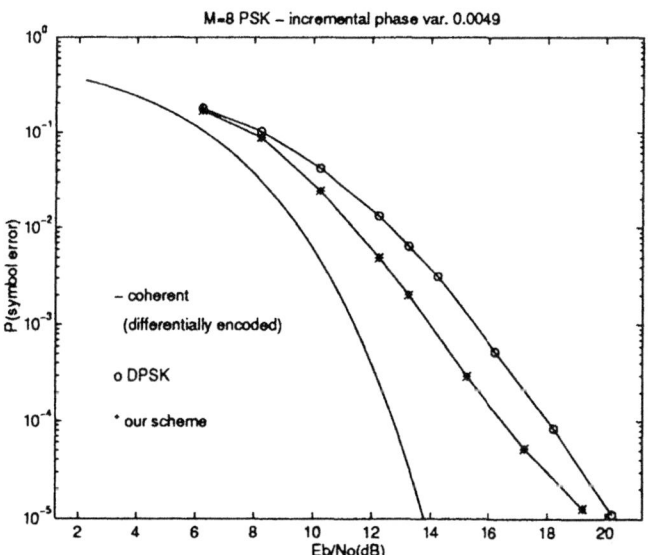

Fig. 6. $P(\epsilon)$ vs. $\frac{E_b}{N_o}$ curves for 8-PSK, $\sigma_w^2 = 0.0049$.

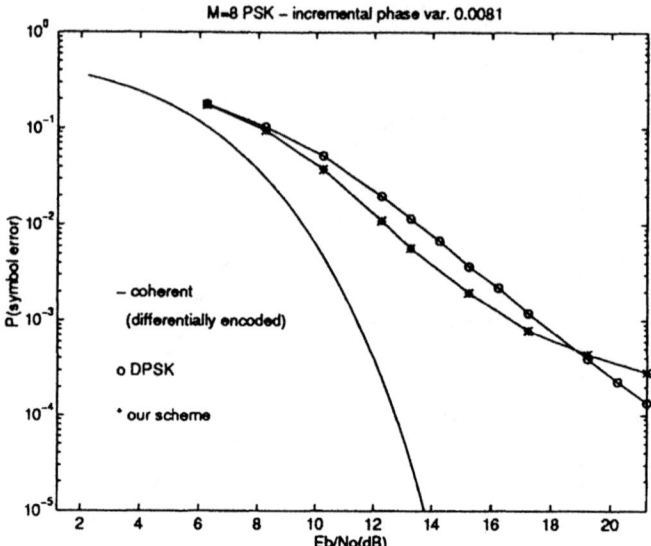

Fig. 7. $P(\epsilon)$ vs. $\frac{E_b}{N_o}$ curves for 8-PSK, $\sigma_w^2 = 0.0081$.

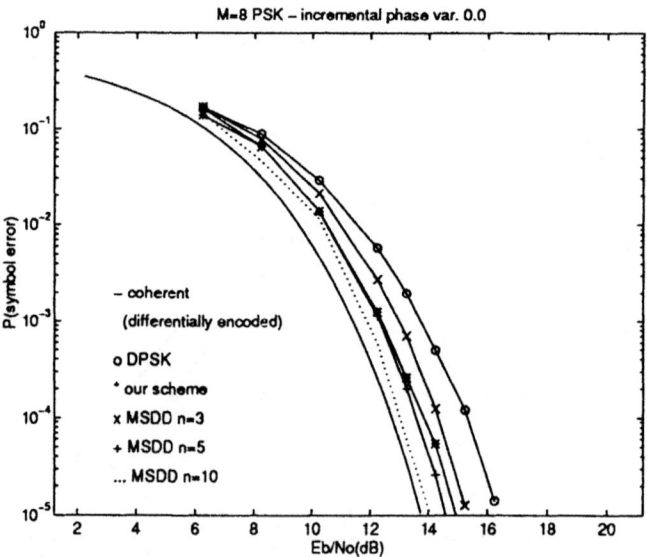

Fig. 8. $P(\epsilon)$ vs. $\frac{E_b}{N_o}$ curves for 8-PSK, $\sigma_w^2 = 0.0$, including MSDD performance ($n = 3, 5,$ and 10).

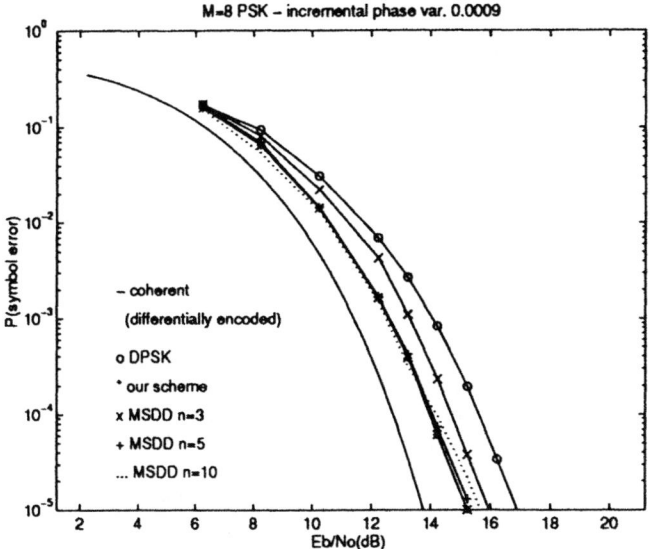

Fig. 9. $P(\epsilon)$ vs. $\frac{E_b}{N_o}$ curves for 8-PSK, $\sigma_w^2 = 0.0009$, including MSDD performance $(n = 3, 5, \text{ and } 10)$.

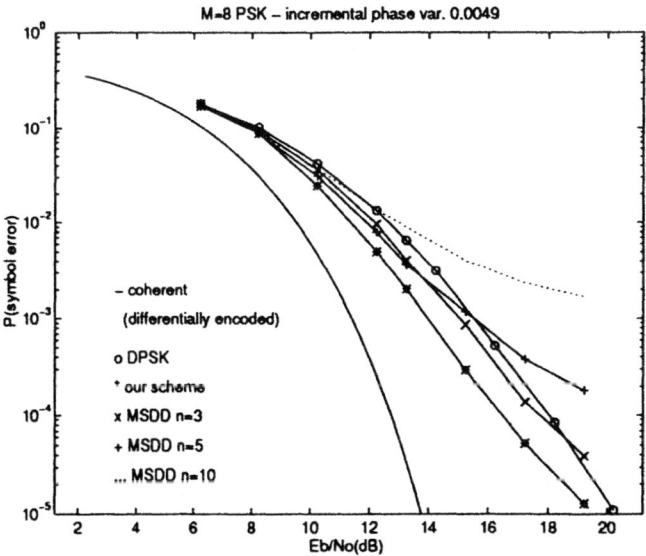

Fig. 10. $P(\epsilon)$ vs. $\frac{E_b}{N_o}$ curves for 8-PSK, $\sigma_w^2 = 0.0049$, including MSDD performance $(n = 3, 5, \text{ and } 10)$.

Comparison of MFSK Variants for a Hybrid DS/FH-CDMA System in Rayleigh Fading

Claude D'Amours[1] and Abbas Yongaçoğlu[2]

[1] Department of Electrical and Computer Engineering,
Royal Military College of Canada, Kingston, Ontario
[2] Department of Electrical Engineering,
University of Ottawa, Ottawa, Ontario

Abstract. This paper compares the spectral efficiency of a hybrid DS/FH-CDMA system employing variants of MFSK modulation. The proposed modulation schemes are: non-coherent MFSK, MFSK-DPSK, and wideband MT-FSK. In each case, the signal to be transmitted is modulated by a high rate BPSK signal (the PN sequence) and then hopped at a rate higher than the symbol rate into different frequency bins that are sufficiently spaced so that the fading process in each frequency bin appears to be independent of the fading in any other frequency bin. The bit error rate performance of each modulation scheme in Rayleigh fading is found by analytical means. Error control coding is also considered. These results are used to find the spectral efficiency of the DS/FH-CDMA system where the multiple access interference is modeled by white Gaussian noise. It will be shown that MFSK with rate 1/2 dual-k coding provides the best results for the cases considered in this paper.

1 Introduction

Code division multiple access (CDMA) systems employ spread spectrum (SS) techniques. There are two types of CDMA systems: direct sequence (DS) CDMA systems and frequency hopped (FH) CDMA systems. In DS-CDMA systems, multiple users may simultaneously access a common channel by employing user specific codes which have low cross-correlation properties. These codes are referred to as pseudonoise (PN) codes as one user's code resembles noise to all other users. In FH-CDMA systems, the carrier frequency of each user is changed at some rate. Low error transmission for a given user requires that the user's signal be hopped into unoccupied frequency bins with very high success.

Hybrid DS/FH-CDMA systems are now being investigated [1-3] as they may combine some advantages of both DS and FH systems while avoiding some of their disadvantages. In this paper, we propose a hybrid DS/FH-CDMA system for use in the base to mobile (forward) link of a land mobile communication system. The channel is modeled by a slowly Rayleigh fading process. The system employs variants of MFSK modulation. The spectral efficiency of the system employing each modulation scheme is determined by analysis and the results are compared.

Section 2 of this paper discusses the different modulation schemes to be used by the hybrid DS/FH-CDMA system. Their bit error rate performances in Rayleigh fading are determined. Diversity combining is also considered. Section 3 demonstrates the performance of these modulation schemes when error control coding is used. We will consider nonbinary BCH codes as well as a nonbinary class of convolutional codes known as dual-k codes. Section 4 describes the hybrid DS/FH-CDMA system and the spectral efficiency of the system employing each modulation scheme with or without coding is determined. Conclusions that are drawn from the results are given in Section 5.

2 Variants of MFSK Modulation and Their Performance in Rayleigh Fading Channels

MFSK modulation and its variants are considered because noncoherent MFSK has been shown to be more power efficient than DPSK for large values of M in AWGN channels [4]. This attribute makes MFSK an attractive modulation scheme for CDMA systems since power efficiency can be traded off against spectral efficiency. Also, it does not require a carrier phase recovery circuit which makes it suitable for mobile communications.

Noncoherent MFSK Modulation. In MFSK, information symbols are represented by specific frequencies, $f_m = f_c + [m - \frac{M+1}{2}]\Delta f$, where $m = 1, 2, \cdots, M$, f_c is the centre frequency, and Δf is the frequency spacing between adjacent symbols. The minimum frequency spacing of adjacent noncoherent MFSK symbols needed for orthogonality is $\Delta f = 1/T_s$ [5].

The optimum receiver of MFSK determines the a posteriori probability that symbol m was transmitted given the received signal. It then chooses the symbol with the highest a posteriori probability. If symbol 1 was transmitted, then assuming that the symbol timing information has been recovered, and given that the signals are noncoherently detected (thus channel phase is irrelevant) the decision variables over one symbol interval are:

$$U_1 = |2aE_s + N_1| \tag{1}$$
$$U_m = |N_m| \tag{2}$$

where a is the magnitude of the complex channel gain and N_m are noise samples at the output of the mth matched filter, and they are complex-valued independent, identically distributed zero mean Gaussian random variables with variance $\sigma = 2E_s N_o$.

Performance of MFSK in Rayleigh Fading. In a Rayleigh fading channel, the channel gain a is time-varying, and its envelope is a stochastic process that has a Rayleigh distribution. If a is constant for all spectral components of the signal, then the channel is referred to as a frequency-nonselective Rayleigh fading channel. The performance of MFSK in a nonselective Rayleigh fading channel is shown in [5] to be:

$$P_M = \sum_{n=1}^{M-1} (-1)^{n+1} \binom{M-1}{n} \frac{1}{nk\bar{\gamma}_b + n + 1} \tag{3}$$

where $\bar{\gamma}_b = \frac{E_b}{N_o} E[a^2]$. The probability of bit error is $P_b = \frac{M}{2(M-1)} P_M$. The bit error rate of MFSK in Rayleigh fading is shown in Figure 1. As we increase from $m = 8$ to $M = 32$, the performance improves marginally.

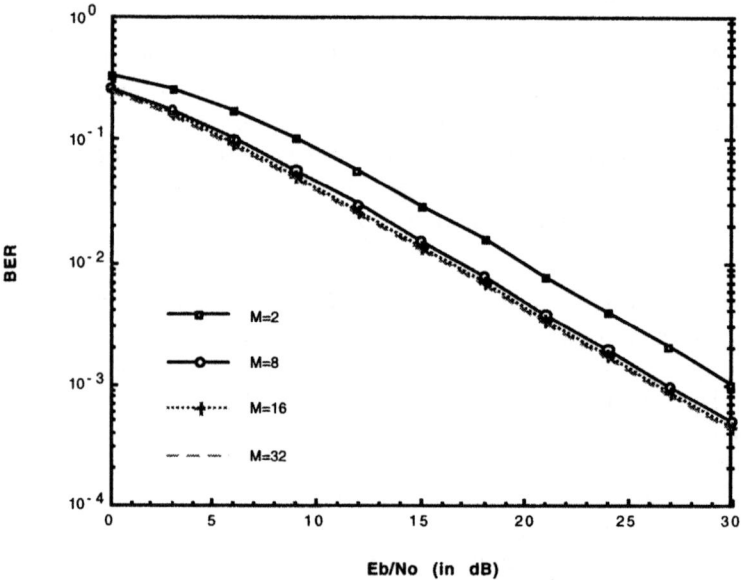

Fig. 1. Bit error rate performance of noncoherent MFSK in flat Rayleigh fading.

Performance of Noncoherent MFSK in Rayleigh Fading with Diversity. Suppose that the same information-bearing noncoherent MFSK signal is transmitted over L frequency-nonselective slowly Rayleigh fading channels. Suppose further that these L channels are sufficiently separated in frequency so that the fading process on each channel is independent of the fading processes on the other $L - 1$ channels. Thus we have an Lth order frequency diversity system.

FSK signals in diversity systems are generally square-law combined, thus assuming symbol 1 was transmitted, the decision variables corresponding to each possible symbol are:

$$U_1 = \sum_{k=1}^{L} |2E_c a_k + N_{k1}|^2 \tag{4}$$

$$U_m = \sum_{k=1}^{L} |N_{km}|^2 \tag{5}$$

where a_k is the complex fading process of the kth channel and $E_c = E_s/L$ is the energy per symbol per diversity channel.

The probability of symbol error is found form the probability distribution functions of these decision variables. It is found to be [5]:

$$P_M = \frac{1}{(L-1)!} \sum_{m=1}^{M-1} \frac{(-1)^{m+1} \binom{M-1}{m}}{(1+m+m\bar{\gamma}_c)^L}$$

$$\times \sum_{i=0}^{m(L-1)} \beta_{im}(L-1+i)! \left(\frac{1+\bar{\gamma}_c}{1+m+m\bar{\gamma}_c}\right)^i \tag{6}$$

where β_{im} is the set of coefficients in the following expansion:

$$\left(\sum_{i=0}^{L-1} \frac{u_1}{i!}\right)^m = \sum_{i=0}^{m(L-1)} \beta_{im} u_1^i \tag{7}$$

Note that when we use $L = 1$ in (6), the equation is the same as (3).

Table 1 provides the average energy per bit to noise spectral density ratio required by noncoherent MFSK to provide a bit error rate of 10^{-3} in Rayleigh fading.

Table 1. $\bar{\gamma}_b$ required to obtain a bit error rate of 10^{-3} for an Lth order diversity system employing MFSK in Rayleigh fading.

L	M=8	M=16	M=32
1	27.0 dB	26.4 dB	26.2 dB
2	17.2 dB	16.7 dB	16.5 dB
3	14.1 dB	13.5 dB	13.3 dB
4	13.5 dB	12.9 dB	12.7 dB

One should note that the gain obtained by employing a fourth order diversity system as opposed to a third order diversity system is small. Further increases in the diversity order will result in very small improvements in power efficiency.

2.1 MFSK-DPSK Modulation and its Performance in Rayleigh Fading

In MFSK-DPSK modulation, $k_f + k_p$ bits are grouped into a symbol. The first $k_f = log_2 M_f$ bits are represented by one of M_f possible frequencies. To ensure orthogonality of the frequency sub-symbols, the minimum frequency separation is $\Delta f = R_b/(k_f + k_p)$. The remaining k_p bits are represented by the differential phase of the symbol. In this paper we will only consider two-phase and four-phase modulation schemes ($k_p = 1$ or 2) because it is well known that PSK signals with more than 4 phases are not power efficient.

At the receiver, the transmitted frequency must first be determined, while information about the phase of each tone is retained in buffers. Once the frequency is determined, the phase for the output of the matched filter of the surviving frequency is passed to a differential detector, where the differential phase sub-symbol is determined. A MFSK-DPSK modulator is shown in Figure 2(a), while the demodulator is shown in Figure 2(b).

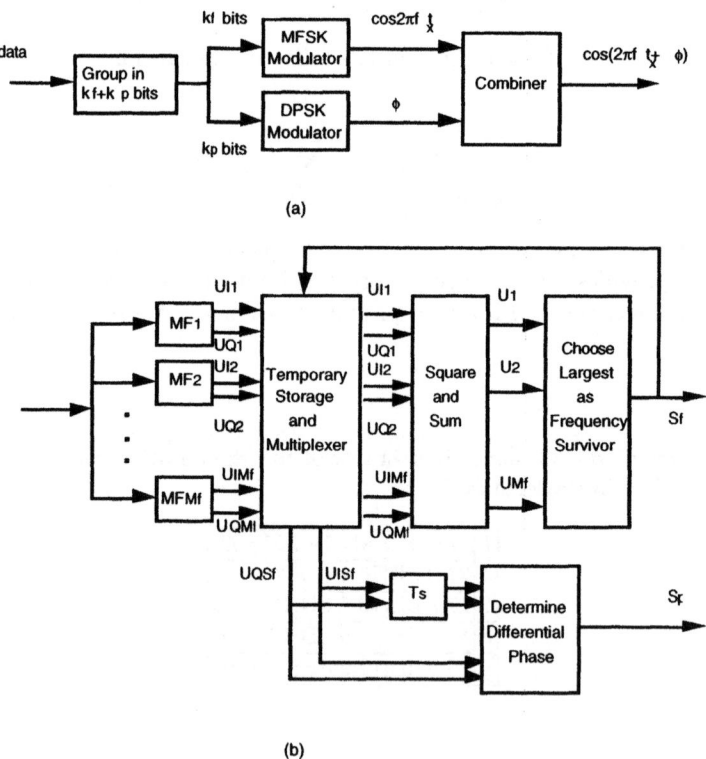

(a)

(b)

Fig. 2. (a) MFSK-DPSK modulator (b) MFSK-DPSK demodulator.

When diversity is employed, the frequency sub-symbol is determined by square-law combining the decision variables from the L MFSK demodulators, while the phase information of each tone from each channel is stored. Once the frequency sub-symbol is determined, the phase information corresponding to the proper frequency is retrieved from each channel and differentially detected. The result is then combined using equal gain combining. The MFSK-DPSK diversity receiver is shown in Figure 3 for $L = 2$.

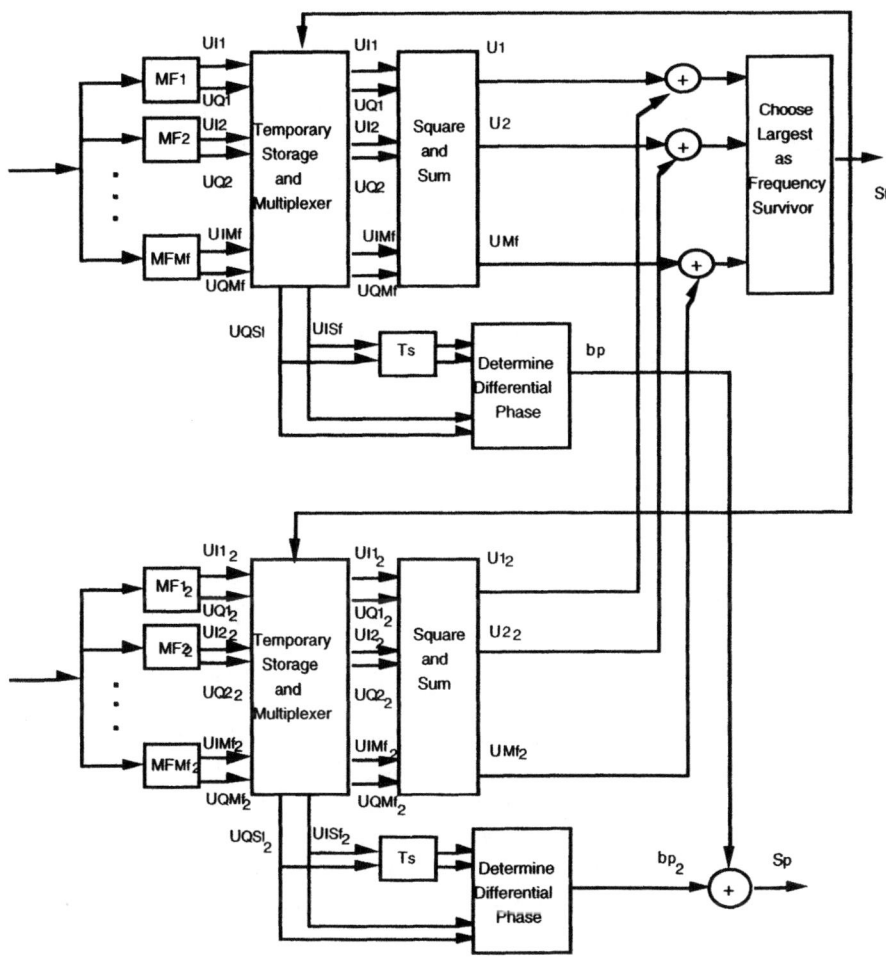

Fig. 3. MFSK-DPSK diversity receiver ($L = 2$).

The average symbol energy to noise spectral density ratio is $\bar{\gamma}_s$. Thus the average bit to noise spectral density ratio is $\bar{\gamma}_b = \bar{\gamma}_s / k_f + k_p$. The frequency sub-symbol is demodulated first. Since the demodulation of this sub-symbol is the same as for MFSK, the sub-symbol error rate in Rayleigh fading for different orders of diversity L is:

$$
P_{Mf} = \frac{1}{(L-1)!} \sum_{m=1}^{M-1} \frac{(-1)^{m+1} \binom{M-1}{m}}{(1+m+m\bar{\gamma}_s/L)^L}
$$
$$
\times \sum_{i=0}^{m(L-1)} \beta_{im}(L-1+i)! \left(\frac{1+\bar{\gamma}_s/L}{1+m+m\bar{\gamma}_s/L} \right)^i \tag{8}
$$

Once the frequency sub-symbol is determined, the received phase of this symbol is retrieved and sent to the differential detector. Assuming that the frequency sub-symbol was correctly detected, the phase sub-symbol error probability is given by P_{DPSK}. For a two-phase system (MFSK-DBPSK), P_{DPSK} is given by [5]:

$$
P_{DPSK} = \left(\frac{1-\mu}{2} \right)^L \sum_{i=0}^{L-1} \binom{L-1-i}{i} \left(\frac{1+\mu}{2} \right)^i \tag{9}
$$

where $\mu = \frac{\bar{\gamma}_s/L}{1+\bar{\gamma}_s/L}$.

For a four phase system (MFSK-DQPSK), P_{DPSK} becomes [5]:

$$
P_{DPSK} = \frac{1}{2} \left[1 - \frac{\mu}{\sqrt{2-\mu^2}} \sum_{i=o}^{L-1} \binom{2i}{i} \left(\frac{1-\mu^2}{4-2\mu^2)} \right)^i \right] \tag{10}
$$

The equations for P_{DPSK} assume that the fading rate is slow compared to the symbol rate.

The probability that the phase sub-symbols are correctly demodulated is dependent on the correct demodulation of the MFSK sub-symbol. If the frequency sub-symbol is correctly detected, then the probability of bit error of the phase sub-symbol is given by P_{DPSK}. However, when the MFSK sub-symbol is incorrectly determined, the incorrect phase information is retrieved from the phase buffers. Therefore, the probability of bit error of this phase sub-symbol is 0.5. This will also affect the next phase sub-symbol since differential detection requires the previous phase to determine the current symbol. Therefore, the bits for the next phase sub-symbol will also have an error probability of 0.5. The probability of bit error for the phase encoded bits is denoted by P_d. It can be found using the Markov chain depicted in Figure 4.

The transition probability matrix for this Markov chain is:

$$
P = \begin{bmatrix} 1-P_{Mf} & P_{Mf} & 0 \\ 0 & P_{Mf} & 1-P_{Mf} \\ 1-P_{Mf} & 0 & P_{Mf} \end{bmatrix} \tag{11}
$$

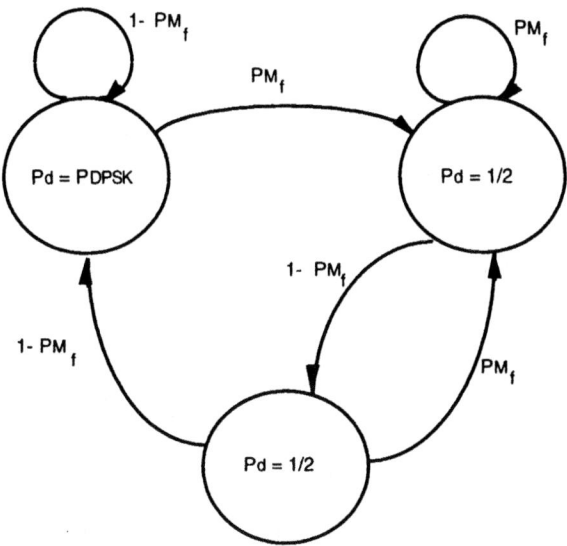

Fig. 4. Markov chain to determine P_d.

The steady state probabilities, π_i, are satisfied by the following equations:

$$\pi = \pi P \tag{12}$$

$$\sum_i \pi_i = 1 \tag{13}$$

Solving these equations, we obtain:

$$\pi_0 = \frac{1 - P_{Mf}}{1 + P_{Mf}} \tag{14}$$

$$\pi_1 = \pi_2 = \frac{P_{Mf}}{1 + P_{Mf}} \tag{15}$$

Thus the probability of bit error in the differentially encoded bit stream is:

$$
\begin{aligned}
P_d &= \pi_0 P_{DPSK} + \frac{\pi_1}{2} + \frac{\pi_2}{2} \\
&= \frac{1 - P_{Mf}}{1 + P_{Mf}} P_{DPSK} + \frac{P_{Mf}}{1 + P_{Mf}} \\
&= \frac{P_{Mf}(1 - P_{DPSK}) + P_{DPSK}}{1 + P_{Mf}}
\end{aligned}
\tag{16}
$$

The overall probability of bit error of MFSK-DPSK modulation is given by:

$$P_{b_{M-D}} = \frac{k_f}{k_f + k_p} \frac{2^{k_f-1}}{2^{k_f} - 1} P_{Mf} + \frac{k_p}{k_f + k_p} P_d \tag{17}$$

Table 2 shows the average bit energy to noise spectral density required for these uncoded modulation schemes to yield a bit error rate of 10^{-3}.

Table 2. $\bar{\gamma}_b$ required to obtain a bit error rate of 10^{-3} for uncoded MFSK-DPSK for different L in Rayleigh fading.

	$L = 1$	$L = 2$	$L = 3$
4FSK-DBPSK	27.4 dB	17.1 dB	14.0 dB
8FSK-DBPSK	27.0 dB	16.5 dB	13.1 dB
16FSK-DBPSK	26.6 dB	16.0 dB	12.6 dB
4FSK-DQPSK	27.3 dB	16.3 dB	13.0 dB
8FSK-DQPSK	26.8 dB	15.8 dB	12.5 dB
16FSK-DQPSK	26.6 dB	15.5 dB	12.2 dB

2.2 Multitone FSK Modulation

Multitone FSK (MT-FSK) is a multifrequency modulation scheme in which energy is transmitted simultaneously over w out of v possible orthogonal frequencies. It can convey at most $log_2 \binom{v}{w}$ bits of information per character. It has the potential to be more bandwidth efficient than MFSK which is a special case of MT-FSK where $w = 1$ and $v = M$. MT-FSK is an application of a more general modulation system described by Slepian called permutation modulation [6].

Permutation Modulation. Suppose we have a finite number of light emitting diodes (LEDs) l_1, l_2, ..., l_v. We also have an alphabet a_1, a_2, ..., a_z that we wish to represent by a combination of illuminated LEDs. If $z = v$, we can represent each letter in the alphabet by illuminating only one LED per symbol. For example $a_1 = l_1$, $a_2 = l_2$, etc. If $z > v$, we can illuminate a pair of LEDs to represent each letter. For example $a_1 = (l_1, l_2)$, $a_2 = (l_1, l_3)$, etc. Thus by using pairs of LEDs to represent each letter, we can represent at most $\binom{v}{2} = v(v-1)/2$ letters using 2 illuminated LEDs. More generally, $\binom{v}{w}$ letters can be represented by only v LEDs by representing each letter by a combination of w illuminated LEDs.

The preceding example was a demonstration of permutation representation. In this example, the medium was LEDs, where the information was conveyed which LEDs were switched on. Permutation modulation is generally implemented using orthogonal waveforms, mainly sinusoids. MT-FSK is a permutation modulation system which employs sinusoids.

Design of Efficient MT-FSK Waveforms. It has been shown that MT-FSK modulation schemes which employ all $\begin{pmatrix} v \\ w \end{pmatrix}$ possible waveforms are not as power efficient as conventional MFSK modulation [7]. However, one does not need to employ all possible waveforms to represent a given alphabet. Suppose only a fraction of these are actually used. Redundancy can be introduced into the modulation scheme, giving us larger distances between symbols, and possibly some power gain. Thus we can sacrifice the bandwidth efficiency of MT-FSK to improve the power efficiency. In this paper, a number of waveforms are designed using the Balanced Incomplete Block (BIB) design which is detailed in [8]. We will denote the signalling sets as (v,w) MT-FSK modulation. The four signalling sets designed for this paper are called (6,3), (8,3), (10,4) and (12,4) MT-FSK. These signalling sets are given in Tables 3-6.

Table 3. (6,3) MT-FSK waveforms.

Symbol	Codeword	Waveform
1	1 1 0 1 0 0	$\cos 2\pi f_1 t + \cos 2\pi f_2 t + \cos 2\pi f_4 t$
2	0 1 1 0 1 0	$\cos 2\pi f_2 t + \cos 2\pi f_3 t + \cos 2\pi f_5 t$
3	0 0 1 1 0 1	$\cos 2\pi f_3 t + \cos 2\pi f_4 t + \cos 2\pi f_6 t$
4	1 0 0 0 1 1	$\cos 2\pi f_1 t + \cos 2\pi f_5 t + \cos 2\pi f_6 t$

Table 4. (8,3) MT-FSK waveforms.

Symbol	Codeword	Waveform
1	1 1 0 1 0 0 0 0	$\cos 2\pi f_1 t + \cos 2\pi f_2 t + \cos 2\pi f_4 t$
2	0 1 1 0 1 0 0 0	$\cos 2\pi f_2 t + \cos 2\pi f_3 t + \cos 2\pi f_5 t$
3	0 0 1 1 0 1 0 0	$\cos 2\pi f_3 t + \cos 2\pi f_4 t + \cos 2\pi f_6 t$
4	0 0 0 1 1 0 1 0	$\cos 2\pi f_4 t + \cos 2\pi f_5 t + \cos 2\pi f_7 t$
5	0 0 0 0 1 1 0 1	$\cos 2\pi f_5 t + \cos 2\pi f_6 t + \cos 2\pi f_8 t$
6	1 0 1 0 0 0 0 1	$\cos 2\pi f_1 t + \cos 2\pi f_3 t + \cos 2\pi f_8 t$
7	1 0 0 0 0 1 1 0	$\cos 2\pi f_1 t + \cos 2\pi f_6 t + \cos 2\pi f_7 t$
8	0 1 0 0 0 0 1 1	$\cos 2\pi f_2 t + \cos 2\pi f_7 t + \cos 2\pi f_8 t$

Table 5. (10,4) MT-FSK waveforms.

Symbol	Codeword	Waveform
1	1 0 0 1 1 0 1 0 0 0	$\cos 2\pi f_1 t + \cos 2\pi f_4 t + \cos 2\pi f_5 t + \cos 2\pi f_7 t$
2	0 0 1 1 0 0 0 1 0 1	$\cos 2\pi f_3 t + \cos 2\pi f_4 t + \cos 2\pi f_8 t + \cos 2\pi f_{10} t$
3	0 1 1 0 1 1 0 0 0 0	$\cos 2\pi f_2 t + \cos 2\pi f_3 t + \cos 2\pi f_5 t + \cos 2\pi f_6 t$
4	1 0 0 0 0 1 0 1 1 0	$\cos 2\pi f_1 t + \cos 2\pi f_6 t + \cos 2\pi f_8 t + \cos 2\pi f_9 t$

Table 6. (12,4) MT-FSK waveforms.

Symbol	Codeword	Waveform
1	1 0 1 0 0 0 1 0 1 0 0 0	$\cos 2\pi f_1 t + \cos 2\pi f_3 t + \cos 2\pi f_7 t + \cos 2\pi f_9 t$
2	1 0 0 1 1 0 0 0 0 0 1 0	$\cos 2\pi f_1 t + \cos 2\pi f_4 t + \cos 2\pi f_5 t + \cos 2\pi f_{11} t$
3	0 0 1 1 0 1 0 1 0 0 0 0	$\cos 2\pi f_3 t + \cos 2\pi f_4 t + \cos 2\pi f_6 t + \cos 2\pi f_8 t$
4	0 1 0 0 1 1 1 0 0 0 0 0	$\cos 2\pi f_2 t + \cos 2\pi f_5 t + \cos 2\pi f_6 t + \cos 2\pi f_7 t$
5	0 0 0 0 1 0 0 1 1 1 0 0	$\cos 2\pi f_5 t + \cos 2\pi f_8 t + \cos 2\pi f_9 t + \cos 2\pi f_{10} t$
6	0 0 0 1 0 0 1 0 0 1 0 1	$\cos 2\pi f_4 t + \cos 2\pi f_7 t + \cos 2\pi f_{10} t + \cos 2\pi f_{12} t$
7	0 1 1 0 0 0 0 0 0 1 1 0	$\cos 2\pi f_2 t + \cos 2\pi f_3 t + \cos 2\pi f_{10} t + \cos 2\pi f_{11} t$
8	1 1 0 0 0 0 0 1 0 0 0 1	$\cos 2\pi f_1 t + \cos 2\pi f_2 t + \cos 2\pi f_8 t + \cos 2\pi f_{12} t$

One can see from inspection of the waveforms or the codewords used to derive them that in (6,3), (10,4), and (12,4) MT-FSK, each waveform shares one and only one frequency tone with every other waveform. In the case of (8,3) MT-FSK, each waveform shares one and only one tone with 6 of the other waveforms and is completely orthogonal to the remaining one.

It has been shown in [9] that these MT-FSK signalling sets are more power efficient than MFSK in AWGN but they perform much worse than MFSK in frequency-nonselective Rayleigh fading. However, by employing wideband MT-FSK, additional diversity can be obtained.

Performance of Wideband MT-FSK in Rayleigh Fading. In wideband MT-FSK, $\Delta f >> R_s$. In this paper, we will choose Δf to be greater than the coherence bandwidth of the channel $(\Delta f)_c$. By doing so, each tone encounters an independent fading process, and frequency diversity can be achieved. Assuming symbol 1 is transmitted, the decision variables U_1 and U_2 for (6,3) MT-FSK are:

$$U_1 = |2a_1 E_t + N_1|^2 + |2a_2 E_t + N_2|^2 + |2a_4 E_t + N_4|^2 \tag{18}$$

$$U_2 = |2a_2 E_t + N_2|^2 + |N_3|^2 + |N_5|^2 \tag{19}$$

where a_i is the time varying complex gain of the ith channel, and $E_t = E_s/w$ is the energy per tone. The symbol error rate can be estimated by the following union bound:

$$P_M < (M - 1)P_2(D) \tag{20}$$

where D is the equivalent diversity order of the modulation scheme, and $P_2(D)$ is given by [5]:

$$P_2(D) = \left(\frac{1-\mu}{2}\right)^L \sum_{i=0}^{L-1} \binom{L-1-i}{i} \left(\frac{1+\mu}{2}\right)^i \tag{21}$$

where $\mu = \frac{E_t}{N_o}/(2 + \frac{E_t}{N_o}$. For (6,3) modulation, $D = w - 1 = 2$. The bit error rate for (6,3) MT-FSK is:

$$P_b < 2P_2(D = 2) \tag{22}$$

Similarly, for (8,3) MT-FSK

$$P_b < 3P_2(D = 2) + P_2(D = 3) \tag{23}$$

$$P_b < 2P_2(D = 3) \tag{24}$$

for (10,4) MT-FSK, and

$$P_b < 4P_2(D = 3) \tag{25}$$

for (12,4) MT-FSK.

Performance of Wideband MT-FSK in Rayleigh Fading with Additional Channel Diversity. Suppose that we had the available bandwidth to transmit a wideband MT-FSK signal over L different channels. Such a scheme would require a minimum bandwidth of $vL\Delta f$. At the receiver, the L symbols are combined using square-law combining. The equivalent decision variables become:

$$U'_m = \sum_{i=1}^{L} U_m^{(i)} \tag{26}$$

It is easy to show that $P_2(D)$ of (21) becomes $P_2(LD)$ and E_t/N_o must be replaced by E_b/N_oL. In other words, we have increased the modulation scheme's inherent diversity order by a factor of L. Table 7 provides the average bit energy to noise spectral density ratio required by MT-FSK to assure a bit error rate of 10^{-3} in Rayleigh fading for different L.

Table 7. $\bar{\gamma}_b$ required to obtain a bit error rate of 10^{-3} for uncoded MT-FSK for different L in Rayleigh fading.

	$L = 1$	$L = 2$	$L = 3$
(6,3) MT-FSK	20.5 dB	16.1 dB	15.0 dB
(8,3) MT-FSK	20.0 dB	15.1 dB	13.8 dB
(10,4) MT-FSK	17.2 dB	14.0 dB	13.4 dB
(12,4) MT-FSK	16.4 dB	13.0 dB	12.4 dB

2.3 Discussion

In this section, we considered a number of modulation techniques. The performance of MFSK, MFSK-DPSK and MT-FSK in Rayleigh fading was examined. Channel diversity was also considered. It can be seen that MT-FSK is more power efficient than MFSK and MFSK-DPSK for $L \leq 2$. For $L = 3$, the best MT-FSK signaling set has roughly the same level of performance as the MFSK-DPSK schemes.

3 Error Control Coding

Error control coding is used to improve the power efficiency of a communications system. Thus coding is quite useful in CDMA systems where power efficiency can be traded off against spectral efficiency. In this section we examine the performance of the above discussed modulation techniques with BCH and RS codes as well as nonbinary convolutional coding.

3.1 Nonbinary BCH and RS Codes

Consider a primitive BCH code defined on GF(M). A codeword from this code will have length equal to $M^x - 1$ where x is an integer. RS codewords are a special class of BCH codes where $x = 1$. Thus RS codewords have length equal to $M - 1$.

Most BCH and RS decoding algorithms do not allow direct soft decision decoding [10]. Some algorithms have been developed to approximate soft decision decoding, and work on new soft decision decoding techniques is being done [11]. However, in this paper, we consider hard decision decoding of BCH and RS codes because this technique is more commonly used.

The number of parity bits in a t-error correcting $(n.m)$ BCH code is $n - m \geq 2xt$, where $x = log_M(n - 1)$. Therefore, the number of errors that a BCH code can correct is:

$$t \leq \frac{n - m}{2x} \tag{27}$$

When a codeword is received with no more than t errors, the decoder will correctly decode it. However, if there are more than t symbol errors, the decoder will add no more than t symbol errors to the codeword. Therefore the symbol error rate after decoding is:

$$P_{M_{dec}} < \sum_{i=t+1}^{n} \frac{t+i}{n} \binom{n}{i} p^i (1-p)^{n-i} \tag{28}$$

where $p = P_M(\frac{m}{n}\bar{\gamma}_b)$ is the code symbol error probability prior to decoding.

Table 8 gives the average bit energy to noise spectral density required by the different modulation schemes to assure a bit error rate below 10^{-3} employing different BCH or RS codes and different L in Rayleigh fading.

Table 8. $\bar{\gamma}_b$ required for an Lth order diversity system employing the different modulation schemes and BCH coding in Rayleigh fading.

Modulation	Code	$L=1$	$L=2$	$L=3$
(6,3) MT-FSK	BCH(255,199)	17.6 dB	14.8 dB	14.2 dB
(10,4) MT-FSK	BCH(255,199)	15.4 dB	14.1 dB	13.5 dB
8FSK	BCH(63,39)	15.7 dB	11.9 dB	10.7 dB
4FSK-DBPSK	BCH(63,39)	16.6 dB	11.8 dB	10.7 dB
(8,3) MT-FSK	BCH(63,39)	15.8 dB	13.4 dB	12.9 dB
(12,4) MT-FSK	BCH(63,43)	14.2 dB	12.9 dB	12.3 dB
16FSK	BCH(255,199)	15.1 dB	10.6 dB	9.5 dB
8FSK-DBPSK	BCH(255,199)	16.4 dB	11.1 dB	9.6 dB
4FSK-DQPSK	BCH(255,199)	16.6 dB	11.0 dB	9.5 dB
16FSK-DBPSK	RS(31,19)	13.9 dB	9.3 dB	8.8 dB
8FSK-DQPSK	RS(31,19)	14.1 dB	9.5 dB	9.0 dB
16FSK-DQPSK	RS(63,47)	14.6 dB	9.3 dB	8.4 dB

The results shown in Table 8 show improvement over the uncoded modulation schemes discussed in Section 2. Generally, the results for MFSK and MFSK-DPSK are better than those seen for MT-FSK. However, since bounds are used to determine the performance of MT-FSK, simulations are required to determine the accuracy of these results.

3.2 Nonbinary Convolutional Codes

In the previous subsection, we discussed error control codes which employed hard decision decoding. Additional power gains may be achieved by employing soft decision decoding. Convolutional codes are well-suited for soft decision decoding by employing the Viterbi algorithm.

There is a class of nonbinary convolutional codes called dual-k codes [12]. A rate $1/2^r$ dual-k encoder takes a k-bit codeword and generates 2^r k-bit codewords. Each of these k-bit codewords are transmitted using an M-ary modulation scheme. Since convolutional coding works best for random error channels, interleaving of the symbol stream is required.

The bit error rate performance of MFSK employing dual-k coding is given by:

$$P_b < \frac{2^{k-1}}{2^k - 1} \sum_{d=4r}^{\infty} w_d P_2(dLD) \tag{29}$$

where w_d are the coefficients in the expansion of the transfer function of the code and provides information about the number of paths which are a distance d from the all-zero path and the fraction of symbol errors associated with each path, $P_2(dLD)$ is given by (21), L is the diversity obtained by hopping, and D is the inherent diversity of the modulation scheme ($D = 1$ for MFSK). In the solution of (29),

$$\mu = \frac{r_c \bar{\gamma}_s / LD}{2 + r_c \bar{\gamma}_s / LD} \tag{30}$$

These equations only apply to MFSK and MT-FSK. In the case of MFSK-DPSK, Viterbi decoding is not practical. Recall that MFSK-DPSK detection requires that a decision be made regarding the frequency sub-symbol in order to retrieve the phase information. In order to perform Viterbi decoding, the phase of each possible frequency tone must be retained in memory so that all possible paths can be verified (since the phase of the previous symbol must be used to determine the phase of the current symbol). This is clearly not practical. Thus MFSK-DPSK is not suited to soft decision decoding.

Table 9 provides the average bit energy to noise spectral density ratio required by MFSK and MT-FSK employing rate $1/2$ dual-k coding in Rayleigh fading. We can see from these results that coded MFSK guarantees a bit error rate of 10^{-3} at a lower average energy per bit to noise spectral density than does coded MT-FSK.

3.3 Discussion

Comparing Tables 8 and 9, we can see that better results are obtained using the dual-k code than the BCH codes that were considered. This is because of the additional advantage of soft decision decoding. However, BCH codes may provide better results if soft decision decoding or even erasure decoding is employed. This is reserved for future studies.

Table 9. $\bar{\gamma}_b$ required for an Lth order diversity system employing the different modulation schemes and rate 1/2 dual-k coding in Rayleigh fading.

Modulation	Code	$L = 1$	$L = 2$	$L = 3$
(6,3) MT-FSK	dual-2	12.4 dB	12.0 dB	12.2 dB
(10,4) MT-FSK	dual-2	11.6 dB	11.4 dB	11.5 dB
8FSK	dual-3	11.9 dB	9.5 dB	9.1 dB
(8,3) MT-FSK	dual-3	11.2 dB	10.7 dB	10.8 dB
(12,4) MT-FSK	dual-3	10.3 dB	10.3 dB	10.4 dB
16FSK	dual-4	11.4 dB	8.8 dB	8.3 dB

4 Hybrid DS/FH-CDMA System for Rayleigh Fading Channels

In this section, the proposed hybrid DS/FH-CDMA system employing the different modulation schemes of Section 2 is discussed and analyzed. The analysis focuses on the system's spectral efficiency in the mobile forward link.

4.1 Hybrid DS/FH-CDMA System Employing MFSK

The proposed hybrid DS/FH-CDMA system in the forward link (base-to-mobile) is pictured in Figure 5. The data bits from the source, at a rate of R_b bits/sec, are grouped into symbols and input into a forward error correction encoder (optional) and interleaved to protect against burst errors. The coded bit rate is R_b/r_c, where r_c is the code rate.

The coded symbol stream is then MFSK modulated. The symbol rate $R_s = R_b/(r_c log_2 M) = R_b/(r_c k)$. The minimum frequency separation is Δf. The MFSK modulated data is then spread by a PSK modulated pseudonoise (PN) sequence of rate R_p chips/sec. Generally $R_p >> R_s$. The bandwidth of the direct-sequence spread signal is $W_{ds} = 2R_p + (M - 1)\Delta f$ [13]. This bandwidth is chosen to be smaller than the coherence bandwidth of the channel.

The signal is then hopped at a rate $R_h = LR_s$ into different frequency slots which have bandwidth greater than W_{ds}. We assume that these frequency slots have bandwidth $W_{ds}(1 + \alpha)$, where α is the excess bandwidth factor of the frequency slot. If there are N_f total frequency slots, then the total bandwidth required is $W_{ss} = N_f W_{ds}(1 + \alpha)$.

At the receiver, the signal is dehopped and despread. The symbols are demodulated at the hop rate, therefore, for orthogonality, $\Delta f = LR_s$. The decision variables are then square-law combined, de-interleaved and decoded.

Spectral Efficiency of Hybrid DS/FH-CDMA Employing MFSK. The effective noise spectral density of a MFSK/DS-CDMA system in an AWGN channel is given in [4,13] as:

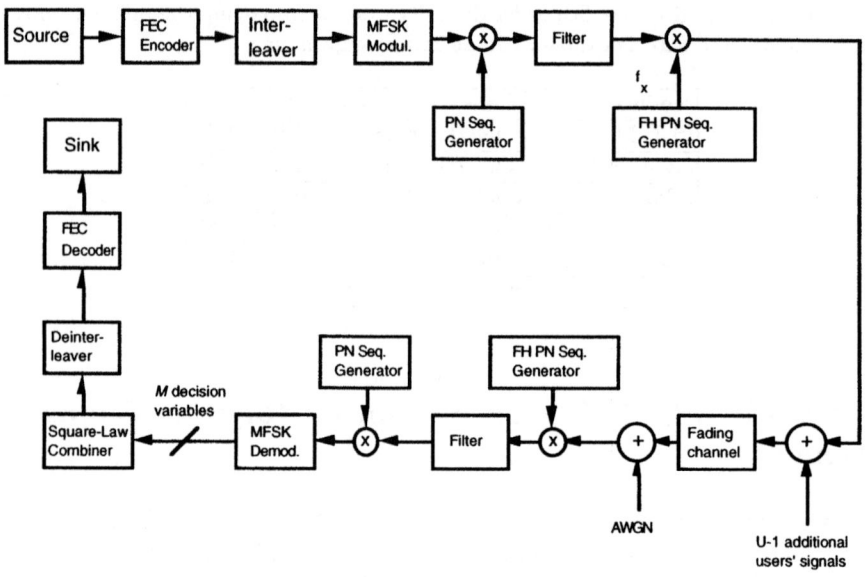

Fig. 5. Hybrid DS/FH-CDMA system in forward link.

$$N_o' = N_o + \frac{2(U-1)E_b R_b R_p}{M^2} \sum_{i=1}^{M}\sum_{j=1}^{M} \frac{1}{4R_p^2 + (\pi \Delta f)^2 (i-j)^2} \qquad (31)$$

where U is the number of simultaneous users accessing the channel. For the hybrid DS/FH-CDMA system employing MFSK and operating in Rayleigh fading, (31) becomes:

$$N_o' = N_o + \frac{2\frac{(U-1)}{N_f}\alpha^2 E_b R_b R_p}{M^2} \sum_{i=1}^{M}\sum_{j=1}^{M} \frac{1}{4R_p^2 + (\pi \Delta f)^2 (i-j)^2} \qquad (32)$$

The average effective noise spectral density ratio becomes:

$$E[N_o'] = N_o[1 + (U-1)E[\alpha^2]\frac{E_b}{N_o}\frac{\beta(M, R_b, R_p, \Delta f)}{N_f}]$$

$$= N_o[1 + (U-1)\bar{\gamma}_b\frac{\beta(M, R_b, R_p, \Delta f)}{N_f}] \qquad (33)$$

where

$$\beta(M, R_b, R_p, \Delta f) = \frac{2R_b R_p}{M^2} \sum_{i=1}^{M} \sum_{j=1}^{M} \frac{1}{4R_p^2 + (\pi \Delta f)^2 (i - j)^2} \tag{34}$$

The effective energy per bit to noise spectral density ratio, $\bar{\gamma}_b{}'$, is given by:

$$\bar{\gamma}_b{}' = \frac{\bar{\gamma}_b}{1 + (U - 1) \frac{\beta(M, R_b, R_p, \Delta f)}{N_f} \bar{\gamma}_b} \tag{35}$$

Rewriting this equation in terms of the number of simultaneous users, we obtain:

$$U = \frac{\bar{\gamma}_b - \bar{\gamma}_b{}'}{\bar{\gamma}_b \bar{\gamma}_b{}' \frac{\beta(M, R_b, R_p, \Delta f)}{N_f}} + 1 \tag{36}$$

The spectral efficiency of the system is defined as the maximum amount of information flow in bps that can be transmitted per unit bandwidth. It is given by:

$$\eta = \frac{U R_b}{W_{ss}} = \frac{R_b}{\bar{\gamma}_b{}' \beta(M, R_b, R_p, \Delta f) W_{ds}(1 + \alpha)} \tag{37}$$

where $\bar{\gamma}_b{}'$ is the equivalent energy per bit to noise spectral density ratio *which produces the highest acceptable bit error rate.*

Spectral Efficiency of DS/FH-CDMA Employing Uncoded MFSK. In a typical land mobile channel, the delay spread is about 5 μs, and thus the typical coherence bandwidth is 200 kHz [5]. Thus one could spread the MFSK signal to a bandwidth $W_{ds} < 200$ kHz. The bandwidth of the frequency slots could be slightly greater than the coherence bandwidth. In this paper, we assume that $W_{ds} = 195$ kHz, and the frequency slots occupy 205 kHz. Therefore $\alpha = 0.05$, and the overall bandwidth $W_{ss} = N_f \times 205$ kHz.

We consider bit rates of 4.8 and 9.6 kbps as these are typical bit rates for voice coders. We can now obtain Δf, R_p, and $\beta(M, \Delta f, R_b, R_p)$. This is shown in Table 10 for 4.8 kbps systems and Table 11 for systems operating at 9.6 kbps..

Using the parameter values in Table 10 and employing the energy per bit to noise spectral density ratios which produce a bit error rate of 10^{-3} from Table 1, we can calculate the spectral efficiency of the system from (37). This is shown in Table 12.

From Table 12, we can see that the spectral efficiency of this system is low. However, the spectral efficiency of CDMA systems can be improved through the use of error control coding.

Table 10. Δf, R_p and $\beta(M, R_b, R_p, \Delta f)$ for hybrid DS/FH-CDMA employing MFSK without coding ($R_b = 4.8$ kbps, $W_{ds} = 195$ kHz).

Modulation	L	R_h (in khops/sec)	Δf (in kHz)	R_p (in kchips/sec)	$\beta(M, R_b, R_p, \Delta f)$
8FSK	1	-	1.6	91.2	0.0261
	2	3.2	3.2	83.2	0.0278
	3	4.8	4.8	76.8	0.0287
16FSK	1	-	1.2	87.6	0.0269
	2	2.4	2.4	79.2	0.0279
	3	3.6	3.6	68.4	0.0287

Table 11. Δf, R_p and $\beta(M, R_b, R_p, \Delta f)$ for hybrid DS/FH-CDMA employing MFSK without coding ($R_b = 9.6$ kbps, $W_{ds} = 195$ kHz).

Modulation	L	R_h (in khops/sec)	Δf (in kHz)	R_p (in kchips/sec)	$\beta(M, R_b, R_p, \Delta f)$
8FSK	1	-	3.2	83.2	0.0557
	2	6.4	6.4	70.4	0.0581
	3	9.6	9.6	57.6	0.0577
16FSK	1	-	2.4	79.2	0.0558
	2	4.8	4.8	57.6	0.0576
	3	7.2	7.2	43.2	0.0525

Table 12. Spectral efficiency (in bps/Hz) of the uncoded hybrid DS/FH-CDMA system employing uncoded MFSK in Rayleigh fading.

	$R_b = 4.8$ kbps			$R_b = 9.6$ kbps		
	$L = 1$	$L = 2$	$L = 3$	$L = 1$	$L = 2$	$L = 3$
8FSK	0.00179	0.0160	0.0317	0.00167	0.0154	0.0316
16FSK	0.00199	0.0179	0.0364	0.00192	0.0173	0.0398

Spectral Efficiency of Hybrid DS/FH-CDMA Employing Coded MFSK. When forward error correction is used, $\Delta f = LR_b/r_c k$, and thus R_p must be decreased so that W_{ds} remains unchanged. Thus the values for $\beta(M, \Delta f, R_b, R_p)$ shown in Tables 10 and 11 must be recalculated for the coded case.

In this paper, we considered many different error control codes. However, in this section, we will consider only one BCH code and the dual-k code. The values for Δf, R_p, and $\beta(M, \Delta f, R_b, R_p)$ are given in Table 13 for 4.8 kbps systems employing different codes and in Table 14 for 9.6 kbps systems employing the different codes.

Table 13. Δf, R_p and $\beta(M, R_b, R_p, \Delta f)$ for hybrid DS/FH-CDMA employing coded MFSK ($R_b = 4.8$ kbps, $W_{ds} = 195$ kHz).

Modulation	L	R_h (in khops/sec)	Δf (in kHz)	R_p (in kchips/sec)	$\beta(M, R_b, R_p, \Delta f)$
8FSK	1	-	2.585	87.87	0.0267
with	2	5.170	5.170	77.55	0.0281
BCH(63,39)	3	7.755	7.755	67.80	0.0277
8FSK	1	-	3.2	83.2	0.0278
with	2	6.4	6.4	70.4	0.0291
dual-3	3	9.6	9.6	57.6	0.0289
16FSK	1	-	1.54	84.57	0.0275
with	2	3.08	3.08	73.8	0.0283
BCH(255,199)	3	4.61	4.61	59.97	0.0286
16FSK	1	-	2.4	79.2	0.0279
with	2	4.8	4.8	57.6	0.0285
dual-4	3	7.2	7.2	43.2	0.0262

Table 14. Δf, R_p and $\beta(M, R_b, R_p, \Delta f)$ for hybrid DS/FH-CDMA employing coded MFSK ($R_b = 9.6$ kbps, $W_{ds} = 195$ kHz).

Modulation	L	R_h (in khops/sec)	Δf (in kHz)	R_p (in kchips/sec)	$\beta(M, R_b, R_p, \Delta f)$
8FSK	1	-	5.17	77.55	0.0505
with	2	10.54	10.54	51.70	0.0541
BCH(63,39)	3	15.51	15.51	31.02	0.0470
8FSK	1	-	6.4	83.2	0.0581
with	2	12.8	12.8	51.2	0.0529
dual-3	3	19.2	19.2	19.2	-
16FSK	1	-	3.08	73.80	0.0565
with	2	6.15	6.15	49.21	0.0548
BCH(255,199)	3	9.23	9.23	27.80	0.0496
16FSK	1	-	4.8	57.6	0.0576
with	2	9.6	9.6	19.2	0.0515
dual-4	3	14.4	14.4	-	-

In Table 14, no value is given for R_p or $\beta(M, \Delta f, R_b, R_p)$ in some cases. This is because R_p is not sufficiently greater than R_h to allow proper operation of the system. The spectral efficiency of the system can be found employing these values and the energy per bit to noise spectral density ratios of Tables 8 and 9. This is shown in Table 15.

We can see that the spectral efficiency is greatly improved when error control coding is employed. We also see that the dual-k codes, by virtue of the soft decision decoder, provide larger spectral efficiencies than the hard decision decoded BCH codes.

Table 15. Spectral efficiency (in bps/Hz) of hybrid DS/FH-CDMA employing coded MFSK.

	$R_b = 4.8$ kbps			$R_b = 9.6$ kbps		
	$L=1$	$L=2$	$L=3$	$L=1$	$L=2$	$L=3$
8FSK-BCH(63,39)	0.0236	0.0539	0.0721	0.0250	0.0512	0.0848
8FSK-dual-3	0.0544	0.0903	0.0997	0.0521	0.0993	–
16FSK-BCH(255,199)	0.0264	0.0723	0.0920	0.0256	0.0743	0.106
16FSK-dual-4	0.0608	0.108	0.132	0.0589	0.120	–

4.2 Hybrid DS/FH-CDMA Employing MFSK-DPSK

The hybrid DS/FH-CDMA system employing MFSK-DPSK operates in the same manner as the system which employs MFSK except that there must be a constraint on the hopping sequence. Hop x of symbol $i + 1$ must fall into the same frequency bin that hop x of symbol i was in since differential detection is to be performed on xth hop of successive symbols and this requires that the fading be dependent.

The idea behind this system is to either decrease the number of tones required by the modulation scheme while maintaining the same symbol rate, or by decreasing the symbol rate (thus decreasing Δf) while employing the same number of tones. In each case, we can increase R_p compared to the previous system discussed.

The equations derived in section 4.1 apply to this system as well except that k must be replaced by $k_f + k_p$, and M is replaced by M_f.

Spectral Efficiency of DS/FH-CDMA Employing Uncoded MFSK-DPSK. The values for R_p , and $\beta(M, \Delta f, R_b, R_p)$ can be calculated in the same manner as the previous section. For sake of brevity, these values are not shown. However, they can be found in [9]. The spectral efficiency of the system employing MFSK-DPSK is also found in the same manner. The results are shown in Table 16.

Compared to DS/FH-CDMA employing uncoded MFSK, this system generally has higher spectral efficiencies. This is due to the improved power efficiency of uncoded MFSK-DPSK compared to uncoded MFSK. However, the results are quite poor, and error control coding is required to further improve the spectral efficiency of the system.

Spectral Efficiency of DS/FH-CDMA Employing Coded MFSK-DPSK. Again, we must recalculate R_p and $\beta(M, \Delta f, R_b, R_p)$ for the coded case. These results can be found in [9]. Only the BCH codes are considered in this section

Table 16. Spectral efficiency (in bps/Hz) of hybrid DS/FH-CDMA employing uncoded MFSK-DPSK.

	$R_b = 4.8$ kbps			$R_b = 9.6$ kbps		
	$L = 1$	$L = 2$	$L = 3$	$L = 1$	$L = 2$	$L = 3$
4FSK-DBPSK	0.00168	0.0172	0.0341	0.00160	0.0164	0.0324
8FSK-DBPSK	0.00180	0.0198	0.0414	0.00176	0.0183	0.0417
16FSK-DBPSK	0.00195	0.0224	0.0455	0.00186	0.0202	0.0463
4FSK-DQPSK	0.00172	0.0215	0.0445	0.00171	0.0201	0.0440
8FSK-DQPSK	0.00193	0.0234	0.0488	0.00186	0.0218	0.0464
16FSK-DQPSK	0.00196	0.0242	0.0506	0.00187	0.0226	0.0489

because soft decision decoding of MFSK-DPSK is impractical. The spectral efficiency of DS/FH-CDMA employing BCH encoded MFSK-DPSK is shown in Table 17.

Table 17. Spectral efficiency (in bps/Hz) of hybrid DS/FH-CDMA employing coded MFSK-DPSK.

	$R_b = 4.8$ kbps			$R_b = 9.6$ kbps		
	$L = 1$	$L = 2$	$L = 3$	$L = 1$	$L = 2$	$L = 3$
4FSK-DBPSK/ BCH(63,39)	0.0200	0.0579	0.0743	0.0191	0.0530	0.0659
8FSK-DBPSK/ BCH(255,199)	0.0204	0.0674	0.0908	0.0199	0.0643	0.0936
16FSK-DBPSK/ RS(31,19)	0.0349	0.0960	0.110	0.0333	0.102	0.118
4FSK-DQPSK/ BCH(255,199)	0.0201	0.0718	0.0970	0.0199	0.0696	0.0970
8FSK-DQPSK/ RS(31.19)	0.0348	0.0960	0.106	0.0332	0.0946	0.100
16FSK-DQPSK/ RS(63,47)	0.0306	0.0777	0.118	0.0287	0.0974	0.112

From Table 17, we can see that the spectral efficiency of DS/FH-CDMA employing coded MFSK-DPSK is much better than when no coding is used. It can also be seen that MFSK-DPSK employing BCH coding provides higher spectral efficiencies than does MFSK employing BCH coding. However, MFSK employing rate 1/2 dual-k coding, due to the soft decision decoding, provides better results than those shown here.

4.3 Hybrid DS/FH-CDMA Employing Wideband MT-FSK

The operation of the hybrid DS/FH-CDMA employing wideband MT-FSK is different than its operation with the previous two modulation schemes. In this case, the minimum frequency spacing is larger than the coherence bandwidth of the channel. Therefore, for a DS/FH-CDMA system employing a (v, w) MT-FSK modulation scheme, the bandwidth of a frequency bin is larger than $v\Delta f$.

The data is MT-FSK modulated. The signal is then spread by a PN sequence of rate R_p using direct-sequence techniques. The null-to-null bandwidth of one spread tone is $2R_p$. We can choose a value for R_p so that $2R_p < (\Delta f)_c$. The signal is then hopped at a rate $R_h = LR_s$ into one of N_f different frequency bins (in this case, N_f is much smaller than when MFSK or MFSK-DPSK are used because of the large frequency bin bandwidth).

At the receiver, the signal is dehopped and despread. The signal is demodulated at the hop rate, and thus we get an Lth order diversity system (not considering the inherent diversity of the modulation scheme).

Performance of DS/FH-CDMA EMploying Uncoded Wideband MT-FSK. The equivalent noise spectral density in a DS/FH-CDMA system employing MT-FSK is given by [9,14]:

$$N_o' = N_o \left[1 + \frac{(U-1)}{N_f} \frac{E_b}{N_o} \frac{R_b}{R_p} \sigma(v, \rho) \right] \tag{38}$$

where $\rho = \Delta f / 2R_p$, and $\sigma(v, \rho)$ is given by:

$$\sigma(v, \rho) = \frac{1}{2v^2} \sum_{i=1}^{v} \sum_{j=1}^{v} \frac{1}{1 + (\pi\rho)^2(i-j)^2} \tag{39}$$

It can be easily shown that the spectral efficiency of this system is given by [9]:

$$\eta = \frac{1}{2\gamma_b'\sigma(v, \rho)[1 + (v-1)\rho]} \tag{40}$$

We continue to employ a channel model with a coherence bandwidth of 200 kHz. Therefore $2R_p < 200$ kHz, and $\Delta f > 200$ kHz. Since R_b does not affect η, any choice of R_b will produce the same results. Assuming $R_b = 9.6$ kbps, then a good choice for R_p is 96 kchips/s and $\Delta f = 211.2$ kHz. Thus $\rho = 1.1$.

It can be shown that $\sigma(v, \rho) = 0.098$ for (6,3) MT-FSK. Similarly, $\sigma(v, \rho) = 0.074$ for (8,3) MT-FSK, 0.06 for (10,4) MT-FSK, and 0.05 for (12,4) MT-FSK. The spectral efficiency of DS/FH-CDMA employing uncoded MT-FSK is given in Table 18.

Comparing these results with those for MFSK or MFSK-DPSK, we see better performance with MT-FSK when $L=1$ or 2. However, when $L=3$, uncoded MFSK-DPSK provides better results.

Table 18. Bandwidth efficiency (in bps/Hz) of hybrid DS/FH-CDMA employing uncoded MT-FSK in Rayleigh fading.

	$L = 1$	$L = 2$	$L = 3$
(6,3) MT-FSK	0.0092	0.0194	0.0250
(8,3) MT-FSK	0.0078	0.0240	0.0324
(10,4) MT-FSK	0.0146	0.0304	0.0349
(12,4) MT-FSK	0.0175	0.0383	0.0439

Performance of DS/FH-CDMA Employing Coded Wideband MT-FSK. Using the values of $\bar{\gamma_b}'$ shown in Tables 8 and 9, we can find the spectral efficiency of DS/FH-CDMA employing coded MT-FSK. These results are shown in Table 19.

Table 19. Bandwidth efficiency (in bps/Hz) of hybrid DS/FH-CDMA employing coded MT-FSK.

	$L = 1$	$L = 2$	$L = 3$
(6,3) MT-FSK/BCH(255,199)	0.0153	0.0272	0.0305
(6,3) MT-FSK/dual-2	0.0452	0.0495	0.0473
(8,3) MT-FSK/BCH(63,39)	0.0204	0.0389	0.0447
(8,3) MT-FSK/dual-3	0.0589	0.0661	0.0646
(10,4) MT-FSK/BCH(255,199)	0.0236	0.0304	0.0342
(10,4) MT-FSK/dual-2	0.0529	0.0554	0.0541
(12,4) MT-FSK/BCH(63,39)	0.0311	0.0439	0.0493
(12,4) MT-FSK/dual-3	0.0712	0.0712	0.0696

Comparing these results to the other modulation schemes, we see that MFSK and MFSK-DPSK provide better results. One should keep in mind that upper bounds are used to produce the bit error rates which are used in the spectral efficiency calculation. Therefore the spectral efficiencies for the coded cases are lower bounds in all cases.

5 Conclusions

We examined the spectral efficiency of DS/FH-CDMA employing three different modulation schemes. Of all the schemes discussed in this paper, MFSK with rate 1/2 dual-k coding provides the best results for all L. The use of dual-k coding with this modulation scheme increases the spectral efficiency from 0.0398 to 0.12 (16FSK). However, MFSK-DPSK provides better results when either no coding or BCH coding is used. MFSK-DPSK is not used with dual-k coding because

Viterbi decoding of this modulation scheme is impractical. One may make a case for the use of MFSK-DPSK with erasure decoded BCH codes. However, the complexity of the modulation scheme compared to MFSK is greater. Therefore the results of this paper suggest that MFSK with dual-k coding is the wisest choice of those presented in this paper for DS/FH-CDMA systems.

References

1. H. Asmer et al., "A hybrid DS/FH spread spectrum system for mobile radio channel: performance and capacity analysis", *Proc. IEEE Veh. Tech. Conf.*, pp. 305-308, 1993.
2. L. Sadiq, A.H. Aghvami, "Performance of an asynchronous hybrid spread spectrum system in the presence of interference, Rician fading, and AWGN", *IEE Proceedings-I*, vol. 138, no. 2, Apr. 1991.
3. Z. Tan, I.F. Blake, "Performance analysis of noncoherent DS-SFH spread spectrum multiple access for indoor wireless communications", *Conf. Record IEEE MILCOM*, pp. 851-855, 1992.
4. J. Wang, A. Yongaçoğlu, "Capacity comparison between direct sequence PSK and MFSK spread spectrum systems in CDMA applications", *Proc. 4th Canadian Conf. Elect. and Comp. Eng.*, Quebec City, Quebec, pp. 53.51-53.54, Sept. 1991.
5. J.G. Proakis, *Digital Communications*, 2nd Ed., New York: McGraw-Hill, 1990.
6. D. Slepian, "Permutation modulation", *Proc. IEEE*, vol. 53, pp. 228-236, Mar. 1965.
7. H.L. Schneider, "Data transmission with FSK permutation modulation", *Bell Syst. Tech. Jour.*, pp. 1131-1138, July-Aug. 1968.
8. F.G. McWilliams, N.J. Sloane, *The Theory of Error Correcting Codes*, Amsterdam: North-Holland Publishing Co., 1977.
9. C. D'Amours, "Hybrid DS/FH-CDMA Systems Employing FSK Based Modulation Schemes", *Ph. D. Dissertation*, University of Ottawa, Sept. 1995.
10. E.R. Berlekamp, et al., "The application of error control to communications", *IEEE Comm. Mag.*, vol. 25, pp. 44-57, Jan. 1987.
11. D.J. Taipale, M.J. Seo, "An efficient soft decision Reed Solomon decoding algorithm", *IEEE Trans. Inf. Theory*, vol. 40, no. 4, pp. 1130-1139, July 1994.
12. J.P. Odenwalder, "Dual-k convolutional codes for noncoherently demodulated channels", *Proc. ITC'76*, pp. 165-174, Sept. 1976.
13. T.T. Ha, *Digital Satellite Communications*, New York: McGraw-Hill, 1986.
14. C. D'Amours, A. Yongaçoğlu, "Hybrid DS/FH-CDMA System Employing MT-FSK Modulation for Mobile Radio", *Proc. IEEE PIMRC Conf.*, Toronto, Ont., pp. 164-168, Sept. 1995.

Finite-Field Wavelet Transforms*

H. Vincent Poor

Department of Electrical Engineering, Princeton University, Princeton, NJ 08544 USA

Abstract. Cyclic wavelet transforms, analogous to cyclic Fourier transforms, are described. Among other things, this construction allows for the development of wavelet transforms over finite fields. Potential applications of this construct in algebraic coding are discussed.

1 Introduction

Over the past decade, multiresolution analysis via wavelet decomposition has emerged as an important tool in the analysis of signals and images (e.g., [6]). Generally speaking, the principal framework within which such techniques have been studied and applied is the same as that used in the discrete-time Fourier analysis of sequences of complex numbers; that is, the sequence to be analyzed is viewed as a mapping from the set of integers $\mathcal{Z}$, to the set of complex numbers $\mathcal{C}$. Of course, Fourier analysis can also be performed on finite-length sequences of complex numbers by viewing them as mappings from a finite cyclic group to $\mathcal{C}$ using the discrete Fourier transform (DFT). The DFT and its extension to the situation in which the complex field is replaced with a finite field are of widespread utility in digital signal processing applications and algebraic coding. In this paper, we discuss an analogous framework for the multiresolution analysis of finite-length sequences of elements from the complex field or a finite field. This framework, which was developed jointly with Giuseppe Caire, Robert Grossman and Sandip Sarkar, is described in more detail in [4, 11]. The potential application of this framework in the develpment of error-control codes with multiple levels of error control is also discussed briefly.

2 Finite-length Wavelet Transforms

An important class of multiresolution decompositions are the so-called *Laplacian pyramid* schemes, in which the resolution of a signal under analysis is successively halved by recursively low-pass filtering the signal and decimating it by a factor of two. The residual (i.e., the error incurred) at each stage of this process is referred to as the *detail* at that stage; and the sequence of details formed by this decomposition is the transform of interest. Suitable choice of the filters used in

---
* This paper was prepared under the support of the U. S. Office of Naval Research under Grant N00014-94-1-0115.

this process renders this transform invertible; and such suitable filters can be characterized through their discrete-time Fourier properties (e.g., [5, 9]).

The Laplacian pyramid can be adapted to define an exact multiresolution wavelet transform for sequences of finite length $N = 2^n$, from an arbitrary field $\mathcal{F}$, where $n > 1$ is an integer. For $j = 1, 2, \ldots, n$, consider matrices H^j and G^j over $\mathcal{F}$ of dimension $2^{n-j} \times 2^{n-j+1}$, satisfying the conditions

$$(H^j)^* H^j + (G^j)^* G^j = N'^{-1} I_{2^{n-j+1}}, \qquad (1)$$

where I_k denotes the $k \times k$ identity matrix, and where $N' \in \mathcal{F}$ is a constant whose choice will be discussed below. (Here, the asterisk denotes the adjoint.)

Within this framework, consider the following algorithm.

Decomposition. Given $c^0 = v$ and an integer $n > 0$, compute $d^1, \ldots, d^n, c^n$ as follows:

$$c^{j+1} = H^{j+1} c^j, \qquad d^{j+1} = G^{j+1} c^j. \qquad (2)$$

Reconstruction. Given a decomposition $\{d^1, \ldots, d^n, c^n\}$, reconstruct the original signal $v = c^0$ by computing, for $j = n - 1, n - 2, \ldots, 0$,

$$c^j = N'[(G^{j+1})^* d^{j+1} + (H^{j+1})^* c^{j+1}]. \qquad (3)$$

The one-to-one mapping $v \leftrightarrow \{d^1, d^2, \ldots, d^n, c^n\}$ defined by decomposition/reconstruction defines a *finite-length wavelet transform*. The vectors d^j and c^j are known as, respectively, the detail and the *lower-resolution* component of c^{j-1}. The condition (1) is known as the *perfect reconstruction* condition.

3 Cyclic Wavelet Transforms

Decomposition/reconstruction subject to the perfect reconstruction condition (1) specifies a finite-length wavelet transform in terms of the matrices $G^1, G^2, \ldots, G^n$ and $H^1, H^2, \ldots, H^n$. As in the case of Fourier analysis, it is of interest to constrain this transform to define a *cyclic* multiresolution analysis of the space of the periodic sequences of period 2^n over $\mathcal{F}$. This can be accomplished by constraining the matrices H^j and G^j to be *2-circulants* [7] for each j; i.e., they are constrained to be of the form:

$$G^j = \begin{pmatrix} g_0^j & g_1^j & g_2^j & \cdots & g_{N_j-1}^j \\ g_{N_j-2}^j & g_{N_j-1}^j & g_0^j & \cdots & g_{N_j-3}^j \\ g_{N_j-4}^j & g_{N_j-3}^j & g_{N_j-2}^j & \cdots & g_{N_j-5}^j \\ \vdots & \vdots & \vdots & \ddots & \vdots \\ g_2^j & g_3^j & g_4^j & \cdots & g_1^j \end{pmatrix}, \qquad (4)$$

and

$$H^j = \begin{pmatrix} h_0^j & h_1^j & h_2^j & \cdots & h_{N_j-1}^j \\ h_{N_j-2}^j & h_{N_j-1}^j & h_0^j & \cdots & h_{N_j-3}^j \\ h_{N_j-4}^j & h_{N_j-3}^j & h_{N_j-2}^j & \cdots & h_{N_j-5}^j \\ \vdots & \vdots & \vdots & \ddots & \vdots \\ h_2^j & h_3^j & h_4^j & \cdots & h_1^j \end{pmatrix}. \qquad (5)$$

where $N_j \triangleq 2^{n-j+1}$. Since a 2-circulant matrix is defined completely by its first row, we can write $G^j = 2 - \text{cir}\{g^j\}$ and $H^j = 2 - \text{cir}\{h^j\}$ where g^j and h^j denote the first rows of G^j and H^j, respectively.

Within this constraint, an interesting interpretation of the algorithm decomposition/reconstruction is possible if we consider the sequences c^j and d^j to be periodic sequences of period equal to their lengths (N_{j+1}). In particular, for matrices satisfying (4) and (5), the $(j+1)^{\text{st}}$ step of Decomposition defines a finite-impulse-response (FIR) filtering of the periodic sequence c^{j-1} with the two FIR filters having impulse response h^j and g^j, followed by a decimation by a factor of two. The periods of the input sequence c^{j-1} is 2^{n-j+1} while the period of the two output sequences c^j and d^j is 2^{n-j}. Similarly, reconstruction can be considered to be interpolation by a factor of two followed by FIR filtering. Note that this filtering and decimation by a factor of two (or interpolation by a factor of two and filtering on the reconstruction side) is completely analogous to the sub-band decomposition scheme for infinite-length sequences described by the Laplacian pyramid [9].

In order to design such transforms, we would like to construct families of sequences $\{g^j, h^j \in \mathcal{F}^{2^{n-j+1}} \mid j = 1, 2, \ldots, n\}$ such that (1) is satisfied for all j with $G^j = 2 - \text{cir}\{g^j\}$ and $H^j = 2 - \text{cir}\{h^j\}$. Any such family defines a *cyclic wavelet transform* (CWT) for the space of periodic sequences of period 2^n over the field $\mathcal{F}$. In the following paragraphs, we consider the design of such cyclic wavelet transforms for the case in which $\mathcal{F} = \mathcal{C}$, the field of complex numbers. The complex field provides a convenient setting withing which to establish the basic ideas of CWT's. These results will then be developed for finite fields in the following section.

To construct the sequences of interest for the complex case, we will first give a result characterizing the 2-circulant matrices satisfying (1) for the case $j = 1$, and then we will give a method to derive a family of sequences $\{g^j, h^j\}$, $j = 2, \ldots, n$, from any two sequences g^1 and h^1 that satisfy the theorem. In the following we will suppress the superscripts 1 on G^1, H^1, g^1, and h^1 for notational convenience, and we define the following discrete Fourier transforms:

$$\gamma_k^j = \sum_{l=0}^{\frac{N}{2}-1} g_{2l+j} \alpha^{2lk}, \quad k = 0, 1, \ldots, \frac{N}{2} - 1, \quad j = 0, 1, \tag{6}$$

and

$$\eta_k^j = \sum_{l=0}^{\frac{N}{2}-1} h_{2l+j} \alpha^{2lk}, \quad k = 0, 1, \ldots, \frac{N}{2} - 1, \quad j = 0, 1, \tag{7}$$

where α is the relevant N^{th} primitive root of unity: $\alpha = \exp\{2\pi i/N\}$.

Theorem 1 (Caire, Grossman & Poor [4]): Consider the cyclic wavelet transform of length $N = 2^n$ over the complex field and let N' be any nonzero complex number. The matrices $G = 2 - \text{cir}\{g_0, g_1, \ldots, g_{N-1}\}$ and $H = 2 - \text{cir}\{h_0,$

$h_1, \ldots, h_{N-1}\}$ satisfy (1) if and only if for each $k = 0, 1, \ldots, \frac{N}{2} - 1$, we have[2]

$$|\gamma_k^0|^2 + |\gamma_k^1|^2 = \frac{1}{N'}, \tag{8}$$

and

$$\eta_k^j = (-1)^j \nu_k \overline{\gamma_k^{1-j}}, \quad j = 0, 1, \tag{9}$$

for some $\nu \in C^{N/2}$ satisfying $|\nu_k|^2 = 1$, $k = 0, 1, \ldots, N/2 - 1$.

Conditions (8) and (9) are the cyclic versions of frequency domain perfect reconstruction conditions characterizing the discrete-time wavelet transform (see, e.g., Eqs. (3.12) and (3.13) of [5]). These conditions allows us to construct sequences whose corresponding 2-circulant matrices satisfy (1) for the case $j = 1$. Given two such sequences, we now wish to construct a family of sequences $\{g^j, h^j \mid j = 1, 2, \ldots, n\}$ that specifies a CWT as described above. Such a construction is given by the following result, which is a straightforward corollary to Theorem 1.

Corollary 1: Suppose $G = 2 - \text{cir}\{g\}$ and $H = 2 - \text{cir}\{h\}$ are $2^{n-1} \times 2^n$ matrices of complex numbers satisfying (1). For each $j = 1, 2, \ldots, n$, define two length-2^{n-j} sequences g^j and h^j by

$$g_{2\ell}^j = \text{DFT}^{-1}\left\{\{\gamma_{2^{j-1}k}^0 \mid k = 0, 1, \ldots, 2^{n-j} - 1\}\right\}_\ell, \tag{10}$$

$$g_{2\ell+1}^j = \text{DFT}^{-1}\left\{\{\gamma_{2^{j-1}k}^1 \mid k = 0, 1, \ldots, 2^{n-j} - 1\}\right\}_\ell, \tag{11}$$

$$h_{2\ell}^j = \text{DFT}^{-1}\left\{\{\eta_{2^{j-1}k}^0 \mid k = 0, 1, \ldots, 2^{n-j} - 1\}\right\}_\ell, \tag{12}$$

and

$$h_{2\ell+1}^j = \text{DFT}^{-1}\left\{\{\eta_{2^{j-1}k}^1 \mid k = 0, 1, \ldots, 2^{n-j} - 1\}\right\}_\ell, \tag{13}$$

for $\ell = 0, 1, \ldots, 2^{n-j} - 1$, where the sequences $\gamma^0, \gamma^1, \eta^0$, and η^1, are defined from g and h as in (6) and (7), and where the operation DFT indicates the discrete Fourier transform of appropriate length (see, e.g., [3]). Then $G^j = 2 - \text{cir}\{g^j\}$ and $H^j = 2 - \text{cir}\{h^j\}$ satisfy (1) for each $j = 1, 2, \ldots, n$.

Corollary 1 defines g^j and h^j to be the sequences obtained by frequency sampling the original sequences g and h with a sampling factor 2^{j-1}. Since the conditions of Theorem 1 are given on the Fourier transform of the even and odd coefficients of g and h, any sequence obtained by such "frequency sampling" of those transforms will obviously satisfy the same conditions. If we look at Decomposition as an FIR filtering followed by a decimation by 2 of a periodic sequence, we see that at each step j the filters constructed in this way have the same frequency characteristics as the two original filters g and h. In this case the frequency sampling procedure does not give any degradation in the filters' frequency responses since, for periodic sequences (and thus for cyclic

---

[2] An overbar on a complex quantity denotes complex conjugation.

convolution), the Fourier transform coincides with the DFT and the frequency response matters only for specific frequency values.

The discrete-time Laplacian pyramid scheme involves the use of a lowpass filter and its complementary bandpass filter to perform the multiresolution analysis of discrete-time sequences [9]. This imposes further conditions on the filters of interest in addition to perfect reconstruction conditions. In the context of cyclic transforms the analogous conditions are the *lowpass condition*

$$\left| \sum_{k=0}^{N-1} h_k \right| = \sqrt{2/N'}, \tag{14}$$

and the complementary *bandpass condition*,

$$\sum_{k=0}^{N-1} g_k = 0. \tag{15}$$

It follows from Theorem 1 that the bandpass condition is equivalent to the condition that $\gamma_0^1 = -\gamma_0^0$. The construction of Corollary 1 assures that this condition holds for all j if it holds for $j = 1$. If the bandpass condition is imposed, then Theorem 1 shows that we also must have $\eta_0^1 = \eta_0^0$, and further that

$$|\gamma_0^0| = |\gamma_0^1| = |\eta_0^0| = |\eta_0^1| = \frac{1}{\sqrt{2N'}}. \tag{16}$$

Thus, from (16) we see that the corresponding lowpass condition (14) is also enforced for each j. (See also [10].)

It is clear from Theorem 1 that, in this $\mathcal{F} = C$ case, the field element N' must be real and positive. Otherwise, the role of N' is not critical in this case, since varying it essentially results in a simple renormalization of the matrices G and H. In particular, there is no lost generality if we simply choose N' to be unity or some other convenient value. As we shall see in the following section, the choice of N' is not arbitrary in the case in which $\mathcal{F}$ is a finite field. This is essentially because the square-root arising in (14) will not be defined for all field elements in a finite field. This point will be discussed further below.

¿From the necessary and sufficient conditions (8) and (9), we see that a cyclic wavelet transform can be designed by first selecting a sequence $g_0, g_1, \ldots, g_{N-1}$ to satisfy (8) (this sequence plays the role of the so-called "mother" wavelet [5]), and then choosing $h_0, h_1, \ldots, h_{N-1}$ from (9). The choice of $g_0, g_1, \ldots, g_{N-1}$ is further reduced to choosing, say, the even-indexed subset $g_0, g_2, \ldots, g_{N-2}$, to satisfy

$$N' |\gamma_k^0|^2 \le 1, \tag{17}$$

and then choosing $g_1, g_3, \ldots, g_{N-1}$ compatibly via the inverse of the relationship (6):

$$g_k = \frac{2}{N} \sum_{l=0}^{\frac{N}{2}-1} \gamma_l^1 \alpha^{-lk}, \quad k = 1, 3, \ldots, N-1. \tag{18}$$

In order to choose the sequence $h_0, h_1, \ldots, h_{N-1}$, it is interesting to rewrite the condition (9) directly as a relationship between the cyclic Fourier transforms of the sequences $g_0, g_1, \ldots, g_{N-1}$ and $h_0, h_1, \ldots, h_{N-1}$, which we denote by $\gamma_0, \gamma_1, \ldots, \gamma_{N-1}$ and $\eta_0, \eta_1, \ldots, \eta_{N-1}$, respectively. In particular, by using the fact that $\alpha^{N/2} = -1$, Eq. (9) can be rewritten straightforwardly as

$$\eta_k = -\nu_k \alpha^k \overline{\gamma_{k-\frac{N}{2}}}, \quad k = 0, 1, \ldots, N-1, \tag{19}$$

where we have used the extension $\nu_k = \nu_{k-\frac{N}{2}}$, $k = \frac{N}{2}, \ldots, N-1$. The corresponding relationship between $g_0, g_1, \ldots, g_{N-1}$ and $h_0, h_1, \ldots, h_{N-1}$, is thus determined by the choice of the sequence $\nu_0, \nu_1, \ldots, \nu_{\frac{N}{2}-1}$. For example, with $\nu_k \equiv 1$, (19) is equivalent to

$$h_k = (-1)^k \overline{g_{[1-k]_N}}, \quad k = 0, 1, \ldots, N-1, \tag{20}$$

where $[x]_N$ denotes x reduced modulo N. (Again, a similarity with discrete-time wavelets is seen when comparing (20) with [5].)

The above principles can be illustrated with some examples.

Example 1 - The Haar Wavelet: ¿From a practical viewpoint, the sequence $g_0, g_1, \ldots, g_{N-1}$ should be chosen to have only a few nonzero elements. As an example, take $N' = 1/2$, and consider the choice

$$g_k = \delta_k, \quad k = 0, 2, \ldots, N-2. \tag{21}$$

This choice is equivalent to $\gamma_k^0 \equiv 1$, which, through (8), imposes the condition $|\gamma_k^1| \equiv 1$, or equivalently, $\gamma_k^1 = \alpha^{\xi_k}$, with $\xi_0, \xi_1, \ldots \xi_{\frac{N}{2}-1}$, taken from the reals. Thus, any sequence of the form

$$g_k = \frac{2}{N} \sum_{l=0}^{\frac{N}{2}-1} \alpha^{-lk+\xi_l}, \quad k = 1, 3, \ldots, N-1, \tag{22}$$

is compatible with the choice (21). Imposition of the bandpass condition restricts only ξ_0 (to be $\frac{N}{2}$). The simplest such sequence results from the choice $\xi_k = \frac{N}{2} + k$, which leads to

$$g_k = -\delta_{k-1}, \quad k = 1, 3, \ldots, N-1; \tag{23}$$

i.e., the mother wavelet in this case is $1, -1, 0, 0, \ldots, 0$.

Using the construct of (20), an example of sequences generating a cyclic wavelet transform are those given by

$$g = \begin{pmatrix} 1 & -1 & 0 & \cdots & 0 & 0 \end{pmatrix}, \tag{24}$$

and

$$h = \begin{pmatrix} -1 & -1 & 0 & \cdots & 0 & 0 \end{pmatrix}. \tag{25}$$

A complete transform is thus specified by (24), (25) and Corollary 1. For example, for the case $N = 8$, we obtain the filters

$$g^1 = (1, -1, 0, 0, 0, 0, 0, 0) \quad h^1 = (-1, -1, 0, 0, 0, 0, 0, 0)$$
$$g^2 = (1, -1, 0, 0) \quad h^2 = (-1, -1, 0, 0)$$
$$g^3 = (1, -1) \quad h^3 = (-1, -1)$$

Note that, in this case, the lower-order filter impulse responses are found by simply taking the first half of that of the preceding filter. In this particular case, this property will hold for any transform length.[3] However, this property will not hold for general CWT's.

Example 2 - Length-4 Transforms: The next level of transform complexity (aside from other choices of the sequence ν_k in the transform of Example 1) arises from setting $g_k = 0$, for $k > 3$. Assuming that the coefficients of the mother wavelet are real, they are related through Theorem 1 by the equations:

$$g_0 g_2 = -g_1 g_3, \tag{26}$$

and

$$(g_0)^2 + (g_1)^2 + (g_2)^2 + (g_3)^2 = \frac{1}{N'}. \tag{27}$$

Note from (26) that a mother wavelet consisting of exactly three consecutive nonzero elements is not allowed in this formulation. Also note that the roles of g_0 and g_2 [resp. g_1 and g_3] are interchangeable. If we assume a normalization such that $g_0 = 1$, and further impose the bandpass condition,

$$g_0 + g_2 = -g_1 - g_3, \tag{28}$$

then, modulo the above noted symmetry, the mother wavelet is given for $N' < (3 + 2\sqrt{2})/8$, by

$$g_0 = 1, \tag{29}$$

$$g_1 = \frac{\zeta - \sqrt{2 - \zeta^2}}{2 - \zeta - \sqrt{2 - \zeta^2}}, \tag{30}$$

$$g_2 = -g_1 \frac{g_1 + 1}{g_1 - 1}, \tag{31}$$

and

$$g_3 = \frac{g_1 + 1}{g_1 - 1}, \tag{32}$$

where $\zeta \triangleq 1 - 2\sqrt{2N'}$. Note that this gives a family of mother wavelets parametrized by N'. (This parametrization results from the choice $g_0 = 1$. Alternatively, we could of course fix N' and consider g_0 to parametrize the family.)

With $N' = 1/2$, (29)-(32) reduces to the previous example, $g_1 = -1$, and $g_2 = g_3 = 0$. For other choices of N' the mother wavelet from (29)-(32) will

---

[3] This transform is analogous to the *Haar wavelet* arising in the discrete-time case.

differ nontrivially from (21), (23). For example, the choice $N' = 1/8$ yields the mother wavelet

$$g_0 = 1; \quad g_1 = \frac{1}{1 - \sqrt{2}}; \quad g_2 = 1; \quad g_3 = \sqrt{2} - 1. \tag{33}$$

Thus, on choosing h from (19), Corollary 1 gives the following $N = 8$ cyclic transform.

$$g^1 = (1, \frac{1}{1 - \sqrt{2}}, 1, \sqrt{2} - 1, 0, 0, 0, 0) \quad h^1 = (\frac{1}{1 - \sqrt{2}}, -1, 0, 0, 0, 0, \sqrt{2} - 1, -1)$$

$$g^2 = (1, \frac{1}{1 - \sqrt{2}}, 1, \sqrt{2} - 1) \quad h^2 = (\frac{1}{1 - \sqrt{2}}, -1, \sqrt{2} - 1, -1)$$

$$g^3 = (2, -2) \quad h^3 = (-2, -2)$$

This example will be discussed again in the following section.

4   Finite-field Wavelet Transforms

We now consider the cyclic wavelet transform described by decomposition/ reconstruction with transform matrices as in (4) and (5) for the case in which $\mathcal{F}$ is a finite field: $\mathcal{F} = \mathrm{GF}(p^r)$, where p is an odd prime and r is a positive integer. We again restrict the data length N to be a power of two, $N = 2^n$; and we assume that $\frac{N}{2}$ divides $p^r - 1$. This latter condition implies the existence of an order-$N/2$ element α of the multiplicative sub-group of $\mathcal{F}$.

Analogously with the complex case, we can again characterize cyclic wavelet transforms by considering the Fourier properties of $g^1, g^2, \ldots, g^n$, and $h^1, h^2, \ldots, h^n$. To do so, we first define polynomials

$$\gamma^m(x) = \sum_{\ell=0}^{\frac{N}{2}-1} g^1_{2\ell+m} x^\ell, \quad m = 0, 1, \tag{34}$$

and let $\eta^m(x), m = 0, 1$, be defined similarly for h^1. We then have the following result, which is essentially the same as Theorem 1.

Theorem 2 [4] : The matrices $2 - \mathrm{cir}\{g^1\}$ and $2 - \mathrm{cir}\{h^1\}$ satisfy (1) for $j = 1$ if and only if, for each $k = 0, 1, \ldots, \frac{N}{2} - 1$, we have

$$\gamma^0(\alpha^{-k})\gamma^0(\alpha^k) + \gamma^1(\alpha^{-k})\gamma^1(\alpha^k) = \frac{1}{N'}, \tag{35}$$

and

$$\eta^m(\alpha^k) = (-1)^m \nu(\alpha^{2k})\gamma^{1-m}(\alpha^{-k}), \quad m = 0, 1, \tag{36}$$

for some rational function $\nu(x)$ of order $\frac{N}{2}$ over $\mathcal{F}$ satisfying $\nu(\alpha^{-k})\nu(\alpha^k) = 1$, $k = 0, 1, \ldots, \frac{N}{2} - 1$.

Again, given two sequences g^1 and h^1 that satisfy this theorem, we want to construct a family of sequences $\{g^j, h^j \mid j = 2, \ldots, n\}$ that specifies a cyclic transform as described above. Analogously with the complex case, such sequences are specified by the following result.

Corollary 2: Suppose $2 - \text{cir}\{g^1\}$ and $2 - \text{cir}\{h^1\}$ satisfy (1). For each $j = 2, \ldots, n$, define a length-2^{n-j} sequences g^j and h^j by

$$g^j_{2\ell+m} = \text{FFFT}^{-1} \left[\{\gamma^m(\alpha^{2^{j-1}k}) \mid k = 0, 1, \ldots, 2^{n-j} - 1\} \right]_\ell \qquad (37)$$

$$h^j_{2\ell+m} = \text{FFFT}^{-1} \left[\{\eta^m(\alpha^{2^{j-1}k}) \mid k = 0, 1, \ldots, 2^{n-j} - 1\} \right]_\ell \qquad (38)$$

for $\ell = 0, 1, \ldots, 2^{n-j} - 1$, and $m = 0, 1$, where the sequences γ^m and η^m are as above, and where the operation FFFT indicates the finite-field Fourier transform [3] of appropriate length. Then $G^j = 2 - \text{cir}\{g^j\}$ and $H^j = 2 - \text{cir}\{h^j\}$ satisfy (1) for each $j = 2, \ldots, n$.

The comments made in the preceding section regarding the design of CWT's over the complex field remain essentially unchanged for this finite-field case. (Of course, the complex DFT's are replaced with their finite-field counterparts.) For example, with regard to the choice of $h_0, h_1, \ldots, h_{N-1}$, in analogy with (19) in the complex case, it is interesting to consider the choice $\nu(x) \equiv 1$, in which case we have

$$h_k = (-1)^k g_{[1-k]_N}, \quad k = 0, 1, \ldots, N - 1 \qquad (39)$$

as before.

An exception regards the nature of the field constant N', which here may be further restricted. For example, the choice of N' is generally constrained if we impose a bandpass condition (15). In particular, this condition together with (35) enforces the condition

$$\gamma^1(1) = -\gamma^0(1) = \pm\frac{1}{\sqrt{2N'}}. \qquad (40)$$

Thus, in order to impose the lowpass condition N' must be constrained to be such that $2N'$ has a square root in $\mathcal{F}$. Note that the corresponding low pass condition is given in this case by

$$\eta^0(1) = \eta^1(1) = \pm\frac{\nu(1)}{\sqrt{2N'}}. \qquad (41)$$

As a specific example, we reconsider Example 2 in this finite-field setting.

Example 3 - A Finite-field CWT: The sequence (29)-(32) is a finite-field mother wavelet for any choice of $n > 1$, and for any choice of N' such that $\sqrt{2N'}$ and $\sqrt{1 + 4\sqrt{2N'} + 8N'}$ exist in the finite field $\mathcal{F}$. An interesting case is that in which $\mathcal{F} = \text{GF}(2^m + 1)$ for an integer $m \geq n - 1$ such that $2^m + 1$ is prime. (These are the so-called *Fermat primes* that also are important in finite-field Fourier analysis.) In this case, exactly half of the nonzero elements of $\mathcal{F}$ - in particular, those elements that are even powers of the primitive element α - have square roots in $\mathrm{GF}(2^m + 1)$ (see, e.g., [2]). Thus, the above conditions imply that N' must be of the form $\frac{\alpha^{2k}}{2}$ for some integer k in order for $\sqrt{2N'}$ to exist, and it must also be of the form $\frac{(1\pm\alpha^\ell)^2}{2}$ for some integer ℓ in order for $\sqrt{1 + 4\sqrt{2N'} + 8N'}$ to exist. Note that the second condition is identical to the first, since all elements of $\mathrm{GF}(2^m + 1)$ can be generated in the form $1 \pm \alpha^\ell$.

As a specific example, consider $\mathrm{GF}(17)$ (i.e., $m = 4$). Here, we have $\alpha = 6$ and $\alpha^2 = 2$, so the possible choices of N' are 1, 2, 4, 8, 9 ($\equiv \frac{1}{2}$), 13 ($\equiv \frac{1}{4}$), $15 \equiv \frac{1}{8}$, and 16. So, for example, $N' = 9$ gives the mother wavelet

$$
\begin{pmatrix} g_0 \\ g_1 \\ g_2 \\ \vdots \\ g_{N-2} \\ g_{N-1} \end{pmatrix} = \begin{pmatrix} 1 \\ 16 \\ 0 \\ \vdots \\ 0 \\ 0 \end{pmatrix}, \tag{42}
$$

which is the $\mathrm{GF}(17)$ equivalent of (24) since $16 = -1$ in $\mathrm{GF}(17)$. In this case, cyclic transforms can be specified for any length up to 32. Thus, for example, together with the choices (37)-(39), we have a complete length-16 transform:

$$
\begin{aligned}
g^1 &= (1, 16, 0, 0, \cdots, 0, 0) \; ; \; h^1 = (16, 16, 0, \cdots, 0, 0, 0) \\
g^2 &= (1, 16, 0, 0, 0, 0, 0, 0) \; ; \; h^2 = (16, 16, 0, 0, 0, 0, 0, 0) \\
g^3 &= (1, 16, 0, 0) \; ; \; h^3 = (16, 16, 0, 0) \\
g^4 &= (1, 16) \; ; \; h^4 = (16, 16)
\end{aligned}
$$

A similar Haar-type transform can be constructed in any finite field.

As another example in $\mathrm{GF}(17)$, the choice $N' = 15$ gives the $\mathrm{GF}(17)$ equivalent of the mother wavelet of (33); namely,

$$
\begin{pmatrix} g_0 & g_1 & g_2 & g_3 & g_4 & \cdots & g_{N-2} & g_{N-1} \end{pmatrix} = \begin{pmatrix} 1 & 10 & 1 & 5 & 0 & \cdots & 0 & 0 \end{pmatrix}. \tag{43}
$$

So, for example, on using the choice (37)-(39), another complete length-16 transform over $\mathrm{GF}(17)$ is thus specified by

$$
\begin{aligned}
g^1 &= (1, 10, 1, 5, 0, \cdots, 0) \; ; \; h^1 = (10, 15, 0, \cdots, 0, 5, 16) \\
g^2 &= (1, 10, 1, 5, 0, 0, 0, 0) \; ; \; h^2 = (10, 15, 0, 0, 0, 0, 5, 16) \\
g^3 &= (1, 10, 1, 5) \; ; \; h^3 = (10, 15, 5, 16) \\
g^4 &= (2, 15) \; ; \; h^4 = (15, 15)
\end{aligned}
$$

5 Multilevel Error Protection

In the preceding sections, we have seen how the general idea of multiresolution analysis can be put into a cyclic framework similar to that of the discrete Fourier transform and the finite-field Fourier transform. One area of application in which these cyclic Fourier transforms have been very useful is in the description, design and analysis of error control codes [3]. Cyclic wavelet transforms have been shown to also play a useful role in error control coding [12, 13], and in particular in the development of codes that allow for different codewords to enjoy different levels of error protection. In this section, we describe this approach briefly.

The basic idea behind this application can be illustrated with the following example. Consider the extended Hamming (8,4) code (see [8]), whose codewords can be thought of as the columns of the matrix

$$
A = \begin{pmatrix}
1 & 1 & 1 & 1 & 1 & 1 & 1 & 1 \\
1 & -1 & 1 & -1 & 1 & -1 & 1 & -1 \\
1 & 1 & -1 & -1 & 1 & 1 & -1 & -1 \\
1 & -1 & -1 & 1 & 1 & -1 & -1 & 1 \\
1 & 1 & 1 & 1 & -1 & -1 & -1 & -1 \\
1 & -1 & 1 & -1 & -1 & 1 & -1 & 1 \\
1 & 1 & -1 & -1 & -1 & -1 & 1 & 1 \\
1 & -1 & -1 & 1 & -1 & 1 & 1 & -1
\end{pmatrix}
$$

and their negatives.

Let us consider the length-8 Haar wavelet of Example 1, and transform the codewords of this code. Taking all three levels of transform, the following is obtained, where the ith column of any matrix denotes the transform of the ith column of the matrix A.

$$
C^1 = \begin{pmatrix}
0 & 2 & 0 & 2 & 0 & 2 & 0 & 2 \\
0 & 2 & 0 & -2 & 0 & 2 & 0 & -2 \\
0 & 2 & 0 & 2 & 0 & -2 & 0 & -2 \\
0 & 2 & 0 & -2 & 0 & -2 & 0 & 2
\end{pmatrix}
$$

$$
D^1 = \begin{pmatrix}
-2 & 0 & -2 & 0 & -2 & 0 & -2 & 0 \\
-2 & 0 & 2 & 0 & -2 & 0 & 2 & 0 \\
-2 & 0 & -2 & 0 & 2 & 0 & 2 & 0 \\
-2 & 0 & 2 & 0 & 2 & 0 & -2 & 0
\end{pmatrix}
$$

$$
C^2 = \begin{pmatrix}
0 & 0 & 0 & 4 & 0 & 0 & 0 & 4 \\
0 & 0 & 0 & 4 & 0 & 0 & 0 & -4
\end{pmatrix}
$$

$$
D^2 = \begin{pmatrix}
0 & 0 & -4 & 0 & 0 & 0 & -4 & 0 \\
0 & 0 & -4 & 0 & 0 & 0 & 4 & 0
\end{pmatrix}
$$

$$
C^3 = \begin{pmatrix} 0 & 0 & 0 & 0 & 0 & 0 & 0 & 8 \end{pmatrix}
$$

$$
D^3 = \begin{pmatrix} 0 & 0 & 0 & 0 & 0 & 0 & -8 & 0 \end{pmatrix}.
$$

The wavelet transform being linear, the negatives of the columns of A are transformed into the negatives of the values shown.

The structure of the transform allows an efficient coding for multilevel error correction. Note that if the detail of a codeword is zero, then it has a non-zero lower-resolution component, and vice versa. Further, at each stage, the number of non-zero coefficients is halved. From this structure, we can devise a nonlinear, multilevel error-protecting code. This code will consist of a base code protecting all codewords uniformly; a second level somewhat like a parity check providing error detection to half of the codewords; and a third level repetition-like code giving double error correcting capabilities to one fourth of the codewords. In particular, the effective information in each codeword (which was four bits) can be encoded as follows.

The first bit is a sign bit. The second bit indicates whether the detail or the lower-resolution component is non-zero. The next two bits contain the information about the transform (there are four patterns in the transform, so two bits suffice). These four bits can be protected by the unique (7,4) code (up to permutations). This gives the base unit error protection to all codewords. Then, the next level of transform requires only one bit of information to encode (as the encoder and decoder both know the transform structure). This bit serves as parity, and is non-zero only for half of the codewords. We send this bit only if it is non-zero. Similarly, the third bit, which is non-zero only for one-fourth of the codewords and is sent only if non-zero, serves as the third layer, forming the repetition code. In this way, the original error protection capability of the code has been non-uniformly distributed onto the codewords as noted above. On assuming that all codewords are equally likely, the rate of this code is 4/7.75, or approximately 0.52. (Recall, the original extended Hamming code had rate 1/2.)

The key features that allow us to contruct the above code is the low density and the alternating nature of the higher-order transforms. Thus, this idea can be repeated for other codes by seeking a suitable CWT that has these properties. Several families of codes that can be treated this way are described in [13]. These include the first-order Reed-Muller codes, the extended Golay code, concatenated codes, and certain dual codes. These desirable properties can be characterized analytically through the notion of *alternating* wavelet transforms (in which, at any level, at most one of the transform coefficients - c^j or d^j - is non-zero) with *dyadic density vectors* (which means that the density of the transformed codewords descreases exponentially as a function of the transform level). Mother wavelets that yield transforms having these properties can be found for the above-noted families of codes [13]. In some case, it is sufficient to consider transforms with real coefficients (as in the above example); however, in other cases, it is necessary to specify transforms in appropriate finite fields.

6 Conclusion

In this paper, we have discussed the design of cyclic wavelet transforms for analyzing finite-length sequences from complex or finite fields. These transforms can be viewed as successive cyclic FIR filtering operations followed by factor-of-two downsampling. It should be noted that these ideas can be extended to decimations of orders other than two and to data lengths other than 2^n. A development of these extensions is found in [11].

An important advantage of the cyclic wavelet analysis over cyclic Fourier analysis is potentially lower computational complexity. If the rows of the matrices G^j and H^j appearing in decomposition/reconstruction each have at most M nonzero elements, then the full decomposition requires at most $2(2M - 1)(N - 1) \sim O(N)$ operations. By comparison, the FFT has $O(N \log_2(N))$ complexity. Thus, if wavelet transforms can be used to perform tasks that are normally performed using FFT's, some practical advantage may be gained if M is small relative to $\log_2(N)$. However, like their discrete-time counterparts, cyclic wavelet transforms may be fundamentally better suited to some tasks than are Fourier transforms.

Potential applications areas for finite-field wavelet transforms are similar to those for the cyclic Fourier transform, or for the discrete-time wavelet transform. As we have noted in Section 5, the multiresolution properties of the finite-field wavelet transform can be exploited to develop useful families of error-correcting codes featuring multilevel error protection. Alternatively, the multiscale/multilocation aspects of the cyclic wavelet transform make then useful in searching for structure in long strings of elements from a finite field. A potential application of this idea is found in biosequence analysis, in which various structural analyses on very long sequences of amino acids (of which there are finitely many) are of interest (see, e.g., [1] for a description of the use of finite-field Fourier analysis in this application).

References

1. D. C. Benson, "Digital signal processing methods for biosequence analysis," *Nucleic Acids Research*, vol. 18, no. 10, pp. 3001 - 3006, 1990.

2. R. E. Blahut, *Theory and Practice of Error Control Codes*. Addison-Wesley, Reading, MA, 1983.

3. R. E. Blahut, *Fast Algorithms for Digital Signal Processing*. Addison-Wesley, Reading, MA, 1984.

4. G. Caire, R. L. Grossman and H. V. Poor, "Wavelet transforms associated with finite cyclic groups," *IEEE Trans. Inform. Theory*, Vol. 37, No. 5, pp. 1157 - 1166, July 1993.

5. I. Daubechies, "Orthonormal bases of compactly supported wavelets," *Comm. Pure Appl. Math.*, Vol. 41, pp. 909-996, 1988.

6. I. Daubechies, "The wavelet transform, time-frequency localization and signal analysis," *IEEE Trans. Inform. Theory*, Vol. 36, pp. 961-1005, Sept. 1990.

7. P. J. Davis, *Circulant Matrices* (Wiley, New York, 1979)

8. F. J. MacWilliams and N. J. A. Sloane, *The Theory of Error-correcting Codes.* (North-Holland: New York, 1977)

9. S. G. Mallat, "A theory of multiresolution signal decomposition: The wavelet representation," *IEEE Trans. Pattern Anal. Machine Intell.*, Vol. 11, pp. 674–693, 1989.

10. D. Pollen, "$SU_I(2, F[z, \frac{1}{z}])$ for F a subfield of C," *J. Am. Math. Soc.*, Vol. 3, pp. 611-624, July 1990.

11. S. Sarkar and H. V. Poor, "Certain generalizations of the cyclic wavelet transform," *Proc. 1995 Conf. Inform. Sci. Syst.*, The Johns Hopkins University, Mar. 22 - 24, 1995.

12. S. Sarkar and H. V. Poor, "Finite-field wavelet transforms and multilevel error protection," *Proc. 1995 IEEE Int'l Symp. Inform. Theory*, Whistler, BC, Sept. 17 - 22, 1995.

13. S. Sarkar and H. V. Poor, "Multilevel error protection: A wavelet transform approach to coding," preprint, January 1996.

Choice of Wavelets for Image Compression

M. K. Mandal, S. Panchanathan and T. Aboulnasr

Dept. of Electrical Engineering, University of Ottawa, Canada - K1N 6N5
E-mail : mandalm@marge.genie.uottawa.ca

Abstract. In wavelet-based image coding the choice of wavelets is crucial and determines the coding performance. Current techniques use computationally intensive search procedures to find the optimal basis (type, order and tree). In this paper, we show that searching for optimal wavelet does not always offer a substantial improvement in coding performance over "good" standard wavelets. We propose some guidelines for determining the need to search for the "optimal" wavelets based on the statistics of the image to be coded. In addition, we propose an adaptive wavelet packet decomposition algorithm based on the local transform gain of each stage of the decomposition. The proposed algorithm provides a good coding performance at a substantially reduced complexity.

1 Introduction

Visual communication is becoming increasingly important with applications in several areas such as, digital television transmission, teleconferencing, multimedia communications, transmission and storage of remote sensing images, image/video databases, educational and business documents, archiving medical images, *etc.* Since digital images and video data are inherently voluminous, efficient data compression techniques are essential for their archival and transmission. The International Standard Organization (ISO) has recently proposed the JPEG standard [12] for still image compression and MPEG standards [5] for video compression. These standards employ discrete cosine transform (DCT) to reduce the spatial redundancy present in the images or video frames. We note that DCT has the drawbacks of blocking artifacts, mosquito noise and aliasing distortions at high compression ratios [3,12].

Recently, discrete wavelet transform (DWT) has emerged as a popular technique for image coding applications [1,3,7]. DWT has high decorrelation and energy compaction efficiency. The blocking artifacts and mosquito noise are absent in a wavelet-based coder due to their overlapping basis functions. The aliasing distortion can be reduced with a proper choice of wavelet filters. In addition, the basis functions are localized in both the spatial and frequency domains. Hence, they are better matched to the human visual system (HVS) characteristics.

In wavelet-based image coding, the coding performance depends on the choice of wavelets. Several wavelets which provide suboptimal coding performance have been proposed in the literature. Recently, a few approaches for selecting the optimal filterbank in an image coder have been proposed in the literature [2,11]. Generally, a wavelet providing optimal performance for the whole image is selected. However, a few cases have been reported where spatially adapted filterbanks were employed. We note that finding the optimal wavelet for a particular image is a computationally intensive task. Tewfik, *et al.* [11] have proposed a technique to find the best wavelet basis by maximizing the L^2-norm of the wavelet approximated signal. Caglar, *et al.* [2] have proposed techniques for designing wavelets which are optimal in the statistical sense. The complexity of these algorithms increases considerably with the filter order.

The performance of wavelet-based coding also depends on the wavelet decomposition structure. Ramchandran, *et al.* [10] have proposed a technique, based on Lagrangian optimization, to find the best basis subtree. This technique minimizes the global distortion for a given bit-rate. However, this algorithm is computationally expensive.

Before employing a computationally intensive procedure for finding the optimal basis, it will be helpful to know, *a priori*, if appreciable gain can be achieved by using the optimal basis over known "good" wavelets. In this paper, we first provide some guidelines to address the following: i) when can one expect appreciable performance gain by using the optimal wavelet rather than standard wavelets ? ii) what should the optimal filter order be for most images? iii) what is a good starting point if using the optimal wavelet is crucial? We then propose an adaptive wavelet packet decomposition algorithm based on the local transform gain at each stage of the decomposition.

2 Choice of wavelet type and order

The subband coding system is based on the frequency selectivity property of the filterbanks. An alias free frequency split and perfect interband decorrelation of coefficients can be achieved only with ideal filterbanks with infinite duration basis functions. Since time localization of the filters is very important in visual signal processing, one cannot use an arbitrarily long filter. In addition, properties such as vanishing moments, phase linearity, time-frequency localization, energy compaction, *etc.*, influence the coding performance. In this section, we will explore the effect of some of these properties on the coding performance.

To determine the effect of the choice of the wavelet filter order, we compare the coding performance of wavelets of various orders within the same family. We then compare the performance of some known wavelet families. Finally, these families are compared to the optimal wavelets found by using a full search method in a multidimensional parameter space. In this study, we have used the following sets of compactly supported orthonormal wavelets.

1. *Daubechies* - 2, 4, 6, 8, 10, 12, 16 and 20 taps (minimum phase) [4]

2. Least asymmetric 8, 10 and 12 taps *Daubechies* wavelets [4]

3. *Coiflet* - 6, 12, 18 and 24 taps [4]

4. *AHQMF* - 4, 6, 12 and 18 taps [Ref. 1, Table-4.11]

5. A dense set of 4 tap wavelets calculated using Pollen's parameterizations [9]

6. A dense set of 6 tap wavelets calculated using Pollen's parameterizations [9]

7. A dense set of 8 tap wavelets calculated using Zou's parameterizations [13]

A N tap two channel wavelet can be chosen using $N/2-1$ free parameters. These free parameters can be used to choose the wavelets which results in a good coding performance. The Daubechies wavelet family was constructed by using all the free parameters to maximize the number of vanishing moments [4]. When the order is greater than seven, more than one wavelet with the same frequency response exist. In this case, if all the zeros of the filter are inside the unit circle, the filter is known as Daubechies minimum phase wavelet. The wavelet that has the least asymmetry (it is not possible to achieve perfect symmetry in two channel filterbanks) is known as Daubechies least asymmetric (Daub-LA) wavelet. In this paper, minimum phase and least asymmetric Daubechies wavelets have been used. Coiflet wavelets are designed by imposing vanishing moments on both the scaling and wavelet functions. The AHQMF wavelets were designed by Akansu and Haddad by optimizing various filter characteristics to achieve a good coding performance [1]. Wavelet parameterizations of Pollen [9], and Zou [13] have been used to systematically generate N tap orthogonal wavelets using the $N/2-1$ free parameters for N = 4, 6 and 8.

In this study, we have used a large number of test images with different levels of activities. An easy way to find the level of activity of an image is to calculate its spectral flatness measure (SFM) which has been defined in the appendix. The images *Airport* and *Mandrill* are highly active, whereas *Barbara, Lena, Girl, Airplane, Sailboat* are moderately active images. *Chest* and *Visualmtf* are images with low SFM. The wavelet coefficients were encoded using a uniform quantization scheme followed by an adaptive arithmetic coder. The quantization step-size (one per band) of each band has been chosen so as to distribute the noise equally throughout the scale space [8]. Peak signal-to-noise ratio (PSNR) has been used as the measure of objective performance; although, the subjective quality of the reconstructed images was verified.

2.1 Effect of filter order for a given wavelet

The order of a wavelet filter is important in achieving good coding performance. A higher order filter can be designed to have good frequency localization which in turn increases the energy compaction. The regularity of wavelets also increases with its order. For example, the regularity of L tap Daubechies wavelet is approximately $0.5*(L/2-1)$. In addition, more vanishing moments can be obtained with a higher order filter. On the other hand, a lower order filter is expected to have a better time localization and therefore preserve the crucial edge information. Fig. 1 summarizes the TFL of various families and orders. It can be seen that as filter order increases, TFL is reduced up to some order. Beyond that, increasing the filter order has a negative effect.

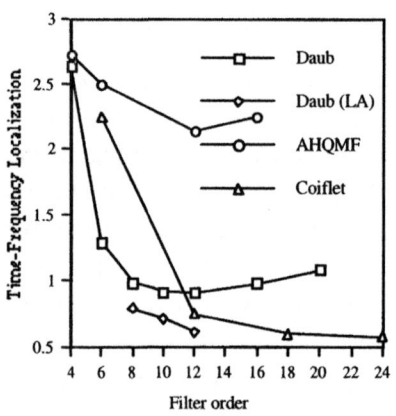

Figure 1 : Time-frequency localizatin
of various wavelet filters

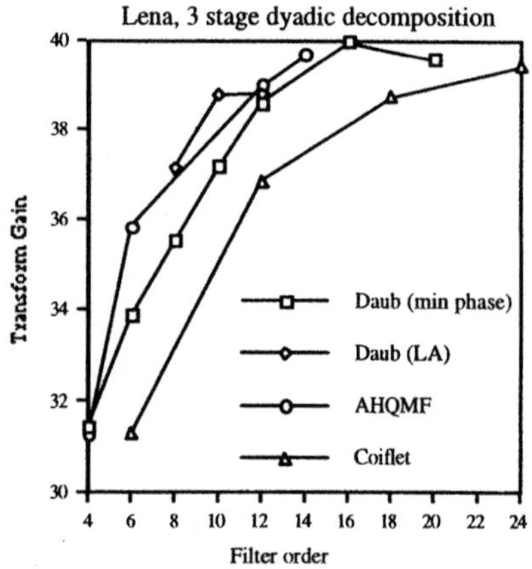

Figure 2 : Comparison of transform coding
gain of various wavelet filters

Wavelet filter performance is also evaluated using the transform coding gain (TCG). TCG is a popular measure of the energy compaction of a transform. The TCG's of various wavelets on Lena image are shown in Fig. 2. It is observed that the TCG's saturate around 10-14 taps for Lena image. Similar behaviour was observed for all the images except, chest and visualmtf, where TCG behaved differently. Fig. 3 shows the coding performance of Daubechies wavelets on the Lena image for a wide range of bit-rates. It is again observed that performance saturates beyond 12 taps. It is known that most natural images are lowpass in nature and hence a regularity of 2 (Daub-10 has a regularity of approximately 2) is sufficient to represent the images well. Thus, using *Daubechies* wavelets of order higher than twelve fails to improve the coding performance, in spite of their higher regularity. However, if wavelets from other families are used, *e.g.*, *Coiflet* which has fewer vanishing moments and less regularity compared to a *Daubechies* wavelet with the same support, the optimal order may change. For *Coiflet* wavelets, performance improvement has been seen up to 18-24 taps (Fig. 4).

The reconstructed images were compared subjectively. It was observed that higher order filters provide a blurred image even if the PSNR is high. On the other hand, lower order filters provide inferior energy compaction and more blockiness. In Daubechies family, the best subjective performance was obtained with 8-12 taps. This can be justified from Fig. 1 which shows the TFL of various wavelets. It is observed that best TFL in Daubechies (min. phase) family is achieved around 10 taps. For *Coiflet* family, however, a longer filter, such as 18-24 tap is expected to provide better results.

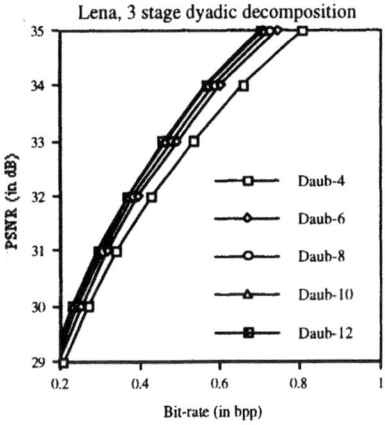

Figure 3 : Objective performance of Daubechies wavelets.

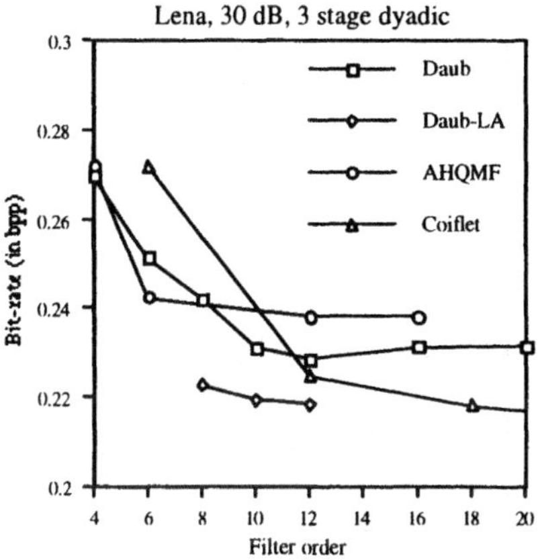

Figure 4 : Relative coding performance of various wavelet families

2.2 Choice of Wavelet function for a given image

As stated earlier, the "best" wavelet basis is dependent on the specific image to be coded. Here, we consider a wide variety of images with different activities. For each image, we find the optimal wavelets with 4, 6 and 8 taps using Pollen's, and Zou's parameterizations. Table 2 provides the parameters of some wavelets used in this paper. Fig. 5 shows the coding performance of various 4-tap wavelets on Lena image. It is observed that the coding performance does depend on the choice of wavelet. The performance curve is symmetric about π as the wavelets corresponding to $\pi - \theta$ and $\pi + \theta$ are the same except for the time reversal. For the images considered, the minimum bit-rate achieved was around $\theta = 1.05$ (we note that Daub-4 has Pollen's parameter, $\theta \approx 1.04$). Table-1 summarizes the improvement provided by optimal 4, 6 and 8 tap wavelets over Daubechies, AHQMF and Coiflet. It is observed that Daub-4 wavelet provides performance very close to that of optimal 4-tap wavelets for all the images. AHQMF-4 also provides similar performance and hence has not been shown here. We also observe that the actual coding performance of the optimal 6-tap wavelet is very close to that of the Daub-6 and AHQMF wavelets (only minor improvement can be achieved using the optimal wavelet). The only exception was the chest image, where Coiflet-6 provides a performance close to the optimal wavelet. For 8-tap wavelets, a performance improvement of 2-10% is possible for most images, using optimal wavelets rather than the Daubechies (min. phase) wavelet.

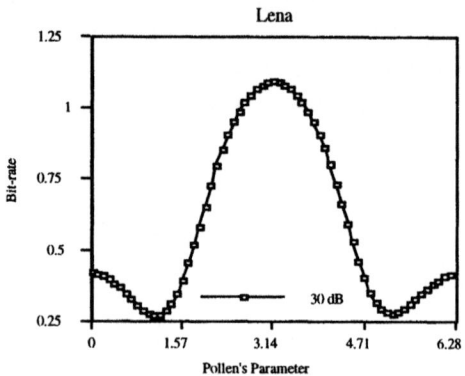

Figure 5 : Bit rate of various 4 tap wavelets on Lena at 30 dB.

Least asymmetric Daubechies wavelets are expected to provide subjectively superior reconstructed images because of their improved phase linearity [4]. Our study shows that they also provide superior energy compaction compared to their minimum phase counterpart (see Table-1 and Fig. 4). In Fig. 3, we observe that the least asymmetric Daubechies wavelets have also a superior TFL. Hence, least asymmetric *Daubechies* wavelets can be considered as a good initial guess for the adaptive search when an optimal basis is required.

Table-1 : Performance improvement (*w.r.t.* bit-rate) in percentage with optimal 4/6/8 tap wavelets over known standard wavelets of same order.

Image	4-tap	6 tap			8 tap	
	Daub-4	Daub-6	AHQMF	Coiflet-6	Daub-8	Daub-8 (LA)
Airport	0.23	0.5	0.5	0.5	1.3	0.6
Mandrill	0.07	1.3	0.2	4.3	1.4	1.2
Airplane	0.33	0.5	1.2	5.4	5.5	0.9
Sailboat	0.75	1.6	0.5	11.1	2.4	1.4
Girl	0.93	7.9	0.5	13.1	7.6	3.2
Lena	1.12	4.7	0.9	13.3	11.4	2.6
Visualmtf	0.91	7.9	2.9	16.7	15.0	7.4
Chest	0.13	12.0	19.4	0.4	31.8	22.2

In Table 1, we observe that for images with high spectral activity (airport, mandrill), the coding performance is fairly insensitive to a relatively broad range of wavelets. In these cases, nearly optimal performance is achieved with *Daubechies* wavelets. Hence, searching for the optimal wavelet is not useful in these cases. For moderate SFM images (which are most common), the performance improvement is marginally superior and still not very encouraging. However, it has been found that for images with low spectral flatness, the choice of wavelet greatly affects the coding performance. It will therefore be well worth the effort to search for the best wavelet for such images.

3 Adaptive wavelet decomposition

The objective of wavelet/subband decomposition is to compact most of the energy in fewer bands. In general, more levels of decomposition are expected to result in a superior coding performance. However, when the number of bands is too large, the performance deteriorates because of two reasons. First, it is uneconomical to individually code small bands with similar statistics. Secondly, the extra information such as tree structure, quantization step sizes, *etc.*, needs to be sent to the decoder. Recently, Coifman, *et al.* [3], have introduced wavelet packets as a family of orthonormal bases. Wavelet packets represent a generalization of the method of multiresolution decomposition and comprise all possible combinations of subband tree decomposition.

For a *d*-level decomposition, 2^{2^d} independent wavelet packet bases are possible. Hence, it is difficult to determine the optimal wavelet packets using exhaustive search procedures. Ramchandran, *et al.* [10] have proposed an optimal technique to find the best basis subtree. In this technique, the image is first decomposed with a regular tree. The complete tree is then pruned using a Lagrangian cost function to find the best basis subtree. Although this algorithm provides an optimal coding performance, it is computationally expensive.

In this section, we propose an adaptive wavelet packet decomposition algorithm using a top-down approach. When an image (or a subimage) is decomposed, four local bands are created at each stage. Let the variance of the four local bands be $\sigma_1^2, \sigma_2^2, \sigma_3^2$ and σ_4^2. Then, the local transform gain is defined as

$$TCG_{local} = \frac{\left(\sigma_1^2 + \sigma_2^2 + \sigma_3^2 + \sigma_4^2\right)/4}{\left(\sigma_1^2 \sigma_2^2 \sigma_3^2 \sigma_4^2\right)^{1/4}}$$

The decomposition will result in appreciable energy compaction if TCG_{local} exceeds a pre-determined threshold. Thus, the decomposition is kept intact if

$$TCG_{local} > 1 + \varepsilon$$

where ε is a small positive number (usually in the range 0.05-0.10). If the above inequality is not satisfied, the decomposition does not result in significant energy compaction and therefore may be considered as non-profitable. In this case, the pre-decomposed band is considered for quantization and coding. If the decomposition is found beneficial (*i.e.*, the inequality is satisfied), the band splitting is maintained and the bands are considered for further decomposition. Among the four bands, the bands that contain appreciable energy are decomposed further. In our simulations we have decomposed the bands that contain more than 5-10% of the total energy of the parent band. This simple top-down splitting technique performs well for most images.

Fig. 6 compares the coding performance of dyadic, optimal [10] and the proposed algorithm. It is observed that the proposed decomposition provides a coding performance superior to that of the dyadic tree. The coding performance of the proposed tree is close to that of the optimal tree at low bit-rates. Table-2 shows the relative performance of dyadic, regular and adaptive tree for three stages of decomposition. The proposed algorithm provides performance very close to that of a regular tree. A regular and a dyadic 3 level decomposition produces 64 and 10 bands, respectively, whereas the proposed algorithm produces about 20-40 bands depending on the image. It is observed that for most images a performance improvement of 3-20% over the dyadic tree decomposition algorithm, can be achieved using the proposed algorithm. We may note that the proposed algorithm provided substantial improvement for both *chest* and *visualmtf* which are low SFM images.

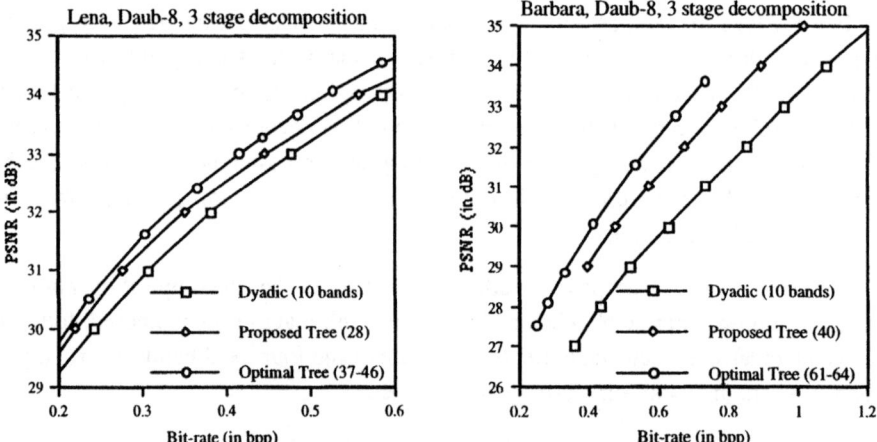

Figure 6 : Performance of the proposed decomposition techniques.

The complexity of the three algorithms is now compared. For simplicity, we consider only the computation required for calculating the wavelet transform. We note that the computational complexity of the wavelet transform for an image size of $N \times N$, employing a simple polyphase decomposition, is approximately [8] :

$$C = \begin{cases} 16N^2L\left(1-4^{-J}\right)/3 & \text{for dyadic decomposition} \\ 4JLN^2 & \text{for regular decomposition} \end{cases}$$

where, L and J are the number of filter tap and decomposition stages, respectively.

Thus, for a 512×512 image decomposed for 3 stages with 8-tap wavelets, the complexity will be approximately 11 and 25 million operations (MOP) for dyadic and regular decomposition, respectively. The proposed wavelet packet algorithm will have a complexity of X MOP, where $11 \leq X \leq 25$.

Table-2 : Bitrates for dyadic, regular and proposed tree for various images at 30 dB PSNR

Image	Dyadic Tree.	Regular Tree	Proposed Tree/bands	% improve-[1] ment
Airport	1.924	1.971	1.924 / 10	0.00
Mandrill	1.377	1.299	1.297 / 31	5.81
Airplane	0.296	0.236	0.235 / 34	20.56
Sailboat	0.481	0.447	0.446 / 43	7.11
Girl	0.098	0.096	0.095 / 28	2.44
Lena	0.245	0.216	0.219 / 28	9.20
Visualmtf	0.342	0.084	0.084 / 22	75.51
Chest (39 dB)	0.251	0.179	0.178 / 34	29.08

[1]improvement in bit-rate over dyadic tree

The optimal algorithm requires the computation of a 3-stage forward DWT and 3-stage inverse DWT to compute the optimal tree. Once the optimal tree is found, the image has to be decomposed again for actual coding. Thus, the total complexity will be approximately $(25 + 25 + Y)$ MOP, where $11 \leq Y \leq 25$. The proposed algorithm, using local transform gain, has a complexity of only one adaptive decomposition, *i.e.*, X MOP. Thus, in the case where $X \approx 18 \approx Y$, the proposed algorithm has less than one-third the complexity of the optimal algorithm. Since, a dyadic decomposition requires a complexity of 11 MOP, the proposed algorithm provides a superior coding performance at a marginal increase in complexity over the dyadic tree.

4 Conclusion

In this paper, we have shown that searching the "optimal" wavelet is warranted only for low activity (SFM) images. Otherwise, *Daubechies* wavelets or *AHQMF's* are close enough to the optimum and not much improvement can be expected by finding the optimal wavelet. The order of the wavelet filter should not in general exceed 12-18 with 8-18 taps generally providing the best time-frequency localization depending on the wavelet family. Least asymmetric Daubechies wavelets provide a better performance due to their superior TFL and phase linearity. Hence, they can be used as an initial guess for an adaptive search algorithm. Finally, a simple adaptive wavelet packet decomposition algorithm has been proposed. The algorithm provides good coding performance at a significantly reduced complexity.

5 Appendix

In this appendix, we define some of the terms used in this paper.

Spectral Flatness Measure

Spectral flatness is a measure of the overall image activity. The spectral flatness of a digital image is defined as the ratio of the arithmetic and the geometric mean of the Fourier coefficients [6]. For two-dimensional digital image, this can be expressed as :

$$SFM = \frac{\frac{1}{MN} \sum_{i=0}^{M-1} \sum_{j=0}^{N-1} |F(i,j)|^2}{\left[\prod_{i=0}^{M-1} \prod_{j=0}^{N-1} |F(i,j)|^2 \right]^{\frac{1}{MN}}}$$

where $F(i,j)$ is the (i,j)-*th* Fourier coefficient of the two-dimensional image. We note that SFM has a dynamic range of [0,1]. Active images (*i.e.*, SFM close to 1) are in general difficult to code.

Transform Coding Gain

A good measure for comparing coding performance of various transforms is the transform coding gain (TCG). For the orthogonal transform, this is defined as [6]

$$TCG = \frac{\sigma_{am}^2}{\sigma_{gm}^2} = \frac{\frac{1}{N} \sum_{i=0}^{N-1} \sigma_i^2}{\left(\prod_{i=0}^{N-1} \sigma_i^2 \right)^{1/N}}$$

where, N is the total number of coefficients to be coded and σ_i is the variance of the *i-th* coefficients. In the following, we rewrite the above equation so that it can be easily used to compute the transform gain of various wavelet filters. Let us assume that there is a total of M bands after the decomposition and the total number of pixels in k-*th* band is P_k. The *transform coding gain* can be rewritten as :

$$TCG = \frac{\sigma_{am}^2}{\sigma_{gm}^2} = \frac{\frac{1}{N} \sum_{k=1}^{M} P_k \sigma_k^2}{\left(\prod_{k=1}^{M} \left(\frac{\sigma_k^2}{P_k} \right)^{P_k} \right)^{1/N}} \qquad \text{where} \quad N = \sum_{k=1}^{M} P_k$$

Time-Frequency Localization

The time and frequency localization of a finite sequence $h[n]$ are estimated [1], respectively, by

$$\sigma_n^2 = \frac{1}{E} \sum_n (n - \overline{n})^2 |h[n]|^2$$

$$\sigma_\omega^2 = \frac{1}{E} \frac{1}{2\pi} \int_{-\infty}^{\infty} (\omega - \overline{\omega})^2 |H(\omega)|^2 d\omega$$

where,

$$\overline{n} = \frac{1}{E} \sum_n n |h[n]|^2 \text{ , the centre of mass of the sequence}$$

$$\overline{\omega} = \frac{1}{2\pi E} \int_{-\infty}^{-\infty} \omega |H(e^{j\omega})|^2 d\omega \text{ , the centre of mass in frequency domain}$$

$$E = \sum_n |h[n]|^2 = \frac{1}{2\pi} \int_{-\infty}^{\infty} |H(\omega)|^2 d\omega \text{ , is the total energy of the sequence}$$

The joint time-frequency localization (TFL) is defined as $\sigma_n \sigma_\omega$. It is well known from the uncertainty principle that the TFL of a continuous function is lower bounded by 0.5. The TFL of any wavelet filter is also lower bounded by 0.5.

References

1. A. N. Akansu and R. A. Haddad: *Multiresolution Signal Decomposition : Transform, Subbands and Wavelets.* Academic Press, Inc. (1992).

2. H. Caglar, Y. Liu, and A. N. Akansu: Statistically optimized PR-QMF design. *Proc. of SPIE : Visual Communications and Image Processing* **1605** (1991) 86-94.

3. R. Coifman, Y. Meyer, S. Quake and V. Wickerhauser: Entropy based algorithms for best basis selections. *IEEE Trans. on Information Theory* **38(2)** (March 1992) 713-718.

4. I. Daubechies: *Ten Lectures on Wavelets.* SIAM, Philadelphia (1992).

5. D. L. Gall: MPEG : A video compression standard for multimedia applications. *Communications of the ACM* **34(4)** (April 1991) 46-58.

6. N. S. Jayant and P. Noll: *Digital Coding of Waveforms: Principles and Applications to Speech and Video.* Prentice Hall, Englewood Cliffs, New Jersey (1984).

7. S. G. Mallat: A theory for multiresolution signal decomposition: the wavelet representation. *IEEE Trans. on Pattern Analysis and Machine Intelligence* **11(7)** (July 1989) 674-693.

8. M. K. Mandal: *Wavelets for Image Compression.* M.A.Sc Thesis, University of Ottawa, Ottawa, Canada (1995).

9. D. Pollen: SU(2, F[z, 1/z]) for F a subfield of C. *J. American Mathematical Society* **3** (July 1990) 611-624.

10. K. Ramchandran and M. Vetterli: Best wavelet packet in a rate distortion sense. *IEEE Trans. on Image Processing* **2(2)** (April 1993) 160-175.

11. A. H. Tewfik, D. Sinha and P. Jorgensen: On the optimal choice of a wavelet for signal representation. *IEEE Trans. on Information Theory* **38(2)** (March 1992) 747-765.

12. G. K. Wallace: The JPEG still picture compression standard. *Communications of the ACM,* **34(4)** (April 1991) 30-45.

13. H. Zou and A. H. Tewfik: Parameterization of compactly supported orthonormal wavelets. *IEEE Trans. on Signal Processing* **41(3)** (March 1993) 1428-1431.

Information in Markov Random Fields and Image Redundancy

Espen Volden, Gérard Giraudon and Marc Berthod

INRIA, 2004 Route des Lucioles,
F-06902 Sophia Antipolis Cedex, France

Abstract. The rate of information transmission of a transmitter-receiver couple was defined by C.E. Shannon in his Information Theory. We use this concept in Computer Vision to define models of image redundancy. Houzelle et Giraudon applied information theory to a simple model considering an image as a set of isolated pixels. We introduce a Markov Random Field model to take into account the spatial neighbourhood of a pixel. We show that we have to determine some parameters of the MRF in order to obtain sufficient statistics from common satellite images, and we propose a measure based on a generalized Ising model. Another model which also takes into account a pixel's spatial context is then proposed. It considers the correspondence between grey level vectors of cliques. We introduce a distance in the grey level space to solve the problem of insufficient statistics. Finally, results for the proposed definitions are presented for some synthetic and a large variety of SPOT XS1, XS2 and XS3 image triples and are compared to the classical correlation coefficient measure.

1 Introduction

When we are exposed to a multi-sensor information fusion problem, a fundamental preliminary issue is to determine whether the information provided by the sensors are essentially the same or substantially different. In other words : are the sensors redundant or complementary ?

This question can be examined either by an a priori or an a posteriori approach. The former uses physical models of the sensors, the physical objects, external conditions and so on, whereas the latter studies the information provided by the sensors, in our case images. We have chosen this second approach and will be speaking of image redundancy rather than sensor redundancy.

Image redundancy should represent the quantity of common information in two images. That is, knowing one image, to what extent are we able to predict the other one. Redundancy and resemblance do not characterize the same properties. For example two random images may look quite the same, but it is difficult to predict one knowing the other, so the redundancy is weak. On the other hand an image and its negative are strongly redundant since it is easy to predict one knowing the other although they do not necessarily resemble. Redundancy is more related to a correlation rather than to a resemblance.

In [4], Giraudon and Houzelle imagine two images being connected by a virtual channel and consider one image as the transmitter and the other as the receiver. They can then apply Shannon's information theory [8] and use the notion of rate of information transmission (called system mutual information in more recent information theory literature) as a measure of image redundancy. Information theory has also been applied to image classification. Maître in [6] defines a *spatial entropy* which is the entropy of the probability distribution of class label configurations on the neighbours of a pixel, and a *global entropy* of the joint probability distribution of a pixel's grey level, its class label and the configuration of class labels on its neighbours. The radiometric entropy (i.e. the entropy of the probability distribution of grey levels), in addition to the global and spatial entropies, are defined globally and for each class. Maître uses these entropies to study the convergence properties of iterative contextual classification algorithms. Lohmann [5] does supervised texture classification based on co-occurrence features. He models the co-occurrence feature vector sampled in some image window as a multinomial distribution. This enables him to compute the mutual information of the co-occurrence feature vectors of the window and of each class, and attribute the class giving highest mutual information to the pixel in the center of the window.

We make the model of Giraudon and Houzelle more explicit. They use information theory to define the entropy of an image and a measure for the redundancy of two images. Their approach amounts to define the quantity of information (i.e. the basic definition of the information theory) as the one produced by the event of a pixel's grey level occurring, knowing the image's histogram. The stochastic model (such a model is necessary to apply information theory) consists in considering an isolated pixel and conferring a probability distribution to it given by the normalized histogram.

Our goal is to use the measure of the redundancy in an image fusion process for classification. In order to obtain a good classification we will require to consider not only isolated pixels but also their neighbourhood. Therefore we need to obtain a definition of redundancy which takes into account the neighbourhood of a pixel. We will develop two models, one based on the theory of Markov Random Fields (MRF), and another using one of the key concepts of this theory : the cliques.

In our first approach we define the information as the one produced by the event of a pixel's grey level occurring, knowing the grey levels of all other pixels. We extend Hamming's definition of the entropy of a Markov chain [3] to a Markov random field and propose a definition of the redundancy of two images in terms of conditional probabilities with respect to the neighbourhood of a pixel. To eliminate the statistical insufficiency we must apply the Hammersley-Clifford theorem [7] and determine parameters of the Markov model. We propose a definition based on the generalized Ising model.

The second model that takes into account the neighbourhood of a pixel, is a quite natural extension of the Isolated-pixels model. Instead of the correspondence pixel to pixel, it considers the one between vectors of neighbouring pixels (i.e. *cliques* in Markovian terminology). We obtain a measure which is applicable to large images, whereas smaller images do not provide enough statistical data.

In order to further increase the statistics we introduce a distance in the grey level space.

We consider images where the sites $s \in S = \{s_1, ..., s_N\}$ have grey levels $x_s \in \Lambda = \{g_1, ..., g_M\}$.

We will use the notation $x = [x_{s_1}, ..., x_{s_N}]$.

2   The Isolated Pixels Model

Every x_{s_i} is considered to be the ith realization of the stochastic variable X_s whose sample space is Λ and whose probability distribution is the normalized grey level occurrence frequency, f_x, (normalized histogram).

$$\forall g \in \Lambda \quad P(X_s = g) = f_x(g) = \frac{1}{N} \sum_{s \in S} \delta(x_s - g)$$

where $\delta(0) = 1$ and $\delta(x) = 0$ for $x \neq 0$.

Considering X_s as an information source, we can apply the information theory and define the quantity of **information** produced by a realization x_s of X_s :

$$I(x_s) = \log \frac{1}{P(X_s = x_s)} = -\log f_x(x_s) . \tag{1}$$

The **entropy** of X_s is defined as the average of the information for all possible values x_s weighted according to the probability of x_s, in other words, it is the mathematical expectation of the information :

$$H(X_s) = \sum_{g \in \Lambda} P(X_s = g) \log \frac{1}{P(X_s = g)} = -\sum_{g \in \Lambda} f_x(g) \log f_x(g) . \tag{2}$$

This is the entropy of X_s when we consider only grey level occurrence probabilities. It is uniquely determined by the image histogram.

If we use the base 2 logarithm, this quantity represents the optimal average number of bits needed to code the grey level of a pixel knowing the image histogram (see [3]).

Similarly for a second image Y : $H(Y_s) = -\sum_{g \in \Lambda} f_y(g) \log f_y(g)$.

Now, let the realization x_{s_i} of the source X_s (transmitter) be the input to an information channel and the realization y_{s_i} of the source Y_s (receiver) be the output of this channel (or vice versa). We can then define the **joint entropy** of the transmitter and the receiver as the mathematical expectation of the information produced by a realization of the variable (X_s, Y_s) :

$$H(X_s, Y_s) = \sum_{g_1 \in \Lambda} \sum_{g_2 \in \Lambda} P(X_s = g_1, Y_s = g_2) \log \frac{1}{P(X_s = g_1, Y_s = g_2)}$$

$$= -\sum_{g_1 \in \Lambda} \sum_{g_2 \in \Lambda} f_{xy}(g_1, g_2) \log f_{xy}(g_1, g_2) \tag{3}$$

where $f_{xy}(g_1, g_2) = \dfrac{1}{N} \sum_{s \in S} \delta(x_s - g_1)\delta(y_s - g_2)$.

Equation (3) expresses the minimum number of bits needed to code the grey level of a site in one image and the grey level of the corresponding site in the other image, knowing their joint histogram.

We also define the **conditional entropy** :

$$H(X_s \mid Y_s) = \sum_{g_1 \in \Lambda} \sum_{g_2 \in \Lambda} P(X_s = g_1, Y_s = g_2) \log \frac{1}{P(X_s = g_1 \mid y_s = g_2)}$$

$$= - \sum_{g_1 \in \Lambda} \sum_{g_2 \in \Lambda} f_{xy}(g_1, g_2) \log \frac{f_{xy}(g_1, g_2)}{f_y(g_2)} \ .$$

$H(X_s \mid Y_s)$ is sometimes called equivocation (Shannon). It measures the average ambiguity of the input when the output is observed. In other words it is the minimum quantity of information required to be added to the output to recover the input. It is zero if we can define a function φ from Λ to Λ so that $\forall s \in S \quad x_s = \varphi(y_s)$.

The **system mutual information** or **rate of information transmission** is defined as the difference between the entropy of the transmitter and the equivocation :

$$R(X_s, Y_s) = H(X_s) - H(X_s \mid Y_s)$$

$$= \sum_{g_1 \in \Lambda} \sum_{g_2 \in \Lambda} P(X_s = g_1, Y_s = g_2) \log \frac{P(X_s = g_1, Y_s = g_2)}{P(X_s = g_1)P(Y_s = g_2)}$$

$$= \sum_{g_1 \in \Lambda} \sum_{g_2 \in \Lambda} f_{xy}(g_1, g_2) \log \frac{f_{xy}(g_1, g_2)}{f_x(g_1)f_y(g_2)} \ . \tag{4}$$

The logarithmic term is known as the mutual information, and the system mutual information is thus the mathematical expectation of the former. It can be easily shown that

$$R(X_s, Y_s) = H(X_s) + H(Y_s) - H(X_s, Y_s) \ .$$

The system mutual information is, therefore, symmetric : $R(X_s, Y_s) = R(Y_s, X_s)$. It measures the quantity of information that has actually been transmitted from the transmitter to the receiver. In other words, it is a measure of the quantity of information that is common to the transmitter and the receiver. It constitutes a measure for the redundancy of two images. More precisely, it represents the reduction in the minimum number of bits needed to code a grey level of a site in an image due to the observation of the grey level of the corresponding site in another image, knowing their joint histogram. We will use the word **redundancy** rather than system mutual information when we are in an image model context.

The definition of image redundancy proposed in (4), only takes into account the two images' joint histogram, (f_{xy}), that is, just the pixel to pixel correspondence and not the spatial neighbourhood. However, our goal is to use the

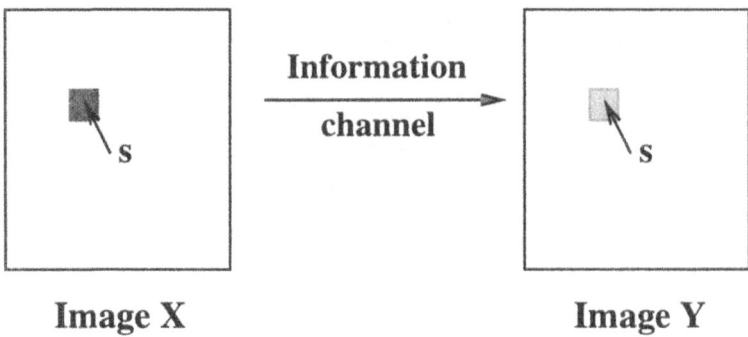

Fig. 1. The isolated pixels model

measure of the redundancy in an image fusion process for classification. In order to obtain a good classification we will require to consider not only the isolated pixels but also their neighbourhood. That is why we need a definition of redundancy that takes into account the neighbourhood of a pixel. The next chapter introduces such a model.

Note that (4) does not take into account the neighbourhood in the grey level space (Λ) either. For example the redundancy of the following two images

$$
\begin{array}{|c|c|}
\hline
23 & 23 \\
\hline
23 & 23 \\
\hline
\end{array}
\qquad
\begin{array}{|c|c|}
\hline
72 & 72 \\
\hline
72 & 73 \\
\hline
\end{array}
$$

is the same as the one of

$$
\begin{array}{|c|c|}
\hline
23 & 23 \\
\hline
23 & 23 \\
\hline
\end{array}
\qquad
\begin{array}{|c|c|}
\hline
72 & 72 \\
\hline
72 & 10 \\
\hline
\end{array}
$$

In chapter 4.2 we define a measure of the redundancy that depends on a distance in the grey level space.

3 The MRF Model

3.1 A General Model

We want to take into account a pixel's neighbourhood and choose naturally a Markov Random Field model. This model was applied to image processing for the first time in the early 1970's [1], and became very popular after the publication of a famous article by the Geman brothers in 1984 [2].

Hamming has defined the entropy of an mth order Markov chain [3]. For a stochastic process $\{t_i\}$, he defines the information produced by the symbol t_i when in the state $(t_{i-m}, \ldots, t_{i-1})$:

$$
I(t_i \mid t_{i-m}, \ldots, t_{i-1}) = \log \frac{1}{p(t_i \mid t_{i-m}, \ldots, t_{i-1})} \qquad (5)
$$

As stated in the introduction, we consider images where the sites $s \in S = \{s_1, ..., s_N\}$ have grey levels $x_s \in \Lambda = \{g_1, ..., g_M\}$. We introduce a homogeneous neighbourhood system $V = \{V_s, \quad s \in S\}$ and use the notation $\nu = |V_s|$ and $s_1, s_2, ..., s_\nu$ for the neighbours of s. To simplify we will consider a homogeneous Markov Random Field.

We can then extend Hamming's definition to our Markov field by defining the **information produced by the grey level of a site knowing the grey levels of the neighbouring sites** :

$$I(x_s \mid x_r, \ r \in V_s) = \log \frac{1}{P(X_s = x_s \mid x_r, \ r \in V_s)} . \tag{6}$$

We define the conditional entropy on Λ as

$$H(X_s \mid x_r, \ r \neq s) = \sum_{g \in \Lambda} P(X_s = g \mid x_r, \ r \in V_s) \log \frac{1}{P(X_s = g \mid x_r, \ r \in V_s)} . \tag{7}$$

Then we define the entropy of X_s as the average for all possible configurations $g = (g_1, ..., g_\nu)$:

$$H(X_s) = \sum_{g \in \Lambda^\nu} P[(X_{s_1}, ..., X_{s_\nu}) = g] H(X_s \mid x_r, \ r \neq s) \tag{8}$$

$$= \sum_{g \in \Lambda^\nu} \sum_{g \in \Lambda} P[(X_{s_1}, ..., X_{s_\nu}) = g, X_s = g] \log \frac{1}{P(X_s = g | x_r, \ r \in V_s)} \tag{9}$$

We choose the normalized frequencies as probability distributions :

$$P[(X_{s_1}, ..., X_{s_\nu}) = g] = f_\nu(g) = \frac{1}{N} \sum_{s \in S} \delta(x_{s_1} - g_1) ... \delta(x_{s_\nu} - g_\nu)$$

$$P[(X_{s_1}, ..., X_{s_\nu}) = g, X_s = g] = f_{\nu+1}(g, g) = \frac{1}{N} \sum_{s \in S} \delta(x_{s_1} - g_1) ... \delta(x_{s_\nu} - g_\nu) \delta(x_s - g) .$$

From now on, functions f will always be normalized, possibly multi-dimensional grey level frequencies. To make the formulas more readable, we will no longer put an index for the dimension of the argument when the arguments are explicit :

$$f(g) = f_\nu(g) \quad , \quad f(g, g) = f_{\nu+1}(g, g) \quad etc.$$

We get the expression for the image entropy :

$$H(X_s) = - \sum_{g \in \Lambda^\nu} \sum_{g \in \Lambda} f(g, g) \log \frac{f(g, g)}{f(g)} . \tag{10}$$

It represents the number of bits required to code a site's grey level when the grey levels of the neighbouring sites are observed, knowing the image's grey level configuration frequency on a site and its neighbours.

For the simple neighbourhood system of four nearest neighbours and $M = |\Lambda| = 256$, we get the following formula :

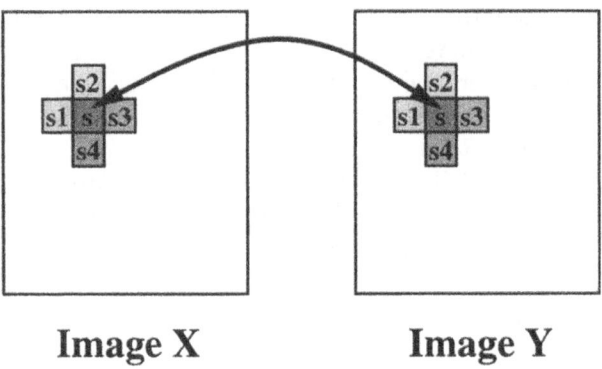

Image X   **Image Y**

Fig. 2. The MRF model

$$H(X_s) = -\sum_{g=0}^{255}\sum_{g_1=0}^{255}\sum_{g_2=0}^{255}\sum_{g_3=0}^{255}\sum_{g_4=0}^{255} f_x(g_1, g_2, g_3, g_4, g) \log \frac{f_x(g_1, g_2, g_3, g_4, g)}{f_x(g_1, g_2, g_3, g_4)} \ .$$

$$(11)$$

We have 256^5 possible realizations of $(X_{s_1}, X_{s_2}, X_{s_3}, X_{s_4}, X_s)$!

The problem is not of computational order as there are only N realizations to be considered. However, for f_x to have a sense as a probability distribution we must have a reasonable number of realizations compared to the number of possible values. For small (compared to 256^5 !), inhomogeneous images, each possible configuration of four grey levels on a site's neighbours will most often occur just once or not at all. A site's grey level will therefore almost always be entirely determined by the knowledge of its neighbours and the entropy will be 0 or at least very low. For the redundancy the problem is even of a higher dimension than for the entropy.

We must therefore reduce the space of possible values. For this we will exploit one of the most important properties of a Markov Random Field that follows from the Hammersley-Clifford theorem [7]. This will enable us to derive a formula for the image entropy which is strictly equivalent to (10) and (11) but contains other conditional probabilities which can be replaced by grey level frequencies on a smaller sample space. A similar expression for the image redundancy is derived easily.

First we associate a clique system to our previously defined neighbourhood system :

$$c = (s_1, \ldots, s_\gamma) \in C \Leftrightarrow s_1, \ldots, s_\gamma \text{ are neighbours two by two}$$

It follows from the Hammersley-Clifford theorem that the conditional probability of a site's grey level with respect to its neighbours is proportional to the exponential of the sum of the potentials of its associated cliques :

$$P(X_s = x_s \mid x_r, \; r \in V_s) = \frac{1}{Z_s} \exp(-U_s(x)) \tag{12}$$

with the local energy given by : $U_s(x) = \sum_{c \mid s \in c} V_c(x_c)$ (13)

where x_c is the restriction of x to c, V_c is called a potential function and Z_s is a normalizing constant. There is, therefore, a one-to-one relation between $P(X_s = x_s \mid x_r, \; r \in V_s)$ and $U_s(x)$.

We can see from (13) and (12) that different grey level configurations on the neighbours of s may cause the same value of $U_s(x)$, that is, the same conditional probability for the grey level x_s of s. We will call such a set of configurations a *state* which will be depicted by α. It is shown in [9] that the entropy can then be written

$$H(X_s) = - \sum_{g \in \Lambda} \sum_{\alpha} p(\alpha, g) \log p(g \mid \alpha) \tag{14}$$

where $p(\alpha, g)$ is the joint probability of a site having grey level g and its neighbours being in state α, and $p(g \mid \alpha)$ is the conditional probability of a site having grey level g knowing that its neighbours are in state α. Approximating these probabilities by normalized frequencies of occurrence of states $f_x(\alpha)$ and of joint occurrences of states and grey levels $f_x(\alpha, g)$ we get the final expression for the image entropy :

$$H(X_s) = - \sum_{g \in \Lambda} \sum_{\alpha} f_x(\alpha, g) \log \frac{f_x(\alpha, g)}{f_x(\alpha)} \; . \tag{15}$$

Comparing the above equation with (10) and (11) we can see that we are now summing over all states α instead of all grey level configurations on four sites.

In a similar way it can be shown [9] that the image redundancy equals

$$R(X_s, Y_s) = \sum_{g_x \in \Lambda} \sum_{g_y \in \Lambda} \sum_{\alpha_x} \sum_{\alpha_y} p(\alpha_x, g_x, \alpha_y, g_y) \log \frac{p(g_x, g_y \mid \alpha_x, \alpha_y)}{p(g_x \mid \alpha_x) p(g_y \mid \alpha_y)} \; . \tag{16}$$

Again we approximate the probability distributions by normalized frequencies of occurrence of states in each image, of joint occurrence of states in the two images and of joint occurrence of states and grey levels in each image and in the two of them to get the final expression for the redundancy of two images :

$$R(X_s, Y_s) = \sum_{g_x \in \Lambda} \sum_{g_y \in \Lambda} \sum_{\alpha_x} \sum_{\alpha_y} f_{xy}(\alpha_x, g_x, \alpha_y, g_y) \log \frac{f_{xy}(\alpha_x, g_x, \alpha_y, g_y) f_x(\alpha_x) f_y(\alpha_y)}{f_{xy}(\alpha_x, \alpha_y) f_x(\alpha_x, g_x) f_y(\alpha_y, g_y)}$$

$$\tag{17}$$

Like in the expression for the entropy, we now sum over all states instead of over all possible configurations. In [9] it is shown that for a first order homogeneous isotropic MRF the number of states is effectively smaller than the number of configurations of grey levels on the neighbours, but that this reduction of the sample space is not sufficient. In order to further reduce the sample space, we will introduce constraints on the potential V which was until now an arbitrary symmetric function (with respect to its arguments). In other words, we will determine parameters of the MRF model. The entropy and the redundancy will therefore be model dependent, and the choice of a specific model should be based on how these measures will be exploited.

3.2   The Generalized Ising Model

A model which is frequently used for the partitioning of homogeneous regions is the Generalized Ising model :

$$V(x_s, x_r) = \beta(1 - 2\delta(x_s, x_r)) \quad \beta \neq 0 \ .$$

With the four nearest neighbours system (first order MRF) this implies

$$\mathcal{A} = \{-\beta, \beta\} \ |\mathcal{A}| = 2$$
$$\mathcal{B} = \{-4\beta, -2\beta, 0, 2\beta, 4\beta\} \ |\mathcal{B}| = 5 \ .$$

We may consider only 5 states instead of 256^4 configurations. These states α_i are given by : i neighbours have the same grey level as s.

The entropy of an image with respect to the Generalized Ising model is :

$$H(X) = -\sum_{g \in \Lambda} \sum_{i=0}^{4} f(\alpha_i, g) \log \frac{f(\alpha_i, g)}{f(\alpha_i)} \tag{18}$$

and the redundancy of two images :

$$R(X, Y) = \sum_{g_x \in \Lambda} \sum_{g_y \in \Lambda} \sum_{i=0}^{4} \sum_{j=0}^{4} f_{xy}(\alpha_i, g_x, \alpha_j, g_y) \log \frac{f_x(\alpha_i) f_y(\alpha_j) f_{xy}(\alpha_i, g_x, \alpha_j, g_y)}{f_x(\alpha_i, g_x) f_y(\alpha_j, g_y) f_{xy}(\alpha_i, \alpha_j)} \tag{19}$$

Equation (19) is a definition of redundancy of two images that takes into account the spatial neighbourhood. The sample space is reduced so that images that are large with respect to the number of grey levels (for example 256×256 pixels and 256 grey levels) provide enough statistical data.

4   The Clique-vector Model

4.1   A General Model

We have just proposed a measure for the redundancy of two images within a Markovian framework. When we determine the form of the potential functions of the Markov field, for example the Generalized Ising model, most images provide enough statistical data.

We will now present another approach and propose a measure which is less model dependent. Actually it is perhaps the most natural extension of the Isolated-pixels model when we want to take into account the neighbourhood of a pixel. Instead of the pixel-to-pixel correspondence, we consider the clique-to-clique correspondence. In the case of only first degree cliques, this model is equivalent to the Isolated-pixels model. We will use here second degree cliques, that is, couples of pixels that are immediate neighbours. This corresponds to a first order MRF as in the previous chapter. Given an image X, we define the

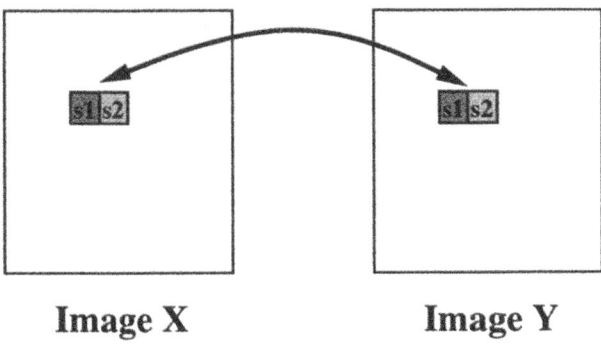

Image X   **Image Y**

Fig. 3. The Clique-vector model

stochastic variable V with sample space Λ^2 and the normalized grey level couple frequency f_x as probability distribution :

$$\forall g \in \Lambda^2 \; : \; P(V = g) = f_x(g) = \frac{1}{|C|} \sum_{c \in C} \delta(\boldsymbol{x}_c - \boldsymbol{g})$$

We define the entropy of the image as the entropy of V :

$$H(X) = H(V) = - \sum_{g \in \Lambda^2} f_x(g) \log f_x(g) \tag{20}$$

Associating another stochastic variable W to an image Y, we define the redundancy of the images X and Y as the system mutual information of V and W :

$$R(X,Y) = R(V,W) = \sum_{g_1 \in \Lambda^2} \sum_{g_2 \in \Lambda^2} f_{xy}(g_1, g_2) \log \frac{f_{xy}(g_1, g_2)}{f_x(g_1) f_y(g_2)} \tag{21}$$

(20) and (21) define an entropy and a redundancy that take into account the spatial neighbourhood of a pixel. They do not depend on a specific Markovian model although choosing vectors (cliques) of two pixels corresponds to an isotropic first order MRF. These measures are meaningful for large images.

To illustrate the importance of the image size, we have applied the above defined measures, and another one which is introduced in chapter 4.2, to random images of different sizes and different number of grey levels (Figs. 4 and 5). Actually, we present the *relative* redundancy which is the redundancy divided by the entropy of the images and expressed in percentage. We will motivate this in chapter 5.2. As it should be very difficult to predict a random image knowing another random image, we want the redundancy to be close to zero. This is the case for the Isolated Pixels and the Generalized Ising models for current image sizes, but not for the Clique-vector model defined in this chapter. This motivates the introduction below of a distance in the grey level space. However, the couple of random images is the worst possible case for the Clique-vector model because the grey level vectors are uniformly distributed. In Sect. 5.2 we will see that for a couple of a random image and a satellite image the problem is greatly reduced and the redundancy is low for all the models.

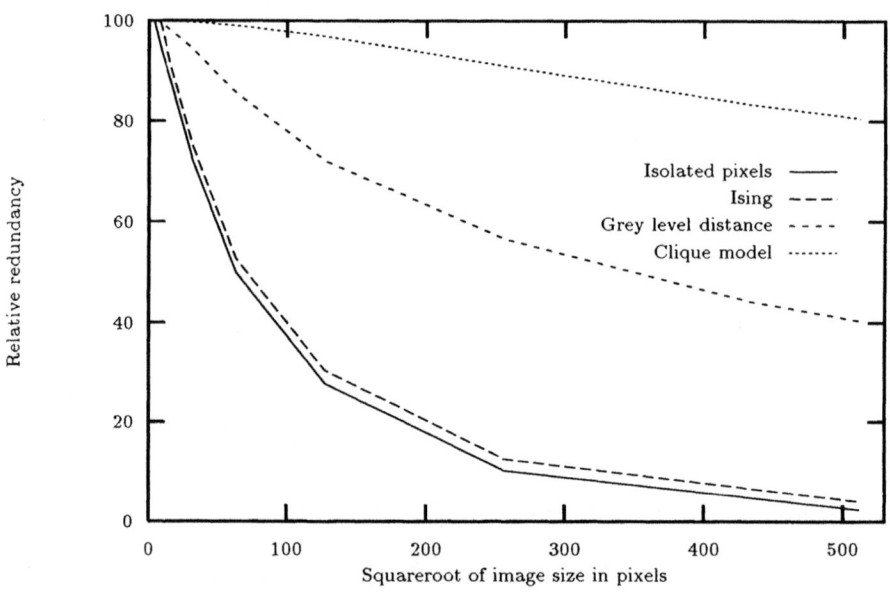

Fig. 4. Relative redundancy as a function of image size for different methods. Random images of 256 grey levels.

4.2 The Grey Level Distance Model

To assure that the measure is meaningful for smaller images, we propose to introduce a distance in the grey level space. We start by representing a clique

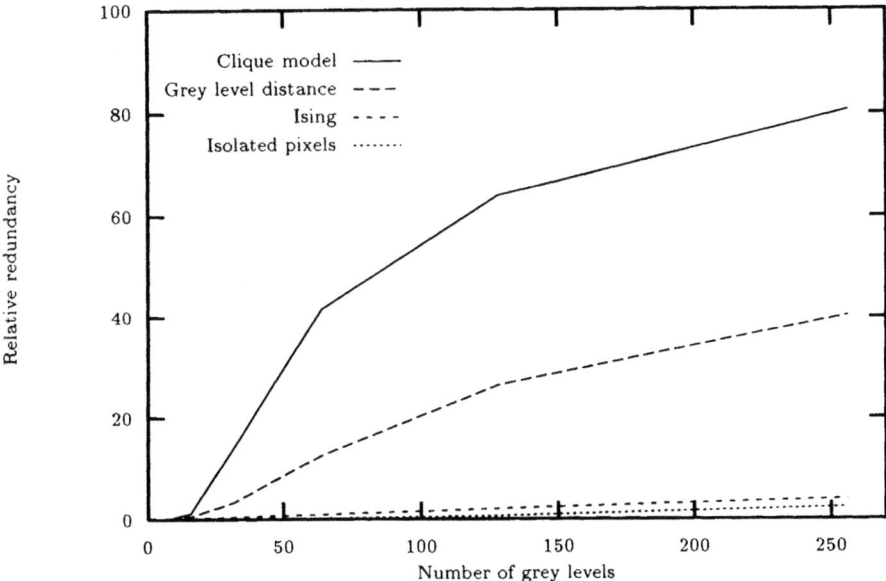

Fig. 5. Relative redundancy as a function of the number of grey levels for different methods. Random images of 512×512 pixels.

by its first and second order moments : the mean and the variance. In order to reduce the sample space we have to reduce their resolution. Since the mean is more robust to noise than the variance, we prefer to keep a higher resolution for the former. When reducing the resolution of the variance, we want a finer discrimination for small variances than for larger ones, which is conform to human vision.

For example we say that the difference between $x_c = (17, 236)$ and $x_c = (16, 237)$ is not very significant and certainly less significant than the difference between $(17, 17)$ and $(16, 18)$. That is, if two cliques of image X of the same value $(189, 54)$ correspond to two cliques of image Y of values $(17, 236)$ and $(16, 237)$, we consider that there is a redundancy, because in both cases a certain gradient in X corresponds to a very strong gradient in Y, even though they are not exactly the same (Fig. 6). The corresponding terms in the expression for the redundancy should therefore provide a contribution to the increase in the latter.

On the contrary, if the same two cliques of X correspond to two cliques of Y of values (17,17) and (16,18), we consider that there is no redundancy, because a certain gradient corresponds to a zero gradient in one case and a small, but non zero gradient in the other (Fig. 7). We do not want this to increase the redundancy.

These presumptions mean that, for example, white additive noise can be eliminated at cliques with a strong gradient, but not at cliques with a weak

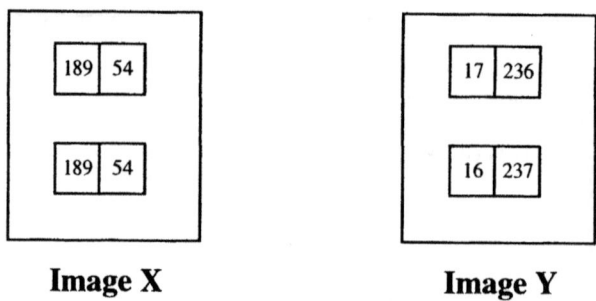

Fig. 6. Increase in the redundancy.

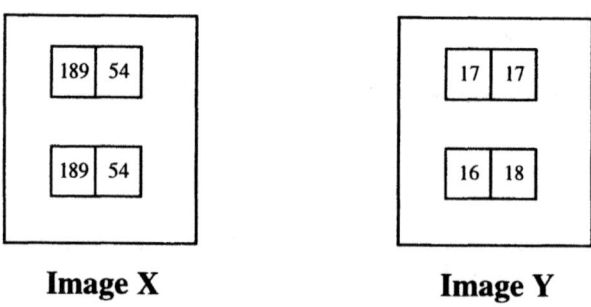

Fig. 7. No increase in the redundancy.

gradient. We do not know if a small difference between two weak gradients is due to noise or not.

We, therefore, propose to replace the grey level couple of the clique by its mean and a concave function of its gradient, for example a logarithm :

$$x_c = (x_{s_1}, x_{s_2}) \longrightarrow z_c = (m_c, d_c)$$

with $m_c = \mathrm{TR}\left(\dfrac{x_{s_1} + x_{s_2}}{2}\right)$ and $d_c = \dfrac{x_{s_1} - x_{s_2}}{|x_{s_1} - x_{s_2}|} \mathrm{TR}\left(\log_2(|x_{s_1} - x_{s_2}| + 1)\right)$

with TR representing rounding a real number to its nearest smaller integer (i.e. truncating).

Using $z_c = (m_c, d_c)$ as the new clique representation we get the following expressions for the entropy and the redundancy :

$$H(X_c) = -\sum_{m=0}^{255} \sum_{d=-7}^{7} f(m, d) \log f(m, d) \tag{22}$$

$$R(X_c, Y_c) = \sum_{m=0}^{255} \sum_{d=-7}^{7} \sum_{m=0}^{255} \sum_{d=-7}^{7} f(m_x, d_x, m_y, d_y) \log \frac{f(m_x, d_x, m_y, d_y)}{f(m_x, d_x) f(m_y, d_y)} \tag{23}$$

where $\quad f(m,d) = \dfrac{1}{|C|} \displaystyle\sum_{c \in C} \delta(m_c - m)\delta(d_c - d)$

$$f(m_x, d_x, m_y, d_y) = \frac{1}{|C|} \sum_{c \in C} \delta(m_{x_c} - m_x)\delta(d_{x_c} - d_x)\delta(m_{y_c} - m_y)\delta(d_{y_c} - d_y)$$

Similarly to (21), equation (23) defines a measure for the redundancy of two images based on a general, but first order, MRF model and therefore takes into account a pixel's spatial neighbourhood. The difference with (21) is that it also takes into account a distance in the grey level space, and thereby reduces the sample space. Consequently it may be applied to smaller images.

5 Results

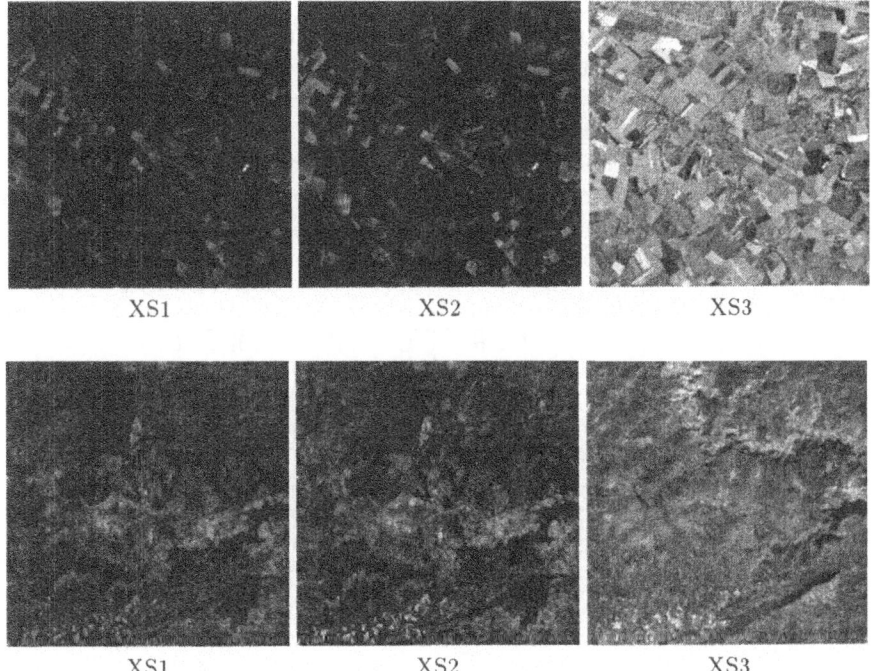

XS1　　　　　　　　XS2　　　　　　　　XS3

XS1　　　　　　　　XS2　　　　　　　　XS3

Fig. 8. Two of the 55 satellite image scenes

5.1 Entropy

We have computed entropies of synthetic and satellite grey level images (Table 1). The 512×512 pixels satellite images (Fig. 8) have been extracted from larger images of 256 grey levels. The entropy can, therefore, be compared to the maximum entropy for a variable with 256 possible realizations, i.e. 8.

However, each extracted 512×512 pixels image typically contains only 70 to 130 grey levels within an interval of length ranging from 80 to 160. For example, an image containing 128 different grey levels cannot have an entropy exceeding 7 for the Isolated-pixels model and the Generalized Ising model. It may then be interesting to compare the entropy of an image to the maximum entropy for the number of grey levels contained in the image, $H_{max} = H_{max}(number\ of\ grey\ levels)$. We have presented the relative entropy, $H_{rel} = H/H_{max}$ in percentage of H_{max} in Table 2.

Actually, the entropy of an image essentially measures two factors : the number of grey levels and the shape of their relative frequency distribution. (Strictly speaking this is valid only for the Isolated-pixels model; the term *number of grey levels* should be replaced by the number of grey level couples for the Clique-vector model and of couples (state, grey level) for the Generalized Ising model.) The relative entropy, H_{rel}, is a measure of the contribution to the entropy due to the shape of the histogram, whereas the maximum entropy, H_{max}, measures the part due to its length (area). When using the terms *number of grey levels* or *length of histogram* we have two choices. We can either use the *number* of grey levels effectively occurring, or the *length of the smallest interval containing them*. We have chosen the length of the interval, considering non occurring grey levels within this interval as being part of the shape of the histogram, and not of its length.

	H Isolated pixels	H Ising	H Cliques	H Grey level distance
XS1	4.49	4.36	7.41	7.22
XS2	4.94	4.79	8.16	7.87
XS3	6.01	5.97	10.24	9.75
Random	8.00	8.00	15.90	11.80
Checkerboard	1.00	1.00	1.00	1.00

Table 1. Mean values for the entropies of 3×55 SPOT images

Note first that for the Isolated Pixels model the entropy of a random image of 256 grey levels is 8. Each grey level has the same probability, and we need $log_2(256) = 8$ bits to code it.

Since the knowledge of the state of the neighbours is of no help for the prediction of the grey level of a pixel in our random image, its entropy is also equal to 8 for the Generalized Ising model.

The Clique-vector model gives an entropy of 15.90 ($H_{rel} = 99\%$), instead of $2 \times 8 = 16$ bits, which we would expect. The reason is that a 512×512 pixels image does not provide enough statistical data. Theoretically, we should have a uniform distribution on the whole sample space for a random image. Since the sample space here is Λ^2 whose cardinal is 256^2, the image must be at least 182×182 pixels (which gives 256^2 second order cliques) so that one could *possibly* have a 100% relative entropy, and must be much larger to guarantee this. We found an entropy of 15.98 for a 1024×1024 pixels random image.

The model introducing a grey level distance gives an entropy of only 11.80 bits (90%) for the random image. This entropy measures the number of bits required to code the sum and a logarithm (base 2) of the difference of two grey levels. The relative entropy is far from 100% (corresponding to 13.09 bits) since neither the sum nor the logarithm of the difference are uniformly distributed. The theoretical value is 12.09 bits.

All models give rather homogeneous values for the entropies of the satellite images.

5.2 Redundancy

After a study of the behavior of the redundancy as a function of the number of grey levels (i.e. the number of significant bits), we decided to use SPOT images whose number of grey levels has been reduced by a factor 6 (≈ 5.5 significant bits of the original 8).

We have computed the redundancies of SPOT XS image couples, firstly for couples of two images of the same scene coming from two different channels (cf. Table (3)). Then we used couples of two images from the same channel with one scene being the one-pixel translation of the other (cf. Table (4)). Finally, Table (5) shows the redundancies for couples of a SPOT XS image and a random image of 128 grey levels.

The values of the redundancy alone have little meaning. They should be compared to the individual entropies. In order to obtain a single meaningful symmetric expression , we have chosen to present below the redundancy of two images divided by the smaller of the individual entropies, given as a percentage of the latter . We refer to this as the *relative redundancy* : $R_{rel}(X,Y) = R(X,Y)/min(H(X),H(Y))$. In this way a 100% relative redundancy will occur when all the information in the image with the smaller entropy is contained in the other image.

The SPOT XS1 and XS2 images, which look very similar, have redundancies between 53% and 65% depending upon the model (cf. Table (3)). The XS1-XS3 and XS2-XS3 couples have much lower redundancies, between 16% and 22%.

For all these SPOT XS image couples, the Generalized Ising model gives the highest values for the redundancy. The three other models give very similar values, except for the XS1-XS2 couples where the Isolated Pixels model gives stronger values.

The Generalized Ising model gives considerably higher redundancies for an image and its one-pixel translation (between 59% and 71%) than the other models (cf. Table (4)). The three other models give very similar values in this case, too (between 44% and 49%).

All models give very low redundancies for a random image and a SPOT XS image (cf. Table (5)). The Isolated Pixels model gives the lowest ones (0.05%), whereas the Clique-vector model gives the highest ones due to insufficient statistical data.

We have also presented the absolute value of the classical correlation coefficient in Tables (3), (4) and (5). The values for an image and its one-pixel translation are very high, as we wanted for our measure. The high values for a couple of SPOT images from different channels, and the low ones for a SPOT image and a random image, correspond also to our intuition of a redundancy measure. However, the intra-scene *variances* of this measure for couples SPOT XS1-XS3 and XS2-XS3 on one hand and XS1-XS2 on the other hand, are very different (Table 6). It is low for XS1-XS2, and extremely high for XS1-XS3 and XS2-XS3. The correlation coefficient is therefore certainly not a good measure of *sensor* redundancy compared to the definitions of redundancy that we have proposed.

6   Conclusion

We have approached the problem of multi-sensor redundancy by defining an image redundancy. Information theory is the basis of our work. A simple model considering an isolated pixel allows a direct application of the definition of redundancy to images.

Our goal is to define a measure for the redundancy that can be exploited in a multi-sensor classification context. This process will not fuse isolated pixels but will take into account their spatial neighbourhood. We, therefore, want a measure for redundancy that do not neglect this. We have shown by two different approaches that the theory of Markov Random Fields constitutes an adequate framework for this. In order to reduce the sample space we must either determine parameters of the Markov field or introduce a distance in the grey level space.

We propose two measures for the redundancy of two images, the Generalized Ising Model (19) and the Grey Level Distance Model (23), and a third one, the Clique-vector Model (21), which should be used only on large (compared to the number of grey levels) images.

The Generalized Ising model gives much higher redundancies for an image and its translation (one pixel) than does the isolated pixels model, which is what we wanted. In an image fusion process, we will often have slight errors in the geometric pixel to pixel correspondence of the images. We do not want these errors to drastically reduce the redundancy.

The Generalized Ising model also gives a higher redundancy for two SPOT images from different channels. This is also what we desired intuitively, as these images are very similar, to the human eye. This model gives higher redundancies between a SPOT image and a random image than does the Isolated Pixels model, but still very low values.

The mean values of the correlation coefficient corresponds also to what we wanted from a redundancy measure. However, the variances for 512×512 pixels image couples within a same larger image are very different for the three XS couples from different channels, and extremely high for two of them. We want a measure that gives rather homogeneous values for image couples of similar, geographically close physical environments taken with the same sensor couple at the same time, for example for SPOT XS1-XS3 image couples of 512×512 pixels taken from a SPOT XS1-XS3 image couple of 2000×2000 pixels of a homogeneous scene. The correlation coefficient satisfies poorly this criterion compared to our proposed redundancy definitions.

The proposed measures of image redundancy can be used in a multi-sensor fusion process. When there are more images available than we want to use in the process, these measures can be of help in selecting the optimal subset of images/sensors according to the priorities (robustness or minimization of the loss of details). They can also be applied within a classification process either by maximization of the redundancy of the classification and the image(s) [10], or locally as measures of the probability of a region belonging to a class. The image redundancy measures also constitute general tools for quantifying the "closeness" of images which in many contexts may be more suitable than the subtraction of two images or their correlation coefficient. This is the case, for example, when image structures above pixel level should not be neglected, or when there are slight errors in the correspondence between the images.

References

1. J. E. Besag. *Spatial Interaction and the Statistical Analysis of Lattice Systems (with Discussion)*. Journal of Royal Statis. Society B., pages 192–236, 1974.
2. S. Geman and D. Geman. *Stochastic Relaxation, Gibbs Distributions and the Bayesian Restoration of Images*. PAMI, 6:721–741, 1984.
3. Richard W. Hamming. *Coding and Information Theory*. Prentice Hall, 1980.
4. Stéphane Houzelle and Gérard Giraudon. *Contribution to multisensor fusion formalization*. Robotics and Autonomous Systems, 13:69–85, 1994.
5. Gabriele Lohmann. *Co-occurrence-based Analysis and Synthesis of Textures*. In 12th International Conference on Pattern Recognition, volume 1, pages 449–453, Jerusalem, October 1994.
6. Henri Maître. *Entropie, information et image - Partie 2*. Technical Report 94 D 006, Ecole Nationale Supérieure des Télécommunications, February 1994.
7. J. Moussouris. *Gibbs and Markov Random System with Constraints*. Journal of Statistical Physics, 10(1):11–33, January 1974.
8. C.E. Shannon. *A Mathematical Theory of Communication*. The Bell System Technical Journal, 27:379–423, July 1948.
9. Espen Volden, Gérard Giraudon, and Marc Berthod. *Modelling image redundancy*. Technical Report 2440, INRIA, December 1994.
10. Espen Volden, Gérard Giraudon, and Marc Berthod. *Image redundancy and Classification*. In Computer Analysis of Images and Patterns, 6th International Conference, CAIP'95, Lecture Notes in Computer Science 970 , pages 206–213, Prague, September 1995.

	Isolated pixels		Ising		Cliques		Grey level distance	
	H_{max}	H_{rel}	H_{max}	H_{rel}	H_{max}	H_{rel}	H_{max}	H_{rel}
XS1	6.81	66	6.81	64	13.62	55	10.56	68
XS2	6.95	71	6.95	69	13.90	59	10.74	73
XS3	7.30	82	7.30	82	14.60	70	11.16	87
Random	8.00	100	8.00	100	16.00	99	13.09	90
Checkerboard	1.00	100	1.00	100	2.00	50	1.58	63

Table 2. Mean values for the maximum and relative entropies of 3×55 SPOT images

	Correlation coefficient	Isolated pixels	Ising	Cliques & Distance
	$\mid \rho \mid$	R_{rel}	R_{rel}	R_{rel}
XS1-XS2	91	58	65	53
XS1-XS3	33	16	21	16
XS2-XS3	39	17	22	17

Table 3. Mean values for the relative redundancies (in %) of 55 SPOT image couples.

	Correlation coefficient	Isolated pixels	Ising	Cliques & Distance
	$\mid \rho \mid$	R_{rel}	R_{rel}	R_{rel}
XS-Translated	91	47	66	48

Table 4. Mean values for the relative redundancies (in %) of 3×55 SPOT images and their one-pixel-translated image.

	Correlation coefficient	Isolated pixels	Ising	Cliques	Grey level distance
	$\mid \rho \mid$	R_{rel}	R_{rel}	R_{rel}	R_{rel}
XS-Random	0.16	0.04	0.55	1.7	0.93

Table 5. Mean values for the relative redundancies (in %) of 3×55 SPOT images and an image of 128 random grey levels

	Correlation coefficient	Isolated pixels	Ising	Cliques	Grey level distance
	$\mid \rho \mid$	R_{rel}	R_{rel}	R_{rel}	R_{rel}
XS1-XS2	5.0	11	17	9.1	9.1
XS1-XS3	153	21	31	15	15
XS2-XS3	238	23	32	18	18

Table 6. Mean values for the intra-scene variances of the relative redundancies of 55 SPOT image couples (in squared %).

Coding of Image Data via Correlation Filters for Invariant Pattern Recognition: Some Practical Results

Jonny Gauvin [1,3], Michel Doucet [1], Denis Gingras [1], and Paul Chevrette [2]

[1] National Optics Institute, Ste-Foy (Qc), Canada
[2] Defense Research Establishment Valcartier, Courcelette (Qc), Canada
[3] Now at Bomem Inc. Quebec (Qc), Canada

Abstract. The National Optics Institute is currently carrying out a project in *automatic target recognition* in order to locate, recognize, and track potential targets (e.g. tanks or troop carriers) monitored in real time by an infrared camera. In this paper, we present an algorithm based on the *Distance Classifier Correlation Filter* (DCCF), a subclass of the *Synthetic Discriminant Functions* family which are often used to solve the target recognition problem. We describe the general approach of DCCF-based template matching and the algorithms used. The effect of the *training set* on the discrimination performance, the optical implementation, and a few practical results are also presented. For a three-class recognition, only 3 misclassifications have been reported on a data bank of more than 200 images.

I   Introduction

Automatic target recognition (ATR) is a complex task due to both the environmental conditions and target aspects changing with time, as well as the requirements of recognizing targets with accuracy without false alarms, irrespective of orientation, or dimension [1]. This recognition problem can be broken down into five main tasks: image preprocessing, segmentation, data fusion, feature extraction, and classification [1]. Only the last two aspects are covered by this paper. They will be solved by using template matching and neural networks.

Template matching is a well known method [2] and is powerful only in a well controlled context involving little object distortion because they are very sensitive to geometric distortion. When the object has to be recognized from many points of view, a large number of matched filters is required since a single filter can only recognize a predetermined view of an object. For this reason, *Synthetic Discriminant Functions* (SDF) based algorithms [3] have been developed to significantly reduce the number of filters required by multiplexing on the same filter several views of the object classes. Unfortunately, only a small number of images can be multiplexed on the same filter without degrading the performance. Many algorithms derived from the original SDF have been developed in order to optimize the performance and the capacity of multiplexing of such filters depending on the kind of invariance they address [3]. The *Distance Classifier Correlation Function* (DCCF) is one of the most recent. This correlation based algorithm is well adapted to solve our ATR problem, since it allows invariance under different orientation of 3D objects.

II DCCF algorithm

In the DCCF solution, the composite correlation filters are optimized for a predetermined range of geometric distortions in order to obtain a similar response for all images of the same class of object. The DCCF-solution involves a set of filters (one for each class) which are optimized with a suitable training set. For classification, an image is compared by correlation to each of the filters in order to determine the best match. This paper does not detail the mathematical derivation of the DCCF algorithm; for a complete mathematical derivation, the reader will refer to Mahalanobis [4]. In this paper we present only the results needed for the generation of the correlation filters required for the classification of objects.

Let us consider a two-class problem involving two different military vehicles (Fig. 1). Each class has a set of images with different points of view of the vehicles to be classified; these pixeleted images have been acquired with an IR video camera. For mathematical convenience the lexicographic representation is used. The lexicographic representation of an $M \times N$ image is a $MN \times 1$ vector containing all the elements of the image taken line by line as shown in Fig. 2. The lexicographic form of an $M \times N$ object $o(n,m)$ is represented by ϱ, a vector having MN elements.

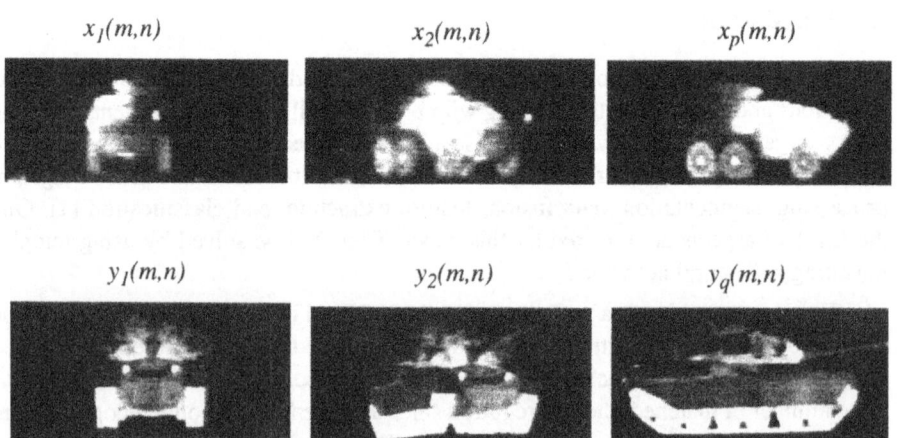

Fig. 1. Typical example of IR images of each class

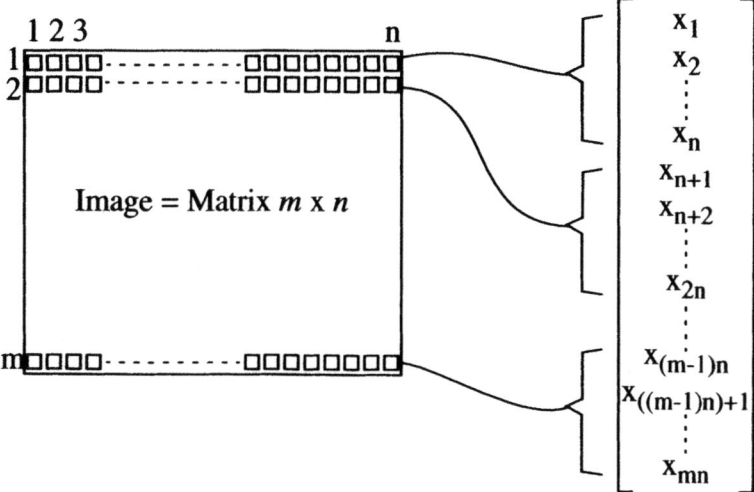

Fig. 2. Lexicographic representation of an image

The DCCF-solution is computed using the Fourier transform (FT) of the images of the training set; by convention the ^ symbol denotes the frequency domain. In a two-class problem, the DCCF-solution consists of a pair of filters. Let $\hat{\underline{x}}_1, \hat{\underline{x}}_2, ..., \hat{\underline{x}}_p$ and $\hat{\underline{y}}_1, \hat{\underline{y}}_2, ..., \hat{\underline{y}}_q$ be the lexicographic representations of the FT of images corresponding respectively to the training set of the first and second class. The means of the classes are defined to be the following vectors:

$$\hat{\underline{m}}_1 = \frac{1}{p}\sum_{k=1}^{p}\hat{\underline{x}}_k \quad , \qquad \hat{\underline{m}}_2 = \frac{1}{q}\sum_{l=1}^{q}\hat{\underline{y}}_l \quad . \tag{1}$$

The DCCF-solution is given by

$$\hat{\underline{h}}_1 = \left(\hat{S}^{-1}(\hat{\underline{m}}_1 - \hat{\underline{m}}_2)\right)\left(\hat{S}^{-1}(\hat{\underline{m}}_1 - \hat{\underline{m}}_2)\right)^{*}\hat{\underline{m}}_1 \quad , \tag{2}$$

$$\hat{\underline{h}}_2 = \left(\hat{S}^{-1}(\hat{\underline{m}}_1 - \hat{\underline{m}}_2)\right)\left(\hat{S}^{-1}(\hat{\underline{m}}_1 - \hat{\underline{m}}_2)\right)^{*}\hat{\underline{m}}_2 \quad . \tag{3}$$

where * symbol refers to the complex conjugate and $\hat{S}$ is a diagonal matrix defined by

$$\hat{S} = \frac{1}{p}\sum_{k=1}^{p}(\hat{X}_k - \hat{M}_1)(\hat{X}_k - \hat{M}_1)^{*} + \frac{1}{q}\sum_{l=1}^{q}(\hat{Y}_l - \hat{M}_2)(\hat{Y}_l - \hat{M}_2)^{*} \quad . \tag{4}$$

In the last expression the $\hat{X}_k$, $\hat{Y}_k$, $\hat{M}_1$, and $\hat{M}_2$ are the diagonals matrices corresponding to the vector $\hat{\underline{x}}_k$, $\hat{\underline{y}}_k$, $\hat{\underline{m}}_1$, and $\hat{\underline{m}}_2$ respectively. The DCCF-filters $\hat{h}_1(m,n)$, $\hat{h}_2(m,n)$ are obtained by rearranging $\hat{\underline{h}}_1$, $\hat{\underline{h}}_2$ into 2D images. The inverse FT $h_1(m,n)$ and $h_2(m,n)$ of the filters $\hat{h}_1(m,n)$, $\hat{h}_2(m,n)$ are used for the classification of the unknown input object $z(m,n)$. The correlation between an object $z(m,n)$ and both 2D functions $h_1(m,n)$ and $h_2(m,n)$ contains the information about the class to which the object belongs. As shown in the Fig. 3, the correlation plane of an input object belonging to the kth class is characterized by a high local maxima while the other correlation plane exhibits a local minimum at the corresponding point. To find such a max-min matching, the absolute maxima are found in both correlation planes and only the one with the best max-min correspondence is retained. The pair of value (d_1,d_2) corresponding to the absolute maximum and its associated minimum are the features used by the post-processing unit to decide whether the input object belongs to one of the classes or if the process is indecisive. It can be noted that this algorithm can be extended to a multi-class problem by taking into consideration all possible combinations of 2 classes. For P classes, $P(P-1)/2$ filter pairs are needed, which produce a signature having $P(P-1)$ scalars (features). Those features are inputs for the neural network used to classify the signatures.

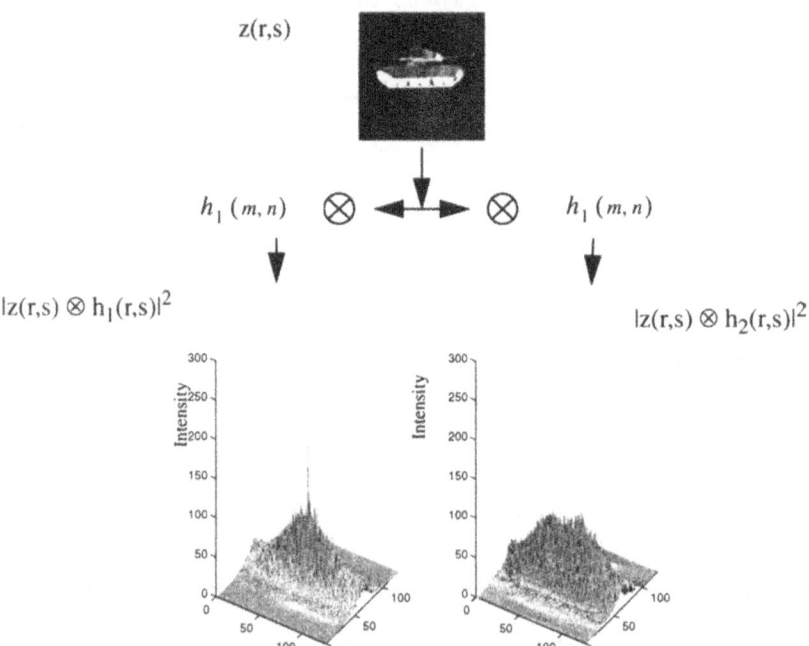

Fig. 3. Typical correlation peaks generated from filter pairs. These correlation planes are used to extract the distance parameters

III Effect of the training set on discrimination

The dimension of the training set (TS) for each class influences the capacity of discrimination of the neural network. If the TS does not contain good representations of the classes, a poor discrimination capability is expected. There are two factors that influence this capacity: the number of objects in the TS and the similarity between each object in the TS. In general, increasing the dimension of the TS in terms of possible number of viewpoints coded on a filter pair, increases the amplitude of cross-correlations and the false alarm rate. On the other hand, if there are not enough images in the TS for a given range of point of view, the similarity between objects outside the TS and the objects of the TS itself becomes too small for proper classification.

Fig. 4 shows the effects of the variation of similarity between the images of the TS on the correlation peaks. In each case, the testing set is composed of 72 views (0 - 360°) of the same target with 5° step between each point of view. We show that for a TS with 10° step, there is a high level of discrimination between the two classes. For a TS with 60° step there is some overlap between correlations peak with both filters therefore producing classification errors. Fig. 5 shows the effect of the discrimination in term of range of points of view. When the range increases, the peak-to-correlation energy ratio (PCE) decrease significantly; the PCE [5] is the ratio between the level of a correlation peak and the total energy contains in the correlation plane.

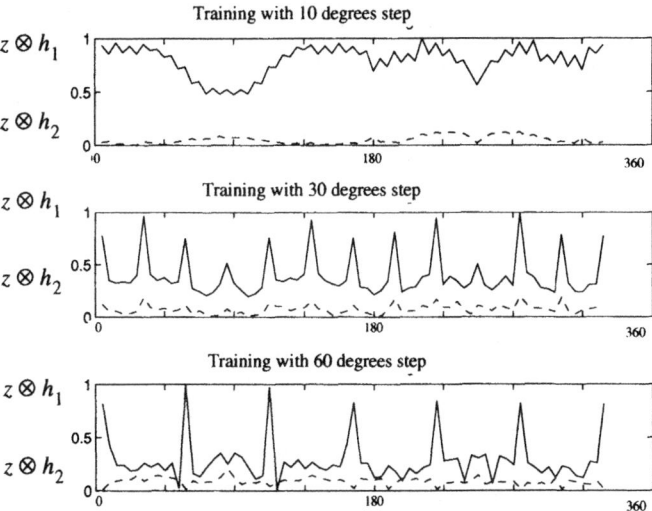

Fig. 4. Effect of the size of the training set on the correlation peak

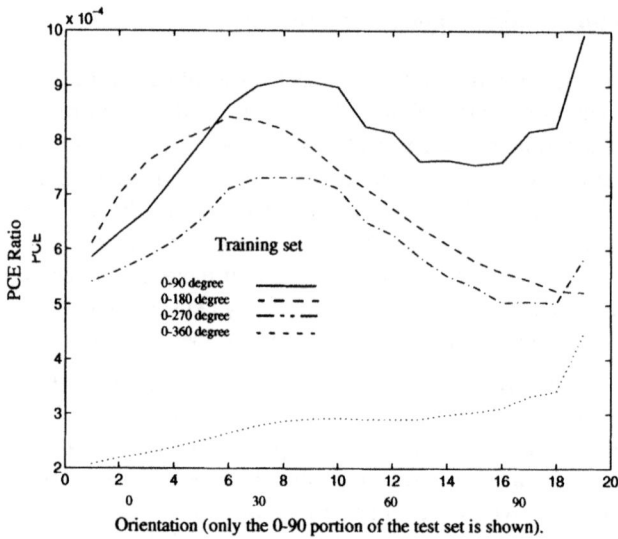

Fig. 5. Effect of the angular range of the training set on the correlation peak

IV Results

In this practical example, the data bank is composed of two classes each of them composed of 19 views of out-of-plane rotation (0 - 90° with constant elevation) acquired by an infrared camera (Fig. 1). The training set used to calculate the DCCF filters is composed of 10 of these images for each class (10° step). Fig. 3 shows a typical correlation output of one set of training images with DCCF filters. The distance parameters (d_1, d_2) are calculated from both correlation planes. Fig. 6 shows the classification map for this two-class problem. This result shows that both classes are well separated even for images outside of the training set.

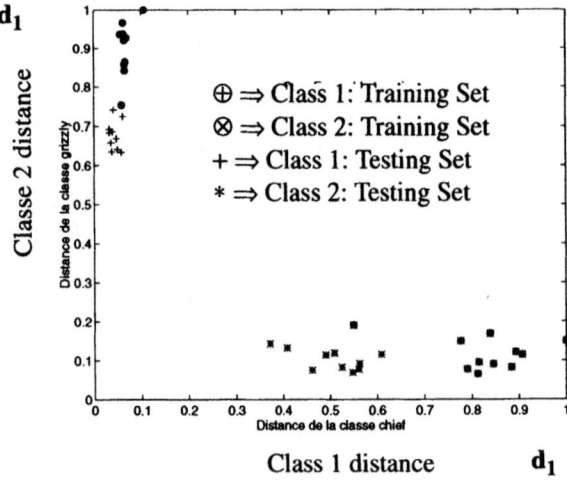

Fig. 6. Classification map for the two-class problem

Fig. 7 shows a more complicated case corresponding to a three-class problem with 0-360° variation in azimuth (with constant elevation). For more robustness, each class is divided into 4 subclasses, one for each 90° range of out-of-plane rotation. This simulation thus requires 12 pairs of composite filters for each combination of the two different classes and the 4 different 90° ranges of orientation. The step in the training sets is 10°. From each pair of filters the distance parameters are extracted to generate a signature of 24 elements. These parameters are used as inputs to a multi-layer perceptron [7] for classification. The multi-layer perceptron is trained using back-propagation learning rules [7] to classify the objects from the DCCF signatures. The neural network has 24 input neurons, 2 hidden layers, and 2 output neurons. Fig. 7 shows the values of the output neurons for the 216 test images. Only 3 misclassification has been obtained; they are enclosed in a circle on the figure.

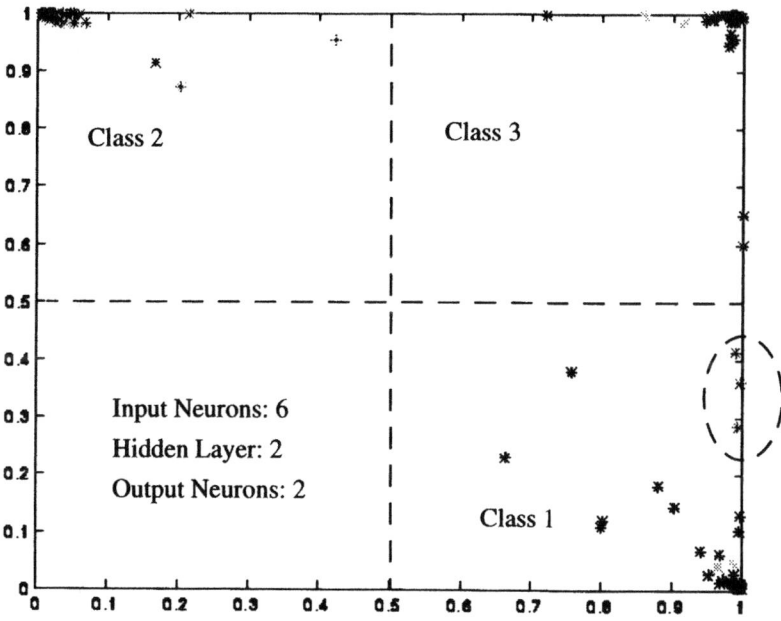

Fig. 7. Multi target classification; three-class problem (0 - 360°)

V Optical implementation

The DCCF is suitable for implementation on real-time optical correlators, since it uses Fourier transform techniques. Fig. 8 shows the basic optical architecture to implement an optical correlator [2]. The major drawback of the optical implementation of the DCCF approach is that the filter requires full complex modulation. Unfortunately, the currently available spatial light modulators (SLM) have a limited coding domain [8], and the DCCF filters have to be modified to fit the limited degrees of freedom availa-

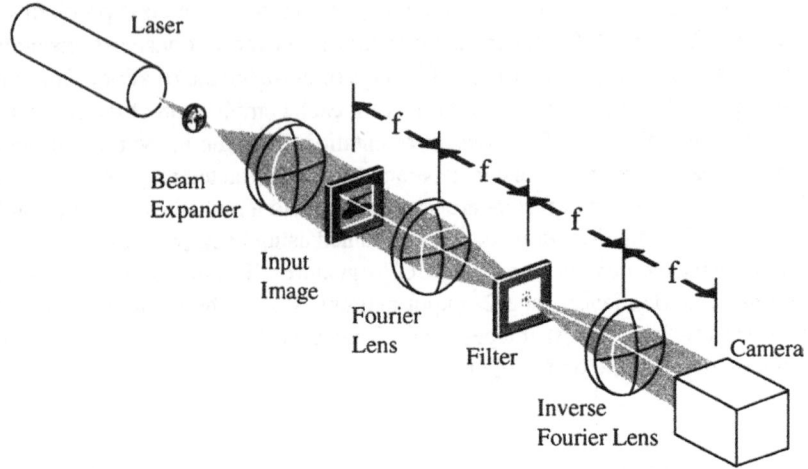

Fig. 8. Optical correlator layout

The Fig. 9 shows the correlation planes obtain optically. In this case, a DCCF filters pairs was optimized to discriminate shuttles from the other types of airplanes.

Airplanes   Shuttle

Fig. 9. Optical results of DCCF filters

ble. This can be achieved by the use a projection technique on the limited complex domain of the SLM, which has been adjusted in complex spiral mode for optimal results[8]. By using this technique, good classification results were achieved even if the number of degrees of freedom used to represent the filters had been reduced to only 32 complex levels. The SLMs used for this experience are the Kodak LC 500 [9].

VI Conclusion

The DCCF template matching has shown good potential for invariant pattern recognition, especially when out-of-plane rotation condition occur. Our simulations have demonstrated good results for the three-class problem using only 24 filters for a 360° off-plane rotation. This algorithm is suitable for optical implementation on correlator, but the currently available spatial light modulators do not yet have sufficient degrees of freedom in terms of complex response to fully implement these filters. The simulations have shown that the performance is sensitive to the training set selection. Further work will focus on the full optical implementation and the characterization of this approach in terms of robustness to noise and clutter.

References

1. Eustache, E., Gingras, D., Poussart, P.: Hybrid system for automatic target recognition. SPIE **2269** (1994) 280-291.
2. Goodman, J. W.: Introduction to Fourier optics. McGraw-Hill Book Company, New York, (1988).
3. Kumar, B. V. K. V. : Tutorial survey of the composite filter designs for optical correlators. Applied Optics **31** (1992) 4773-4801.
4. A. Mahalanobis et al.: Quadratic distance classifier for multi-class SAR ATR using correlation filters. SPIE **1875** (1993) 84-95.
5. Kumar, B. V. K. V., Hassebrook, L. : Performance measures correlation filters. Applied Optics **29** (1990) 2997-3006.
6. M.L Minsky, M.L., Papert, S.A.: Perceptrons. MIT Press, Cambridge, (1990).
7. Kröse, B. A., van der Smagt, P. P.: An introduction to neural networks, The University of Amsterdam, 1993.
8. Laude, V., Réfrégier, P.: Multicriteria characterization of coding domains with optimal Fourier spatial light modulator filters. Applied Optics **33** (1994) 4465-4471.
9. Bergeron, A. et al., Phase calibration and applications of a liquid-crystal spatial light modulator. Applied Optics **34** (1995) 5133-5139.

Improving Myoelectric Signal Classifier Generalization by Preprocessing with Exploratory Projections

Peter J. Gallant, Evelyn L. Morin and Lloyd E. Peppard

Department of Electrical and Computer Engineering
Queen's University, Kingston, Ontario, Canada K7L 3N6

Abstract. We present an approach to classifying myoelectric signals (MES) that involves distinct feature extraction and classification stages. The extraction of MES features is accomplished by a self-organizing neural network composed of BCM neurons that performs Exploratory Projection Pursuit. This approach produces a meaningful representation of the input data using a limited number of features which are usefully differentiable for classification. Classification is accomplished by a simple backpropagation network. This forms the basis for a myoelectric control system that exhibits a lower initial state selection error than other neural network based approaches which have recently been reported. This system offers enhanced functionality for a myoelectric prosthesis and simplifies user training through adaptation to the individual MES characteristics of each user.

1 Introduction

1.1 Objectives

The overall objective of this work is to develop a method of processing and classifying myoelectric signals so that a greater number of functions of a prosthetic device[1] can be controlled by the myoelectric signal (MES) associated with a voluntary muscle contraction. The usefulness of a myoelectric control system is closely related to the number of distinct *muscle contraction classes* which can be resolved from the input myoelectric signal and the degree to which the identified class membership of the signal corresponds to the desired signal generated by the user. Recent insight into the nonrandom behaviour of the myoelectric signal at the onset of a muscle contraction allows the application of advanced pattern analysis techniques to myoelectric control [1]. Further advances in neurophysiology and neural computation have suggested that advanced statistical analysis procedures can be carried out by artificial neural networks (ANN) which are modeled after functioning biological neurons [2] [3].

---

[1] The term "prosthetic device" as used here will refer to an electrically-powered artificial limb or a functional neuromuscular stimulation (FNS) system used to stimulate muscle activity in functionally compromised limbs (e.g. due to spinal cord injury).

In this work, a strategy for myoelectric control is developed which increases the functionality of myoelectric control systems by enhancing the reliability of control decisions and responsiveness to user actions. The adaptive nature of the control strategy will shift more of the burden of training from the user to the control system, thereby decreasing user training time and fatigue, which should lead to better patient acceptance of the prosthetic device.

1.2 The Steady-State Myoelectric Signal

DeLuca [4] describes the myoelectric signal as the electrical manifestation of the neuromuscular activation associated with a contracting muscle. Figure 1 shows two samples of the myoelectric signal taken when a muscle has reached a steady-state contraction level at: (a) a low-intensity contraction level (approximately 10 percent of maximum contraction level); and (b) a medium intensity level (50 percent of maximum value). Traditional *level-coded* approaches to myoelectric control were limited to an analysis of the signal variance under steady-state contraction conditions. This typically limited the classifier to at most three distinct control levels and did not allow for the identification of the type of muscle contraction giving rise to the observed variance of the MES.

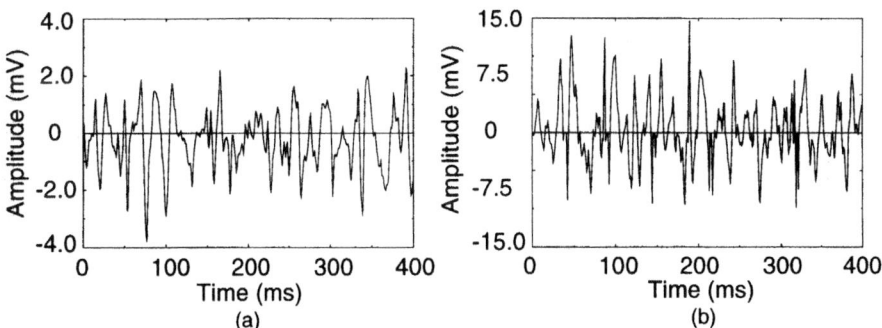

Fig. 1. Two samples of the myoelectric signal at: (a) 10 percent of maximum contraction level; and (b) 50 percent of maximum contraction level. Note the differing amplitude scales.

Because of the small magnitude of the action potential (1 - 300 μV as observed using bipolar skin surface electrodes), the myoelectric signal is normally amplified differentially to reduce the common-mode noise component. As typical MES bandwidths lie well below 1 kHz, the amplified signal is bandlimited by a low-pass filter to approximately 1 kHz. The filtering process does not significantly affect the information contained in the myoelectric signal and minimizes aliasing when sampled at the Nyquist rate.

1.3 The Transient Myoelectric Signal

The amplitude of the myoelectric signal has traditionally been regarded as a summation of numerous motor unit action potentials which has a normal distribution for steady-state muscle contractions. However, recent studies by Hudgins indicate that there is considerable deterministic structure in the myoelectric signal within the first 200 ms after the onset of a voluntary muscle contraction [1] [7]. Although a physiological explanation for this phenomenon has not yet been suggested, an analysis of variance (ANOVA) study on several amputee and non-amputee subjects suggests that this transient MES contains sufficient structure to allow for identification of the muscle action causing the observed MES and differentiation between different types of muscle contraction actions [7] . Figure 2 shows two samples of a 200 ms transient MES generated by a voluntary elbow flexion activity from a below elbow amputee (BEA) subject[2]. The data were acquired using a bipolar pair of stainless-steel electrodes placed on the skin surface, such that one active electrode was located over the distal third of the short head of the biceps brachii and the other over the distal third of the triceps brachii. This arrangement is designed for maximum pickup of myoelectric signals from active muscles in the upper arm and has been used in other studies such as those done by Graupe and Cline [8] and Kelly *et al.* [9]. Similarity between the samples is evident despite some noise interference.

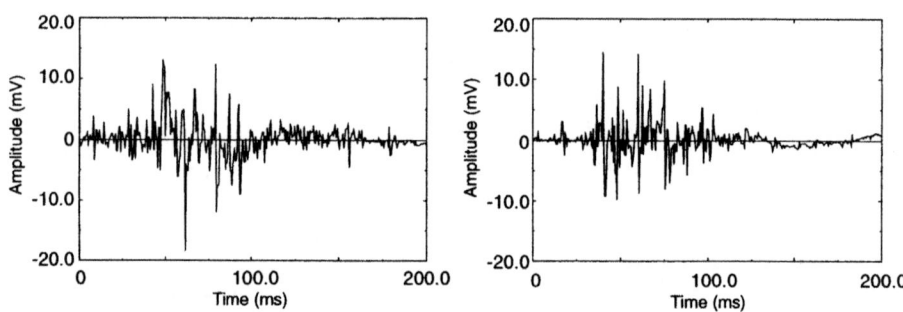

Fig. 2. The transient MES generated during the first 200 ms after a voluntary elbow flexion activity performed by a below elbow amputee subject.

Another observation made by Hudgins, which is critical to the current work, is that the structure of the observed transient MES appears to be distinct for each different muscle contraction activity. Figure 3 shows the structure of the transient MES generated by four different muscle contraction activities by a single normally limbed subject (subject A).

---

[2] The data for these trials were obtained from Dr. Bernard Hudgins of the Institute of Biomedical Engineering at the University of New Brunswick.

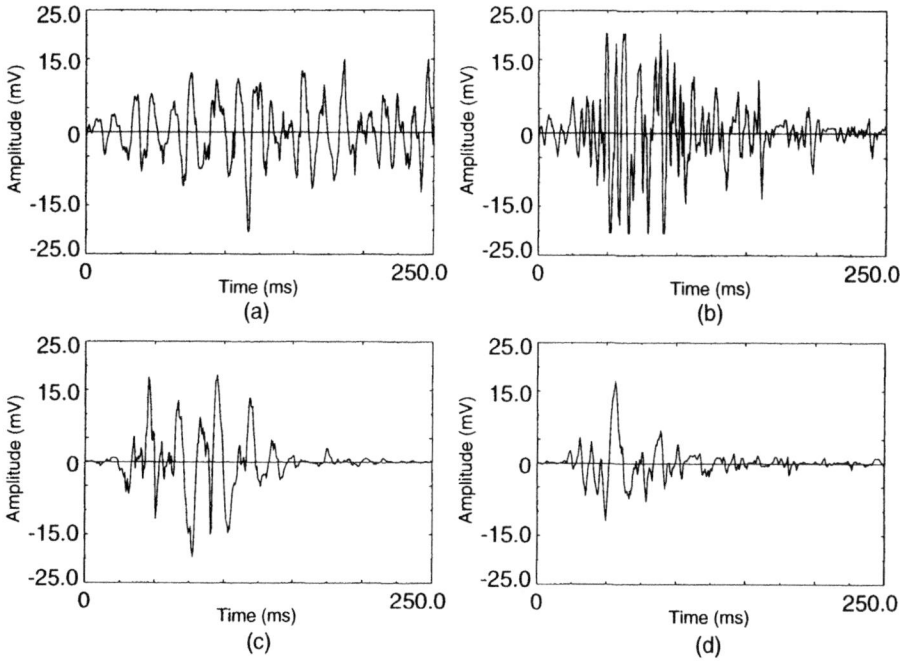

Fig. 3. Transient MES structure for four different voluntary muscle contractions: (a) elbow extension; (b) elbow flexion; (c) wrist flexion; (d) wrist pronation.

2 Myoelectric Control and Neural Networks

2.1 Myoelectric Signal Processing for Control Applications

Hudgins' identification of nonrandom structure in the myoelectric signal at the onset of a muscle contraction suggests that pattern-recognition techniques may be usefully applied to the myoelectric control problem. If the myoelectric signal patterns caused by a variety of voluntary muscle motions can be reliably identified, they can be used to control prosthetic devices. Inherent to the process of pattern recognition is some form of *classification decision* in which input data are assigned to a limited number of distinct *classes*. Data samples within the same class are assumed to have one or more specific *features* in common which would cause them to be placed in the same class. This basic classification-process model is used in several approaches which have been developed to analyze the structure of the myoelectric signal. Of relevance to the current work are approaches which take advantage of the unique computational abilities of *artificial neural networks* (ANN) to perform the classification task. ANN-based classifiers are able to classify input data into distinct classes by "learning" the optimum set of boundary definitions given the desired class membership of each input case. This adaptive behaviour makes the neural network a powerful paradigm for classification applications where the relationship between the inputs and the respective desired output classes is complex or difficult to resolve.

2.2 A Multi-State Control System Based on Transient Analysis

A recent demonstration by Hudgins of nonrandom structure in the *transient* myoelectric signal (during the first 200 ms after the onset of a muscle contraction task) offers the potential for using a new method of identification of muscle contraction actions from a single-site myoelectric signal [1] [7] . Initial work by Hudgins on pattern classification based on the transient MES demonstrated that at least four distinct muscle contraction tasks can be reliably identified.

The acquired myoelectric signal was converted to a digital signal which was monitored until its mean absolute value (MAV) exceeded a predetermined threshold, indicating the onset of a new muscle contraction. A set of features was then extracted for several time segments of the transient myoelectric signal to form the input representation to the classifier network. Five features were taken from each of five 40 ms time segments over the 200 ms duration of the transient MES. The feature set included the mean-absolute value (MAV) of the signal, the difference in the MAV in adjacent time segments, the number of zero crossings and slope changes in the signal per time segment, and a fractal measure of the waveform length.

The analysis of variance (ANOVA) performed by Hudgins on these features extracted from signal measurements recorded from several subjects indicated that they each contribute some useful information which can be used in the classification stage to determine the type of voluntary muscle contraction initiated by the user. Hudgins also reported that this approach resulted in a classification error rate of approximately 15 percent, and that the observed initial state selection error rate for all patients who used the system for the first time was over 40 percent. The need to improve the classification performance of the myoelectric control system is a prime motivator of the current work.

2.3 Neural Network Based Myoelectric Signal Processing Methods

Artificial neural networks have recently been applied to the problem of multifunction myoelectric control [9] [1] . In addition, neural-network-based signal analysis has been shown to be useful in medical diagnoses of heart conditions [10] and neuromuscular disorders [11] . In both applications, the objective is to classify the input signal into one of several classes based on a set of features extracted from the input data. These approaches have significantly influenced the direction of the present work.

3 Myoelectric Control System Design

3.1 Myoelectric Control: A Classification Task

Myoelectric control applications require that some form of classification decision be applied to the input data. The classification maps the input signal into a number of distinct *control classes*, each representing a different type of voluntarily actuated muscle activity. A relationship can then be developed between the different muscle activities (represented by the various identified classes) and the appropriate control actions desired by the user. For example, the function of opening the hand of a myoelectric limb could be assigned to the muscle task of a strong contraction of an upper arm muscle (such as for flexion of the elbow) if this is convenient for the patient.

3.2 Dimensionality Reduction, Feature Extraction and Feature-Based Representation

Feature Extraction and Projection. The inherent sparsity of high dimensional input spaces requires an extremely large number of training exemplars to adequately populate the input space. With a limited amount of training data and a limit on computational resources for training a classifier network, it is probable that large subspaces of a high dimensional input space are devoid of exemplars for which the output of the classifier is known. In this situation, there is an increased likelihood of unexpected classifier outputs and misclassification errors. To address this problem, it is necessary to reduce the dimensionality of the input data presented to the classifier. The lower-dimensional representation must retain the most *meaningful* features of the input data. The methodology behind this approach is summarized by Figure 4. The original m-dimensional input data are reduced to an n-dimensional representation by extracting meaningful features from the higher-dimensional input representation. This can also be regarded as performing a projection from R^m into R^n according to a defined *projection index*. An appropriately chosen projection index can emphasize certain distinct qualities, or *features*, present in the m- dimensional input representation in the n-dimensional representation (which is also called the *feature-based representation*). The classification decisions, which assign membership to one of p classes, are then generated by assigning decision boundaries within the feature-based representation. The main question then becomes: which features of the high-dimensional input representation should be retained in the lower-dimensional (feature-based) representation, given that classification must be performed on the lower-dimensional representation?

Development of the Input Representation. The present work uses a time-power spectrogram as the input representation to the feature-extraction network. This scheme was chosen due to its relative simplicity[3] and its good performance in a voice recognition application [22]. The 200 ms transient MES immediately following the detected onset of a voluntary muscle contraction was divided into a number of adjacent time slices (typically 10 time slices of 20 ms duration each). The power spectral density (PSD) of each time slice was generated by a fast Fourier transform (FFT) technique (typically an 8- or 16- point FFT PSD was generated). The amplitude of the PSD, representing the shape of the power

---

[3] Other feature extraction and representation schemes, such as principal component analysis or orthogonal wavelet transforms, could be employed and are currently being evaluated.

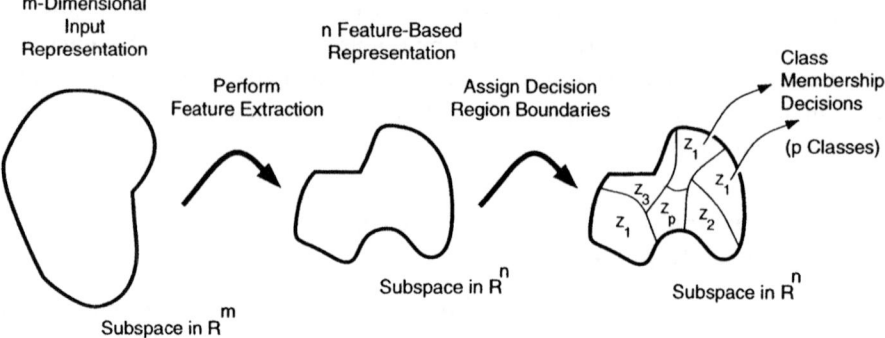

Fig. 4. Methodology behind feature-based classification: the m-dimensional representation is reduced to an n-dimensional representation by a feature extraction process. Decision boundaries can then be developed for the n-dimensional representation to delineate class membership

spectrum for each time slice, forms a matrix of values representing the input spectrogram. The dimensionality of the input spectrographic representation used in the current work is 320 dimensions (20 time slices, 16-point FFT PSD). It is suggested as a part of the current work that an input representation based on the shape of the PSD over several adjacent time slices is an attractive alternative to the set of time-domain features suggested by Hudgins.

Generating Feature-Based Representations. A critical consideration in the design of a feature-based classifier is the choice of a suitable projection index which can be applied to the input representation to form a feature-based representation that is most suitable for classification. With this goal in mind, a feature extraction method is required which results in a low-dimensional projection that possesses properties useful to the classification task, such as tight *clustering* of data associated with a particular class. Though there are a large number of possible projection indices from which to choose [12] [13] [3] , the choice is influenced largely by the requirement that the resulting low-dimensional subspace should contain information that emphasizes significant differences between individual data clusters. This suggests that the projection index should be chosen such that the statistical distributions of data points in the resulting lower-dimensional projection (the extracted features) are non-normal.

Introduction to Exploratory Projection Pursuit. Exploratory Projection Pursuit (EPP) is a method of statistical analysis proposed by Huber [12] which performs feature extraction (or dimensionality reduction) while at the same time emphasizing clustering in the resulting lower-dimensional representation. EPP has been the subject of much attention and has recently been suggested by Ripley as a potentially useful method of reducing the dimensionality of data prior to classification [14] . The EPP technique is useful for classification applications because it attempts to produce "interesting" projections of high-dimensional data in terms of a lower-dimensional representation. The projections are said to be *interesting* if they are able to represent subtle underlying structures in apparently unstructured data, as is the case for the MES. Interesting projections are found by finding an extremum (usually the minimum) of an objective function known as a *projection index*.

Principal component analysis (PCA) falls into the domain of Class II objective functions and can sometimes find clusters in the input data despite being sensitive to outliers. Huber generally prefers objective functions of Class III.

Several heuristics guide the search for an appropriate objective function which emphasizes departure from normality:

1. A multivariate distribution is normal (and therefore not interesting) if and only if all of its one-dimensional projections are normal.
2. If the projection that is least normal is still close to normality, it is not useful to look at any other projection as all other projections will be uninteresting.
3. For most high-dimensional collections of point data, most low-dimensional projections are approximately normal [15].

3.3   Neural Networks for Feature Extraction

Introduction. Due to the generally large computational requirements of classification problems, *neural networks* have become popular as classifiers due to their inherent massive parallelism and powerful computational capabilities. The ability of networks of simple processing elements to perform classification tasks after undergoing "training" by learning algorithms such as backpropagation is well known. More recently, there has been significant interest in the mathematical analysis of the dynamic learning mechanisms of neural networks. In 1982, Oja made the remarkable observation that a feedforward neural network trained with a Hebbian learning rule was able to perform a simple principal component analysis (PCA), extracting the first principal component of the input data distribution [16] . This work was later extended so that neural networks could be made to perform a complete PCA [17] [18].

This early work suggested that even the simplest neural-processing elements possess the ability to perform useful computation, and inspired others to investigate the computational power of individual neurons. Of importance to the current work is an investigation made by Intrator into the computational abilities of a model of neurons, found in the visual cortex, known as Bienenstock, Cooper, and Munro (BCM) neurons [19]. Intrator suggests the possibility that the observed behaviour of the BCM neuron is such that it is actually performing a form of EPP, and that the EPP algorithm can be implemented using artificial neural networks performing the *BCM synaptic weight update rule* [3] . The BCM rule was originally developed as a model of synaptic plasticity in the visual cortex [20] . A description of the function of a single BCM neuron in this domain forms the basis for a complete development of the synaptic weight update rule and the neural network architecture required to perform the EPP feature extraction.

Chinese Character Recognition via Orthogonal Moments

Simon X. Liao[1] and Miroslaw Pawlak[2]

[1] The University of Winnipeg, Winnipeg, Manitoba, Canada, R3B 2E9
[2] The University of Manitoba, Winnipeg, Manitoba, Canada, R3T 5V6

Abstract. To select a suitable feature vector extracted from the interested character for the purpose of classification is essential in the design of a character recognition system. Moment descriptors have been developed as features in pattern recognition since Hu[14] first introduced the moment method. Describing a character with moments means that global properties of the character are used rather than local properties. This nature makes the method of moments a proper candidate for Chinese character recognition system. In this paper, new Legendre moment spaces for Chinese character recognition are proposed which provide significant improvements in terms of Chinese character recognition.

I Introduction

One of the difficult problems in the design of a character recognition system is to select a suitable feature vector extracted from the interested character for the purpose of classification. Various character recognition techniques have been utilized to abstract characterizations for efficient character representations, see [11] and the references cited therein. Such characterizations are defined by measurable features extracted from the characters. Therefore, the effectiveness of the technique for an application is dependent on the ability of a given technique to uniquely represent the character from the available information. Since no one single technique will be effective for all recognition problems, the choice of character characterization is driven by the requirements of a specific recognition task.

Moment descriptors of various forms have been developed following Hu[14] as features in pattern recognition [1][15][24][28][29]. Describing image with moments instead of other traditional methods of image processing means that global properties of the image are used rather local properties. Based on Hu's *Uniqueness Theorem*[14], the double moment sequence is uniquely determined by an image function $f(x, y)$, and conversely, $f(x, y)$ is uniquely determined by the moment sequence. This nature makes the method of moments a proper candidate in character recognition.

neuron may represent a one-dimensional feature of the input data extracted by the projection of the input vector according to the weight vector developed by the BCM rule.

The BCM theory can now be used to generate an objective function formulation of EPP. Consider the activity of the BCM neuron as a decision problem: whether or not the BCM neuron should become active in response to a given input pattern. This decision depends on the value of the synaptic weights and, ultimately, on the value of the modification threshold (θ_M for notational simplicity) which determines the weight values. For each input pattern $x_i \epsilon R^N$ of the set of n input patterns $\Omega = (x_1, ..., x_n)$, there are two decisions in the decison space $D = [\text{no firing, firing}] = [0,1]$. With each decision to fire for a specific input there is an associated cost function known as the *loss function* L_θ which depends on the value of the modification threshold θ_M. The average loss over all inputs is the measure of the *risk* R_θ of the firing decision. Thus, the family of loss functions generated by applying all possible firing decisions over all input vectors yields a measure of the risk as:

$$L_\theta : \Omega \times D \to R_\theta \tag{7}$$

For a particular value of $\theta = \tilde{\theta}$, there is an optimal decision $\delta_{\tilde{\theta}}$ in the set of all possible decisions $(\Omega \times D)$ that minimizes the risk:

$$R_{\tilde{\theta}}\left(\delta_{\tilde{\theta}}\right) = \min_\theta \left\{ \sum_{i=1}^n P\left(x_i\right) L_\theta\left(x_i, \delta\left(x_i\right)\right) \right\} \tag{8}$$

where $P\left(x_i\right)$ is the probability of input pattern x_i occurring. When the threshold represents a vector in R_N, R_θ can be regarded as a projection index in terms of the Exploratory Projection Pursuit theory [2] [12]. In terms of the specific $\widehat{\phi}$ function which has been defined for the BCM neuron, it is possible to calculate a specific loss function associated with a particular decision; this function L_m is related to the value of the modification threshold and the nonlinear activity of the neuron for the input pattern under consideration:

$$L_m\left(x, \delta_m\right) = -u \int_{\theta_M}^{\sigma(\mathbf{x}\cdot\mathbf{m})} \widehat{\phi}\left(s, \theta_M\right) ds \tag{9}$$

where u is a proportionality constant related to the BCM learning rate. Assuming that the sigmoidal nonlinearity has a linear region which is large enough to span most reasonable values of $(\mathbf{x} \cdot \mathbf{m})$, the loss function over the space of reasonable values of synaptic activity becomes

$$L_m\left(x, \delta_m\right) = -\frac{u}{3}\left\{(\mathbf{x}\cdot\mathbf{m})^3 - E\left[(\mathbf{x}\cdot\mathbf{m})^2\right](\mathbf{x}\cdot\mathbf{m})^2\right\} \tag{10}$$

and the risk associated with this loss function is:

$$R_\theta\left(\delta_\theta\right) = -\frac{u}{3}\left\{(\mathbf{x}\cdot\mathbf{m})^3 - E\left[(\mathbf{x}\cdot\mathbf{m})^2\right]\right\} \tag{11}$$

Intrator suggests that if the risk is rewritten in the following form:

$$\frac{R_\theta\left(\delta_\theta\right)}{E^2\left[\left(\mathbf{x}\cdot\mathbf{m}\right)^2\right]} = -\frac{u}{3}\left\{\frac{E\left[\left(\mathbf{x}\cdot\mathbf{m}\right)^3\right]}{E^2\left[\left(\mathbf{x}\cdot\mathbf{m}\right)^2\right]} - 1\right\} \tag{12}$$

the term:

$$\frac{E\left[\left(\mathbf{x}\cdot\mathbf{m}\right)^3\right]}{E^2\left[\left(\mathbf{x}\cdot\mathbf{m}\right)^2\right]} \tag{13}$$

is proportional to the skewness of the input distribution. Thus an analysis of the risk associated with a particular projection can be regarded as a measure of *interestingness* according to the EPP criteria, which emphasizes deviation from normality in the distribution of the projected data. This result also resembles Huber's example of standardized absolute cumulants, which is a Class III objective function that was deemed to be a useful objective function for the EPP algorithm. Having shown that the risk associated with a particular projection of the input data onto the weight vector is a useful objective function in terms of the projection pursuit criterion, it is necessary to adapt the weight vector such that the risk is minimized over all possible projections. The minimization of the risk function R_θ with respect to the i'th synaptic weight of the BCM neuron involves a gradient descent of the risk with respect to the weight value:

$$\frac{\partial m_i}{\partial t} = -\frac{\partial}{\partial m_i}R_\theta\left(\delta_\theta\right) = uE\left[\phi\left(\mathbf{x}\cdot\mathbf{m},\theta_M\right)x_i\right] \tag{14}$$

For the nonlinear version of the BCM neuron, the BCM learning rule becomes

$$\frac{\partial m_i}{\partial t} = -\frac{\partial}{\partial m_i}R_\theta\left(\delta_\theta\right) = uE\left[\phi\left(\sigma\left(\mathbf{x}\cdot\mathbf{m}\right),\theta_M\right)\sigma'\left(\mathbf{x}\cdot\mathbf{m}\right)x_i\right] \tag{15}$$

where the sigmoidal nonlinear function σ is assumed to be continuous and differentiable to yield σ'.

Neural Network Implementation of Exploratory Projection Pursuit
In order to perform effective classification, it is necessary to extract a number of features from the input data. What is required is a method of performing several projections in parallel to extract the feature set to be used for the classification stage of the system. Intrator suggests that several BCM neurons connected in a mutually-inhibiting configuration can carry out EPP in parallel [2] . Consider a network of 3 BCM neurons connected as a lateral inhibition network. The lateral inhibitory connections are necessary to ensure that the 3 BCM neurons will each extract different features using a different EPP projection index.

In this configuration, all nodes receive inhibitory signals from all other nodes; this tends to inhibit the output activity of the target neuron. The inhibited activity of the k'th BCM neuron in a network composed of Q BCM neurons is:

$$\tilde{c}_k = c_k - \eta\sum_{j\neq k}^{Q}c_j \tag{16}$$

where η represents the magnitude of the weight value associated with the laterally inhibitory connections. The modification threshold of the k'th neuron based on the inhibited activity becomes:

$$\widetilde{\theta_M^k} = E\left[\widetilde{c}_k{}^2\right] \tag{17}$$

The BCM weight update equation is modified to include the inhibitory effects of the *(Q-1)* neurons that send similarly-weighted inhibitory signals to the target neuron in response to being exposed to input pattern x_i:

$$\frac{\partial m_k}{\partial t} = \frac{\partial R_\theta}{\partial m_k} = -\mu\left[1 - \eta\left(Q - 1\right)\right] E\left[\phi\left(\widetilde{c}_k, \widetilde{\theta_M^k}\right) x_i\right] \tag{18}$$

In the case where the non-linearity of the neuron is considered, the weights are updated according to the relationship:

$$\frac{\partial m_k}{\partial t} = \frac{\partial R}{\partial m_k} = -\mu E\left[\phi\left(\widetilde{c}_k, \widetilde{\theta_M^k}\right)\left(\sigma'\left(\mathbf{x}\cdot\mathbf{m_k}\right) - \eta\sum_{j\neq k}\sigma'\left(\mathbf{x}\cdot\mathbf{m_j}\right)\right) x_i\right] \tag{19}$$

These new equations governing the plasticity of synaptic weights for a lateral inhibition network of BCM neurons suggest that Q simultaneous exploratory projections of the input data can be conducted in parallel by Q BCM neurons. Two useful properties emerge from the BCM neural network:

1. **Selectivity**: each BCM neuron in the network becomes selectively tuned to a *specific*, interesting feature in the input data according to the EPP criteria; and
2. **Dispersion**: each BCM neuron becomes selectively tuned to a different feature in the input data. The loose coupling of the neurons by laterally inhibitory signals prevents two or more nodes from having the same level of output activation by a process known as *dispersion* [21] . This observation has been empirically confirmed in the course of the current work for the myoelectric signal processing application, and by Intrator and Tajchman in the processing of speech signals [22] .

3.4   Control System Structure

Figure 5 shows the complete structure of the proposed myoelectric control system. The MES was acquired by Dr. Bernard Hudgins at the University of New Brunswick using a bipolar electrode pair placed at an appropriate location on the skin surface of the patient (typically over the biceps brachii and triceps brachii on the upper arm). The signal was then amplified differentially and low-pass filtered to 1 kHz by an external analog circuit before being sampled digitally at a sampling rate of 2 kHz. The digital data were used to compute a moving mean absolute value (MAV) that was compared with a threshold value in order to detect and signal the onset of a voluntary contraction. When a voluntarily-generated contraction was detected, the next 200 ms of incoming digital data

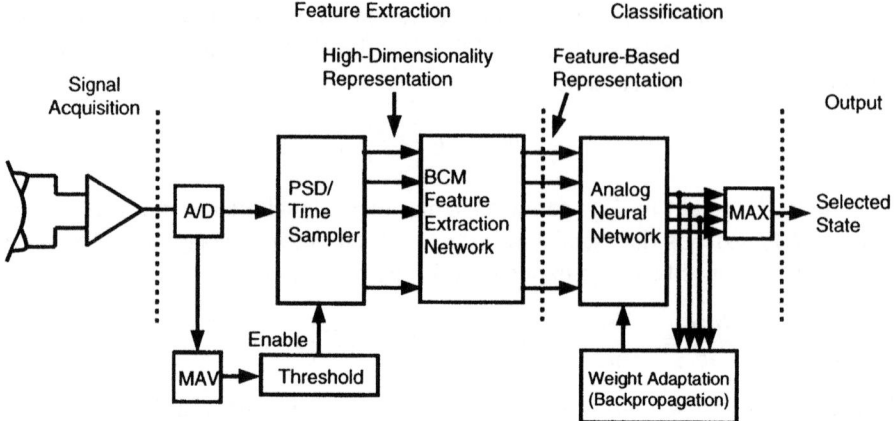

Fig. 5. Overall structure of the proposed myoelectric control system.

were processed to form the input representation, consisting of power spectral densities of several time slices of the sampled MES (typically 16-point power spectra of 20 time slices, forming a 320-dimensional input representation). The resulting spectrogram was presented to the feature extraction network composed of several BCM neurons. When this network had been trained appropriately, the activity levels of the BCM neurons in the network (resulting from the linear projection of the spectrograph onto the weights) represent a projection of the input data that is interesting according to the Exploratory Projection Pursuit criteria. The lower-dimensional representation, given by the output activity of the BCM neurons, was then mapped into an associated class label that represented the type of voluntary muscle contraction being initiated by the user. This mapping was accomplished by a three-layer feedforward neural network trained with the backpropagation learning rule. This network was trained, prior to normal operation, in a supervised learning mode. The output of the system corresponded to the class label (representing a certain voluntary muscle contraction activity) for which the output of the output-layer neuron representing that class label was a maximum over all output neurons in the network

4   System Simulation and Evaluation

4.1   Effect of Neural Network Architecture on Classifier Performance

A series of simulations were conducted to determine the effectiveness of the proposed classifier in terms of the correct classification rate for each control decision for a group of amputee and non-amputee subjects. A range of neural network architectures were tested to determine two key parameters: the number of BCM neurons to include in the feature extraction network (corresponding to the degree of dimensionality reduction from the 320-dimensional spectrographic input representation) and the number of neurons to include on the hidden layer of the classifier network. This second parameter is critical as neural networks often exhibit a behaviour that is consistent with *Ockham's Razor*: more complex networks with more hidden-layer neurons tend to memorize the training data set and often fail to generalize well on previously unseen data. In order to test the classifier, simulations were performed on myoelectric signal data from the normal (NOR), below-elbow amputee (BEA), and above-elbow amputee (AEA) subjects performing the four contraction tasks shown in Figure 3. In each simulation, the number of BCM neurons in the feature extraction network was varied while the number of hidden layer neurons in the classifier network remained constant at 5 neurons. Figure 6 shows the results of these simulations.

The effect of increasing the number of BCM neurons in the feature extraction network on the classification performance for the NOR subjects is negligible. The apparent robustness of the classifier with respect to data from the NOR subjects is not entirely unexpected; there appears to be sufficient information in the input representation of the transient MES of these subjects to allow for good classifier performance. In data from BEA and AEA subjects, however, a significant decrease in performance is observed as the number of BCM neurons is increased.

The apparent correlation between degraded classifier performance and an increased number of BCM neurons (corresponding to a larger number of extracted features forming the input representation to the backpropagation network) leads to the formulation of the following hypothesis: *beyond a certain limit, the inclusion of additional BCM neurons in the feature extraction network has a "distracting" effect on the backpropagation (classifier) network, resulting in degraded classifier performance.* We define the term *distraction* here to mean the degradation in classifier performance due to the inclusion of additional features with relatively low incremental information content.

Although it appears that the performance decrease has a more marked effect on the data from BEA subjects, the small number of amputee subjects for which data were available makes it difficult and probably premature to suggest that the type of amputation plays a role in the degree of sensitivity to the *distracting* effect of the presence of additional BCM neurons in the feature extraction network (i.e. the additional BCM neurons do not necessarily contribute additional information which is useful for classification, and the classifier network must nevertheless expend computational effort on dealing with this useless information).

To gather further support for the hypothesis concerning the role of additional BCM neurons as distractors to the classifier, a similar series of simulations was conducted in which the number of BCM neurons was varied while the number of neurons in the hidden layer of the classifier was set to 8 (see Figure 7). In this trial the performance of one of the normally limbed subjects (subject A) was degraded by approximately 10 percent compared to the previous trial with 5 neurons on the classifier hidden layer. In addition, the impact of an increased number of BCM neurons did not affect the classification performance of subject I (AEA) as it did in the previous trial. However, the effect of an increased number

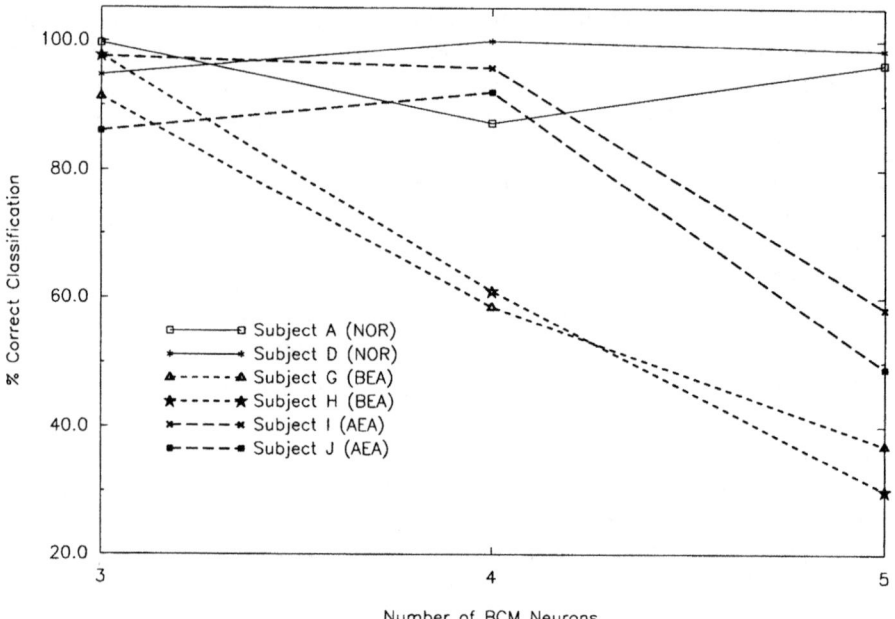

Fig. 6. Classifier performance for various numbers of extracted features (BCM neurons) for a classifier with 5 hidden neurons.

of BCM neurons had a more severe effect on the other AEA subject (subject J), and appears to have a generally consistent effect on both of the BEA subjects.

It is clear that classifier performance depends greatly on the complexity of the classifier, which can be controlled by altering the number of neurons on the hidden layer of the backpropagation classifier network. A large number of trials were conducted to determine the sensitivity of the classifer network performance to the number of hidden layer neurons involved. It was found that architectures involving between 5 and 11 hidden-layer neurons demonstrated good training set classification performance and generalization performance using the cross-validation technique while still requiring reasonable training times. Classification results for both the 5 neuron and 11 neuron classifier are shown in Table 1.

4.2   Classification Results

Simulation results indicated that very good classification performance (exceeding 90 percent) is possible using judiciously chosen network architectures and simulation parameters. Table 1 shows the performance of two candidate architectures for the proposed myoelectric control system, and compares these to results reported by Hudgins using MES data from the same set of subjects. It should be noted that direct comparison of these results is difficult due to several technical factors, including the fact that Hudgins' results reflect a *holdout approach* to classifier testing in which training and testing data sets are generated

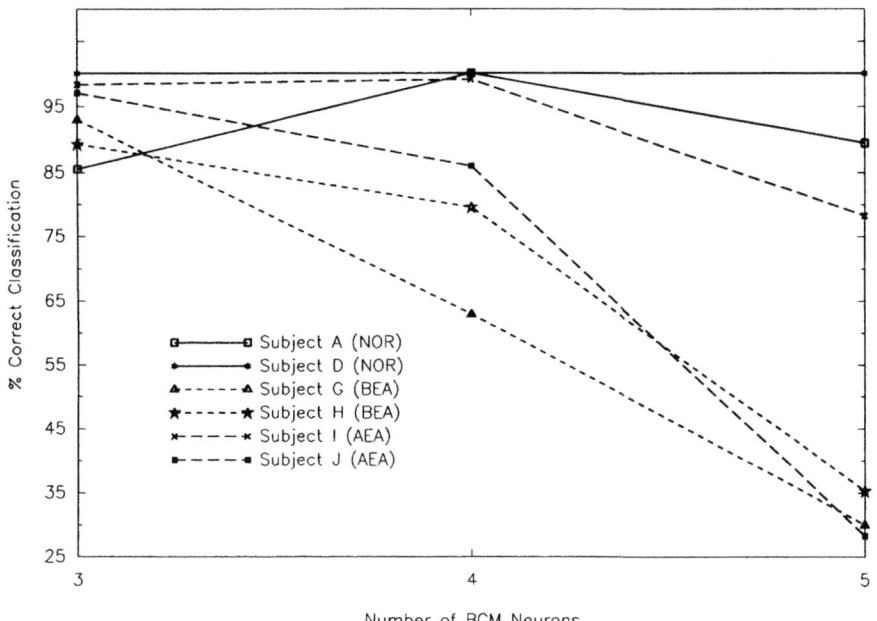

Fig. 7. Classifier performance for various numbers of extracted features (BCM neurons) for a classifier with 8 hidden neurons.

Subject	Type	Cross-Validation Method		Holdout Method
		3 BCM / 5 Hidden	3 BCM / 11 Hidden	Hudgins' Back-prop 30:8:4
A	NOR	99.7	85.5	98.0
B	NOR	100.0	100.0	88.0
C	NOR	95.0	100.0	96.0
D	NOR	94.7	100.0	81.0
E	NOR	87.0	100.0	90.0
F	NOR	99.8	100.0	90.0
G	BEA	91.3	93.0	90.0
H	BEA	97.7	89.3	78.0
I	AEA	97.5	98.3	90.0
J	AEA	86.0	97.0	86.0

Table 1. Classification results for various subjects for four contraction types. Subject types: NOR - *normal subject*, AEA - *above-elbow amputee*, BEA - *below-elbow amputee*.

by dividing the available sample data into approximately $\frac{1}{2}$ training data and $\frac{1}{2}$ testing data, whereas the results from the current work use the cross-validation approach to classifier testing due to the small number of MES exemplars which were available for classifier training and testing. This problem was especially severe for the AEA subjects (subjects I and J), since many of the recorded data samples for these users were corrupted by amplifier saturation or by a lack of sufficient data to form a complete 200 ms window after the detected onset of the contraction activity. These problems are likely associated with the difficulties experienced by the AEA subjects in producing four distinct muscle contraction activities. Because of these considerations, the number of available MES sample cases ranged from 20 cases for subject J to approximately 80 cases for several of the normally limbed subjects. It is prudent to consider the effects of the limited number of test data when interpreting the classification results. Instances resulting in 100 percent classification *should be considered to be optimistic* because of the limited number of available cases, despite the use of the multiple cross-validation technique. Regardless, it is evident that the classification performance of the proposed myoelectric control system compares favourably with that proposed by Hudgins.

5 Conclusions

This work has resulted in the development of a new approach to myoelectric control that involves the sampling and representation of a transient myoelectric signal as a time-frequency spectrograph, preprocessing of this representation by a network of BCM neurons performing a form of Exploratory Projection Pursuit, and classification into a control decision using a neural network-based classifier. Simulation results indicate that this approach results in a high correct classification rate for both amputee and non-amputee subjects. In addition, an hypothesis relating the increase in dimensionality of the feature-based representation to an observed degradation in classifier performance has been presented. It is suggested that the presence of additional BCM neurons in the feature extraction network serves as a distraction to the classifier network. Empirical evidence from a series of simulations supports this claim, and forms the basis for establishing the upper bound on the search for the appropriate topology (in terms of the number of neurons required) for the BCM feature extraction network.

Future work involving this approach is focused on further simplifying the proposed myoelectric control system by incorporating the function of the BCM feature extraction neurons into the first layer of the classifier network. In addition, alternative feature extraction algorithms and classifier architectures are also being investigated.

References

1. B. Hudgins, P. Parker, and R. Scott, "The Recognition of Myoelectric Patterns for Prosthetic Limb Control," in *Ann. Intl. Conf. IEEE Engineering in Biology Society*, vol. 13, pp. 2040–2041, 1991.

Function of the BCM Neuron Model. It is useful to provide a review of the BCM weight update rule, based on an analysis originally suggested by Intrator [2]. This analysis will show that the behaviour of the BCM rule can be regarded as an objective function of the type suggested by Huber as being compatible with the premise of the EPP approach which uses functions which emphasize departure from normality.

Adopting the notation used by Intrator, the input to a cell containing N synapses can be expressed as a vector $\mathbf{x}$, where $\mathbf{x} = (x_1, ..., x_N)$ and the vector of synaptic weights associated with each input is $\mathbf{m} = (m_1, ..., m_N)$. Assuming that the output of the neuron is linear, activity of the BCM neuron is given by:

$$c = \mathbf{x} \cdot \mathbf{m} \tag{1}$$

In order to enhance the arbitrary mapping functionality of the network and to increase the robustness of the BCM neuron to outliers in the input data, a nonlinear model of the BCM neuron is often used. The activity of the neuron in this case becomes

$$c = \sigma(\mathbf{x} \cdot \mathbf{m}) \tag{2}$$

where σ is a continuous, differentiable and monotonically increasing function such as the *sigmoid function*

$$\sigma(\alpha) = \frac{1}{1 + e^{-\alpha}} \tag{3}$$

Following the approach proposed by Intrator [3] , the modification threshold θ_M is expressed as:

$$\theta_M = \bar{c^2} = E\left[\sigma^2(\mathbf{x} \cdot \mathbf{m})\right] \tag{4}$$

For the nonlinear neuron model, the $\widehat{\phi}$ function determining the plasticity of synaptic weights in the BCM model is defined as:

$$\widehat{\phi}(c, \theta_M) = c^2 - \frac{4}{3}c\theta_M \tag{5}$$

The BCM synaptic weight update equation of the ith synapse is given by:

$$\frac{\partial m_i}{\partial t} = \mu\widehat{\phi}(\mathbf{x} \cdot \mathbf{m}, \theta_M)x_i \tag{6}$$

in proportion to the value of presynaptic input x_i consistent with Hebb's classical weight update rule. The variable μ, known as the *BCM learning rate*, is used to control the rate at which the synaptic weights are updated. The activity (output) of the BCM neuron can be considered to be the projection of the input vector $\mathbf{x}$ on the synaptic weight vector $\mathbf{m}$. Thus, the output of the BCM neuron can be regarded as a single-dimensional projection of the input data according to the values of the synaptic weight which are dictated according to the BCM weight update rule. Because of the selectivity property of the BCM rule, the synaptic weights can become selectively tuned to input patterns in the data which cause activation of the BCM neuron. In this case, the output of the BCM

2. N. Intrator, "A Neural Network for Feature Extraction," in *Advances in Neural Information Processing Systems*, vol. 2, pp. 719–726, 1990.

3. N. Intrator, "Feature Extraction Using an Unsupervised Neural Network," *Neural Computation*, vol. 4, pp. 98–107, 1992.

4. C. DeLuca, "Physiology and Mathematics of Myoelectric Signals," *IEEE Trans. Biomedical Engrg.*, vol. 26, pp. 313–325, June 1979.

5. P. Parker and R. Scott, "Myoelectric control of prosthesis," *CRC Crit. Rev. Biomed. Engrg.*, vol. 13, no. 4, pp. 283–310, 1986.

6. P. Parker, J. Stuller, and R. Scott, "Processing for the multistate myoelectric channel," *Proc. IEEE*, vol. 65, no. 5, pp. 662–674, 1977.

7. B. Hudgins, *A New Approach to Multifunction Myoelectric Control.* PhD thesis, University of New Brunswick, 1991.

8. D. Graupe and W. Cline, "Functional separation of EMG signal via ARMA identification methods for prosthesis control purposes," *IEEE Trans. Sys. Man Cyber.*, vol. SMC-5, no. 2, pp. 252–259, 1975.

9. M. Kelly, P. Parker, and R. Scott, "The Application of Neural Networks to Myoelectric Signal Analysis: A Preliminary Study," *IEEE Trans. Biomedical Engrg.*, vol. 37, pp. 221–230, Mar. 1990.

10. E. Pietka, "Feature extraction in the computerized approach to the ECG analysis," *Pattern Recognition*, vol. 24, no. 2, pp. 139–146, 1991.

11. C. Schizas, C. Pattichis, I. Schofield, P. Fawcett, and L. Middleton, "Artificial Neural Net Algorithms in Classifying Electromyographic Signals," in *Proc. First IEE Conf. on Artificial Neural Networks*, vol. 1, pp. 134–138, 1989.

12. P. Huber, "Projection Pursuit (with discussions)," *Annals of Statistics*, vol. 13, no. 2, pp. 435–475, 1985.

13. J. Friedman and J. Tukey, "A Projection Pursuit Algorithm for Exploratory Data Analysis," *IEEE Trans. on Computers*, vol. C-23, pp. 881–890, Sept. 1974.

14. B. Ripley, "Statistical aspects of neural networks," *in press*, 1992.

15. P. Diaconis and D. Freedman, "Asymptotics of graphical projection pursuit," *Annals of Statistics*, vol. 12, pp. 793–815, 1984.

16. E. Oja, "A simplified neuron model as a principal component analyzer," *J. Math. Biology*, vol. 15, pp. 267–273, 1982.

17. T. Leen, "Dynamics of learning in recurrent feature-discovery networks," in *Advances in Neural Information Processing Systems* (R. Lippman, J. Moody, and D. Touretzky, eds.), vol. 3, pp. 70–76, 1991.

18. T. Sanger, "An optimality principle for unsupervised learning," in *Advances in Neural Information Processing Systems* (D. Touretzky, ed.), vol. 1, pp. 11–19, 1989.

19. E. Bienenstock, L. Cooper, and P. Munro, "Theory for the Development of Neuron Selectivity: Orientation Specificity and Binocular Interaction in Visual Cortex," *Journal of Neuroscience*, vol. 2, pp. 32–48, Jan. 1982.

20. M. Bear, L. Cooper, and F. Ebner, "A Physiological Basis for a Theory of Synapse Modification," *Science*, vol. 237, pp. 42–48, 1987.

21. J. Marshall, "Self-organizing neural networks for perception of visual motion," *Neural Networks*, vol. 3, pp. 45–74, 1990.

22. N. Intrator and G. Tajchman, "Supervised and unsupervised feature extraction from a cochlear model for speech recognition," in *Neural Networks for Signal Processing - Proceedings of the 1991 IEEE Workshop* (B. Juang, S. Kung, and C. Kamm, eds.), pp. 460–469, IEEE Press, 1991.

II Moments

Geometric Moments

The two-dimensional geometric moments of order (p, q) of the image function $f(x, y)$, which is assumed that $f(x, y)$ is a piecewise continuous function, are defined as

$$M_{pq} = \int_{-\infty}^{+\infty} \int_{-\infty}^{+\infty} x^p y^q f(x, y) dx dy, \tag{1}$$

where $p, q = 0, 1, 2, ..., \infty$.

The central moments of $f(x, y)$ are defined as

$$\mu_{pq} = \int_{-\infty}^{+\infty} \int_{-\infty}^{+\infty} (x - \bar{x})^p (y - \bar{y})^q f(x, y) dx dy, \tag{2}$$

where

$$\bar{x} = \frac{M_{10}}{M_{00}}, \qquad \bar{y} = \frac{M_{01}}{M_{00}}.$$

The central moments μ_{pq} defined in Eq. (2) are invariants under the translation of coordinates[14]. Also, the central moments μ_{pq} can be expressed in term of the moments M_{pq} defined in Eq. (1)[14][29].

For a digital image, the double integrations in M_{pq} are usually approximated by double summations,

$$M_{pq} = \sum_{i=1}^{M} \sum_{j=1}^{N} h_{M_{pq}}(x_i, y_j) f(x_i, y_j), \tag{3}$$

where

$$h_{M_{pq}}(x_i, y_j) = \int_{x_i - \frac{\Delta x}{2}}^{x_i + \frac{\Delta x}{2}} \int_{y_j - \frac{\Delta y}{2}}^{y_j + \frac{\Delta y}{2}} x^p y^q \, dx dy \tag{4}$$

represents the double integration of $x^p y^q$ over the pixel $[x_i - \frac{\Delta x}{2}, x_i + \frac{\Delta x}{2}] \times [y_j - \frac{\Delta y}{2}, y_j + \frac{\Delta y}{2}]$. $\Delta x = x_i - x_{i-1}$ and $\Delta y = y_j - y_{j-1}$ are the sampling intervals in the x and y directions, respectively. See [16][17] for more discussions.

Similarly, μ_{pq} can be written as

$$\mu_{pq} = \sum_{i=1}^{M} \sum_{j=1}^{N} h_{\mu_{pq}}(x_i, y_j) f(x_i, y_j), \tag{5}$$

where

$$h_{\mu_{pq}}(x_i, y_j) = \int_{x_i - \frac{\Delta x}{2}}^{x_i + \frac{\Delta x}{2}} \int_{y_j - \frac{\Delta y}{2}}^{y_j + \frac{\Delta y}{2}} (x - \bar{x})^p (y - \bar{y})^q dx dy, \tag{6}$$

$$\bar{x} = \frac{M_{10}}{M_{00}} \qquad \bar{y} = \frac{M_{01}}{M_{00}}.$$

The summation limits M and N are the dimensions of the digitized image $f(x_i, y_j)$, in which i and j are the discrete locations of the image pixels.

Legendre Moments

The nth - order Legendre polynomial is

$$P_n(x) = \sum_{j=0}^{n} a_{nj} \cdot x^j$$

$$= \frac{1}{2^n n!} \cdot \frac{d^n}{dx^n}(x^2 - 1)^n. \tag{7}$$

The Legendre polynomials $\{P_m(x)\}$ [7] are a complete orthogonal basis set on the interval [-1, 1]:

$$\int_{-1}^{+1} P_m(x)P_n(x)dx = \frac{2}{2m+1}\delta_{mn}, \tag{8}$$

where δ_{mn} is the Kronecker symbol.

The Legendre moment of $f(x, y)$ with order (m, n) is defined by

$$\lambda_{mn} = \frac{(2m+1)(2n+1)}{4} \int_{-1}^{+1} \int_{-1}^{+1} P_m(x)P_n(y)f(x,y)dxdy, \tag{9}$$

where $m, n = 0, 1, 2, ..., \infty$.

If an image function $f(x, y)$ is digitized into its discrete version $f(x_i, y_j)$ with an $M \times N$ array of pixels, the double integration of (9) can be approximated by double summation:

$$\lambda_{mn} = \frac{(2m+1)(2n+1)}{4} \sum_{i=1}^{m} \sum_{j=1}^{n} h_{\lambda_{mn}}(x_i, y_j)f(x_i, y_j), \tag{10}$$

where

$$h_{\lambda_{mn}}(x_i, y_j) = \int_{x_i - \frac{\Delta x}{2}}^{x_i + \frac{\Delta x}{2}} \int_{y_j - \frac{\Delta y}{2}}^{y_j + \frac{\Delta y}{2}} P_m(x)P_n(y)dxdy. \tag{11}$$

III Character Recognition via Central Moments

Historically, Hu[13][14] published the first significant paper on the utilization of moment invariants for image analysis and object representation in 1961. Hu demonstrated the utility of moment techniques through a simple pattern recognition experiment. The first two moment invariants were used to represent several known digitized patterns in a two-dimensional feature space. The experiment was performed by using a set of 26 capital English letters as input patterns. In the two-dimensional feature space, all the points representing each of the characters were fairly distinct except those of M and W.

Chinese character recognition is believed to be a typical practical problem that depends on general shapes rather than details of the image. Compared with the set of English letters, the Chinese character set is larger, and in terms of character recognition, is more difficult to classify.

In this paper, we use two sets of Chinese characters as the testing images. The first set is composed of five Chinese characters, which are named C_1 to C_5 and shown in Fig. 1 from left to right. The reason to use these five Chinese characters is that they are very similar to each other. Actually, among more than 60,000 Chinese characters, one cannot find another set of five (even set of three or set of four) in which all individuals are so similar to each other. The second set is composed of 90 randomly selected Chinese characters, which are named S_1 to S_{90} and shown in Fig. 2 from left to right and top to bottom. Each Chinese character consists of 24×24 pixels and has 32 graylevels. All characters have the gray level 11 and the background has the value 21.

Fig. 1. Five Chinese characters used for testing. From left to right: C_1, C_2, C_3, C_4, and C_5.

Similar to Hu's experiment, a simulation program of a character recognition model using only two central moment invariants is employed. The following two moments invariants,

$$X_1 = \sqrt{\mu_{20} + \mu_{02}} \tag{12}$$

and

$$X_2 = \sqrt{(\mu_{30} - 3\mu_{12})^2 + (3\mu_{21} - \mu_{03})^2}, \tag{13}$$

where μ_{pq} was defined in (5), are used to compute the representations of all known characters in the image plane (x, y). A point (X_1, X_2) in a two dimensional space represents a Chinese character.

The values of X_1 and X_2 for two sets of Chinese characters are given in Table 1 and Table 2, respectively. Fig. 3 and Fig. 4 display those results in the two-dimensional (X_1, X_2) plane.

Characters	X_1	X_2
C_1	3.5328	0.0933
C_2	3.5440	0.0818
C_3	3.5433	0.0616
C_4	3.5574	0.0254
C_5	3.5559	0.0794

Table 1. Values of the five Chinese characters in the central moment feature space.

Sample	X_1	X_2	Sample	X_1	X_2	Sample	X_1	X_2
S_1	3.6184	0.2566	S_{31}	3.6512	0.2411	S_{61}	3.5429	0.0565
S_2	3.5487	0.0162	S_{32}	3.5564	0.1498	S_{62}	3.5485	0.1071
S_3	3.5474	0.0029	S_{33}	3.4875	0.0674	S_{62}	3.5618	0.0980
S_4	3.5504	0.0769	S_{34}	3.4743	0.1785	S_{64}	3.5798	0.1332
S_5	3.5551	0.0811	S_{35}	3.5606	0.0598	S_{65}	3.5286	0.1251
S_6	3.5339	0.2560	S_{36}	3.5578	0.1511	S_{66}	3.4988	0.1657
S_7	3.5092	0.1201	S_{37}	3.5509	0.0610	S_{67}	3.5440	0.0818
S_8	3.5725	0.1149	S_{38}	3.5382	0.1388	S_{68}	3.5354	0.0843
S_9	3.6230	0.0114	S_{39}	3.5589	0.0596	S_{69}	3.5327	0.2650
S_{10}	3.5499	0.1434	S_{40}	3.5840	0.0639	S_{70}	3.5091	0.0972
S_{11}	3.5749	0.0771	S_{41}	3.5689	0.0613	S_{71}	3.5424	0.2206
S_{12}	3.5591	0.0477	S_{42}	3.4838	0.1280	S_{72}	3.5314	0.1080
S_{13}	3.5577	0.2699	S_{43}	3.5259	0.0878	S_{73}	3.5216	0.2923
S_{14}	3.4994	0.1735	S_{44}	3.5056	0.1187	S_{74}	3.5340	0.0785
S_{15}	3.5576	0.1663	S_{45}	3.5435	0.1615	S_{75}	3.5996	0.0398
S_{16}	3.4926	0.1638	S_{46}	3.5750	0.0304	S_{76}	3.5075	0.1668
S_{17}	3.5146	0.1091	S_{47}	3.5053	0.3824	S_{77}	3.5067	0.0730
S_{18}	3.5050	0.0607	S_{48}	3.5609	0.1297	S_{78}	3.5466	0.0675
S_{19}	3.5035	0.0953	S_{49}	3.6086	0.1638	S_{79}	3.5668	0.0329
S_{20}	3.5339	0.0733	S_{50}	3.5615	0.0674	S_{80}	3.6527	0.1474
S_{21}	3.5569	0.0480	S_{51}	3.5390	0.1286	S_{81}	3.5889	0.0140
S_{22}	3.5958	0.1324	S_{52}	3.4933	0.1307	S_{82}	3.5776	0.1211
S_{23}	3.5467	0.0199	S_{53}	3.4940	0.1729	S_{83}	3.4619	0.2198
S_{24}	3.5022	0.1514	S_{54}	3.5621	0.0845	S_{84}	3.5986	0.0416
S_{25}	3.5264	0.1419	S_{55}	3.5675	0.0680	S_{85}	3.5646	0.1387
S_{26}	3.5378	0.0837	S_{56}	3.5750	0.1012	S_{86}	3.5460	0.0678
S_{27}	3.5033	0.1533	S_{57}	3.5387	0.2538	S_{87}	3.6025	0.0930
S_{28}	3.5615	0.0590	S_{58}	3.5175	0.2213	S_{88}	3.5371	0.1616
S_{29}	3.5787	0.1090	S_{59}	3.5182	0.1422	S_{89}	3.5542	0.2515
S_{30}	3.5322	0.0200	S_{60}	3.5571	0.2579	S_{90}	3.5860	0.1156

Table 2. Values of the 90 Chinese characters in the central moment feature space.

Fig. 3 and Fig. 4 show that for both sets of characters, all characters are very close to each other in the two dimensional (X_1, X_2) space. ¿From the classification point of view, this disadvantage certainly will limit the use of the central moments on the character recognition tasks.

IV Character Recognition via Legendre Moments

Similar to the two classification measures defined in (12) and (13), the following two measures

$$Y_1 = \sqrt{\lambda_{20} + \lambda_{02}} \tag{14}$$

and

$$Y_2 = \sqrt{(\lambda_{30} - 3\lambda_{12})^2 + (3\lambda_{21} - \lambda_{03})^2} \tag{15}$$

are employed in our new recognition model. Where λ_{mn} are the Legendre moments defined in (9).

Characters	Y_1	Y_2
C_1	2.0204	1.6965
C_2	2.1636	0.2535
C_3	2.1899	1.7548
C_4	2.2032	2.1281
C_5	2.1653	3.2418

Table 3. Values of the five Chinese characters in the Legendre moment feature space.

Using the same two sets of Chinese characters, the distributions of all characters in the two dimensional space (Y_1, Y_2) are illustrated in Fig. 5 and Fig. 6. To make the results from two different methods comparable, The same scale is used in Fig. 3 and Fig. 5. For the same reason, the scale of Fig. 3 is the same as that of Fig. 5.

Table 3 and Table 4 display the values of Y_1 and Y_2 which represent the two sets of Chinese characters in the two dimensional (Y_1, Y_2) space.

From Table 3 and Fig. 5 we can see that the five characters are well separated in the two dimensional (Y_1, Y_2) space. In other words, on this particular Chinese character recognition task, using Legendre moments is superior.

Table 4 and Fig. 6 show that most of those randomly selected 90 Chinese characters are well separated in the two-dimensional (Y_1, Y_2) space. However, it is observed that two characters, S_{44} and S_{57}, are very close to each other in the Legendre moment feature space. Although the results shown in Fig. 6 are indeed better than those of Fig. 4, yet the Legendre moment two-dimensional feature space cannot be used as a successful technique to recognize a specific Chinese character from the whole Chinese character set.

One option to improve the Legendre moment technique is to add a new feature to the Legendre moment feature space. We use the following equation as the third feature:

$$Y_3 = \sqrt{(\lambda_{20} - \lambda_{02})^2 + 4\lambda_{11}^2}. \tag{16}$$

In this new three-dimensional Legendre moment feature space, characters S_{44} and S_{57} have Y_3 values 0.6933 and 1.6876, respectively. Therefore, all Chinese characters shown in Fig. 2 can be separated successfully. Table 5 displays the values of Y_1, Y_2, and Y_3 for all 90 Chinese characters.

V Conclusions

In this paper, we proposed new Legendre moment feature spaces for Chinese character recognition. Compared with the Central moment feature space proposed by Hu[14], the new method provided significant improvements in terms of Chinese character recognition.

Sample	Y_1	Y_2	Sample	Y_1	Y_2	Sample	Y_1	Y_2
S_1	1.5346	4.7334	S_{31}	1.5260	5.6012	S_{61}	1.7716	0.8286
S_2	1.9007	2.0506	S_{32}	1.8379	2.9933	S_{62}	1.8951	1.6680
S_3	1.8135	0.6016	S_{33}	1.8906	2.4768	S_{63}	1.9745	2.6982
S_4	1.7271	2.5868	S_{34}	1.8940	3.7551	S_{64}	1.9486	1.4085
S_5	1.9632	1.1077	S_{35}	1.7915	1.8526	S_{65}	1.6608	2.6182
S_6	1.8792	4.8260	S_{36}	2.0227	3.3827	S_{66}	1.6585	3.6280
S_7	1.9411	2.6879	S_{37}	2.1192	1.8604	S_{67}	2.1636	0.2535
S_8	1.9408	2.0360	S_{38}	1.9756	1.8860	S_{68}	1.7241	1.7722
S_9	1.9248	2.2374	S_{39}	1.7896	1.7319	S_{69}	1.7064	5.7469
S_{10}	1.7645	3.9552	S_{40}	1.9206	2.5398	S_{70}	1.7558	0.5146
S_{11}	1.7005	1.2156	S_{41}	1.8540	1.5533	S_{71}	2.0121	4.0992
S_{12}	1.8596	1.1031	S_{42}	1.9207	2.1761	S_{72}	2.0239	2.3041
S_{13}	1.7067	5.0808	S_{43}	1.8654	2.8219	S_{73}	1.8900	6.0365
S_{14}	2.0086	3.8532	S_{44}	1.8708	2.4153	S_{74}	1.8122	1.1317
S_{15}	2.0362	3.8763	S_{45}	1.9319	1.4443	S_{75}	1.9149	1.5990
S_{16}	1.6787	3.5999	S_{46}	1.4783	0.0995	S_{76}	1.8685	0.6289
S_{17}	2.0319	2.7340	S_{47}	1.9144	6.2822	S_{77}	1.7135	1.2120
S_{18}	1.7548	1.2384	S_{48}	1.7970	2.6013	S_{78}	1.7870	1.0894
S_{19}	1.9031	1.9106	S_{49}	1.5632	3.6298	S_{79}	1.7331	0.7432
S_{20}	2.0153	2.7606	S_{50}	2.2657	1.8740	S_{80}	1.5097	1.7610
S_{21}	1.5705	1.3187	S_{51}	1.7894	2.3395	S_{81}	1.7795	0.0637
S_{22}	1.7817	1.9157	S_{52}	1.7126	3.1672	S_{82}	1.4753	1.7641
S_{23}	1.6696	0.9155	S_{53}	1.6793	4.4645	S_{83}	1.5775	2.8834
S_{24}	2.0691	1.5532	S_{54}	2.2200	1.7281	S_{84}	1.6220	3.5224
S_{25}	2.1259	3.5769	S_{55}	1.7195	0.8730	S_{85}	1.8122	1.6799
S_{26}	1.6907	1.2791	S_{56}	2.1327	0.4946	S_{86}	2.0188	2.5025
S_{27}	2.0025	2.3081	S_{57}	1.8733	2.3991	S_{87}	2.0721	1.3334
S_{28}	1.8056	0.5816	S_{58}	2.1976	4.2732	S_{88}	1.8996	2.1385
S_{29}	1.6444	2.3904	S_{59}	1.9050	2.0736	S_{89}	1.6954	6.1273
S_{30}	2.0144	1.8385	S_{60}	1.5251	6.0759	S_{90}	1.9719	3.3317

Table 4. Values of the 90 Chinese characters in the Legendre moment feature space.

Two sets of Chinese characters, one including five similar Chinese characters and the other containing 90 randomly selected characters, are used as the input patterns in this research. For the set of five, the experiment demonstrated that all five representations in the two-dimensional Legendre moment feature space are well separated. The performance of recognizing 90 randomly selected Chinese characters with the Legendre moment feature space is much more refined than that of the Central moment feature space as well. However, the distance in the two-dimensional feature space between two characters, S_{44} and S_{57}, is quite small. To improve the recognizing ability, we added one new feature to the two-dimensional Legendre feature space. The new three-dimensional Legendre feature space is able to separate all 90 randomly selected Chinese characters successfully.

Since the most complicated Legendre polynomials involved in the method is $P_3(x)$, with the development in the VLSI moment generator chip computes area[4], the real-time Chinese character recognition becomes possible.

Because of some technical reasons, we could not obtain the whole set of Chinese characters and test all of them individually. However, with a possible fourth feature being added to the three-dimensional Legendre moment feature space, we are very optimistic to say that this character recognition method can be applied successfully to the whole set of Chinese characters.

Acknowledgments

This research was supported by the University of Winnipeg Major Research Grant and the NSERC Grant A8131.

References

1. Y.S. Abu-Mostafa and D. Psaltis, *Recognitive aspects of moment invariants*, IEEE Trans. Pattern Anal. Machine Intell., Vol. PAMI-6, pp. 698-706, Nov. 1984.
2. Y.S. Abu-Mostafa and D. Psaltis, *Image normalization by complex moments*, IEEE Trans. Pattern Anal. Machine Intell., Vol. PAMI-7, pp. 46-55, Jan. 1985.
3. F.L. Alt, *Digital pattern recognition by moments*, J. Assoc. Computing Machinery, Vol. 9, pp. 240-258, 1962.
4. R.L. Andersson, *Real time gray scale video processing using a moment generating chip*, IEEE Journal of Robotics and Automation, Vol. RA-1, No. 2, June 1985.
5. S.O. Belkasim, M. Shridhar and M. Ahmadi, *Pattern recognition with moment invariants: a comparative study and new results*, Pattern Recognition, Vol. 24, No. 12, pp. 1117-1138, 1991.
6. J.F. Boyce and W.J. Hossack, *Moment invariants for pattern recognition*, Pattern Recognition Lett., Vol. 1, no. 5-6, pp. 451-456, July 1983.
7. R. Courant and D. Hilbert, *Methods of Mathematical Physics*, Vol. I. New York: Interscience, 1953.
8. P.J. Davis and P. Rabinowitz, *Methods of Numerical Integration*, Academic Press, New York, 1975.
9. R.O. Duda and P.E. Hart, *Pattern Classification and Scene Analysis*, Wiley, New York, 1973.
10. K.J. Dudani, K.J. Breeding, and R.B. McGhee, *Aircraft identification by moment invariants*, IEEE rans. Comput. Vol. C-26, pp. 39-46, Jan. 1977.
11. V.K. Govindan and A.P. Shivaprasad, *Character recognition – a review*, Patt. Recogn., Vol. 23, no. 7, pp. 671-683, 1990.
12. M. Hatamian, *A Real-time two-dimensional moment generating algorithm and its single chip implementation*, IEEE Trans. Acoust. Speech Signal Process. ASSP-34, 1986, pp. 99-126.
13. M.K. Hu, *Pattern recognition by moment invariants*, proc. IRE 49, 1961, 1428.
14. M.K. Hu, *Visual problem recognition by moment invariant*, IRE Trans. Inform. Theory, Vol. IT-8, pp. 179-187, Feb. 1962.
15. A.K. Jain, *Fundamentals of Digital Image Processing*, Prentice-Hall, 1989.
16. S. X. Liao, *Image Analysis by Moments*, Ph.D. dissertation, The University of Manitoba, 1993.

17. S. X. Liao and M. Pawlak, *On image analysis by moments*, IEEE Trans. PAMI, Vol. 18, No. 3, pp. 254-266, March, 1996.
18. Chin Lu, *A survey on Chinese computing research*, Hong Kong Computer Journal, Vol. 9, No. 12, December 1993.
19. M. Pawlak, *On the reconstruction aspects of moment descriptions*, IEEE Symposium on Information Theory, San Diego, January 1990.
20. M. Pawlak, *On the reconstruction aspects of moment descriptors*, IEEE Trans. Information Theory, Vol. 38, No. 6, pp. 1698-1708, November, 1992.
21. M. Pawlak and X. Liao, *On Image Analysis via Orthogonal Moments*, Vision Interface '92, pp. 253-258, May, 1992.
22. M. Pawlak and X. Liao, *On image analysis by orthogonal moments*, 11th IAPR International Conference On Pattern Recognition, pp. 549-552, Aug.30 - Sept.3, 1992.
23. M. Pawlak and S. X. Liao, *On digital approximation of moment descriptors*, Machine GRAPHICS & VISION Vol. 3, Nos. 1/2, pp. 61-68, 1994.
24. R.J. Prokop and A.P. Reeves, *A survey of moment-based techniques for unoccluded object representation and recognition*, Graphical Models And Image Processing, Vol. 54, No. 5, September, pp. 438-460, 1992.
25. William H. Press, Brian P. Flannery, Saul A. Teukolsky, and William T. Vetterling, *Numerical Recipes in C*, Cambridge University Press, 1988.
26. T.H. Reiss, *The revised fundamental theorem of moment invariants*, IEEE Trans. Pattern Anal. Machine Intell., Vol. PAMI-13, No. 8, August 1991, pp. 830-834.
27. F.W. Smith and M.H. Wright, *Automatic ship photo interpretation by the method of moments*, IEEE Trans. Comput., Vol. C-20, pp. 1089-1094, Sept. 1971.
28. M.R. Teague, *Image analysis via the general theory of moments*, J. Optical Soc. Am., Vol. 70, pp. 920-930, August 1980.
29. C.H. Teh and R.T. Chin, *On image analysis by the methods of moments*, IEEE Trans. Pattern Anal. Machine Intell., Vol. PAMI-10, pp. 496-512, July 1988.
30. Yu. V. Vorobyev, *Method of Moments in Applied Mathematics*, New York, Gordon and Breach Science Publishers, 1965.

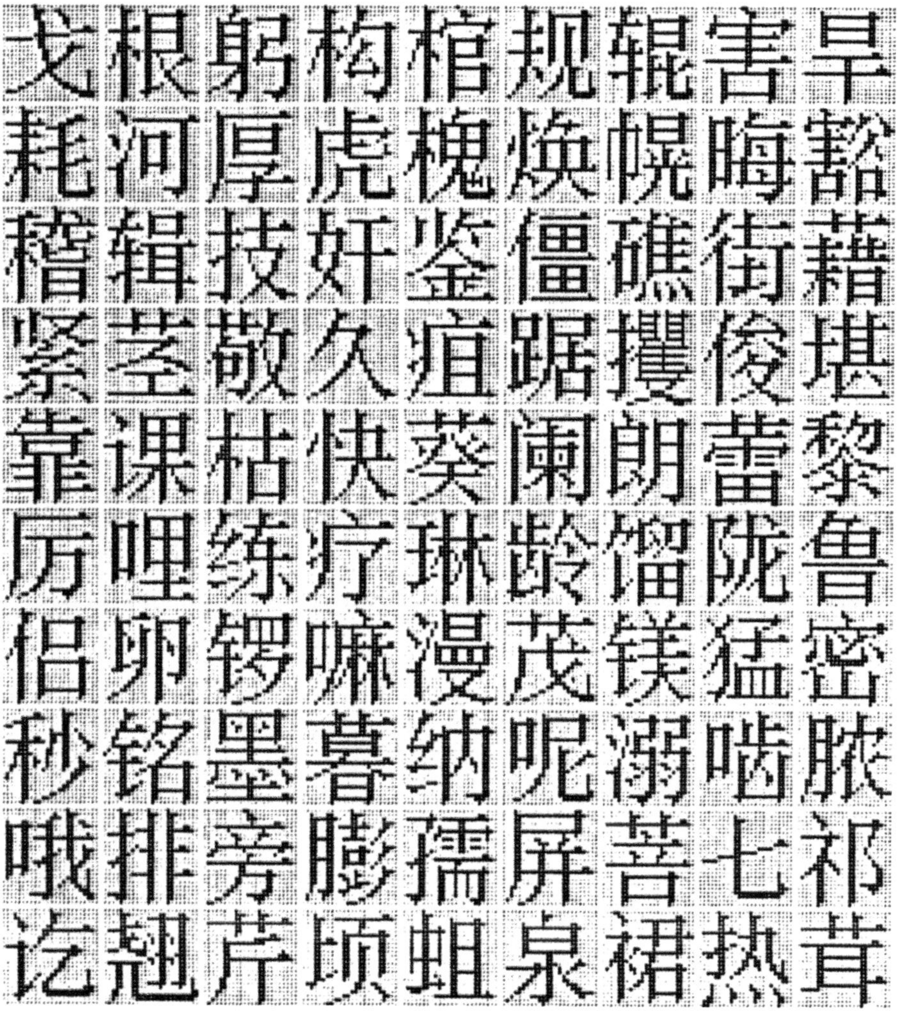

戈 根 躬 构 棺 规 辊 害 旱
耗 河 厚 虎 槐 焕 幌 晦 豁
稽 辑 技 奸 鉴 僵 礁 街 藉
紧 茎 敬 久 疽 踞 攫 俊 堪
靠 课 枯 快 葵 阑 朗 蕾 黎
厉 哩 练 疗 琳 龄 馏 陇 鲁
侣 卵 锣 嘛 漫 茂 镁 猛 密
秒 铭 墨 暮 纳 呢 溺 啮 脉
哦 排 旁 膨 孺 屏 菩 匕 祁
讫 翘 芹 顷 蛆 泉 裙 热 茸

Fig. 2. Ninety randomly selected Chinese characters.

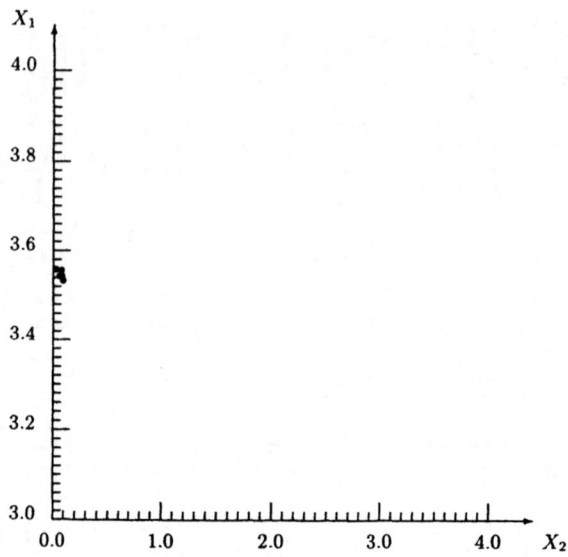

Fig. 3. Representations of the five Chinese characters in the central moment feature space.

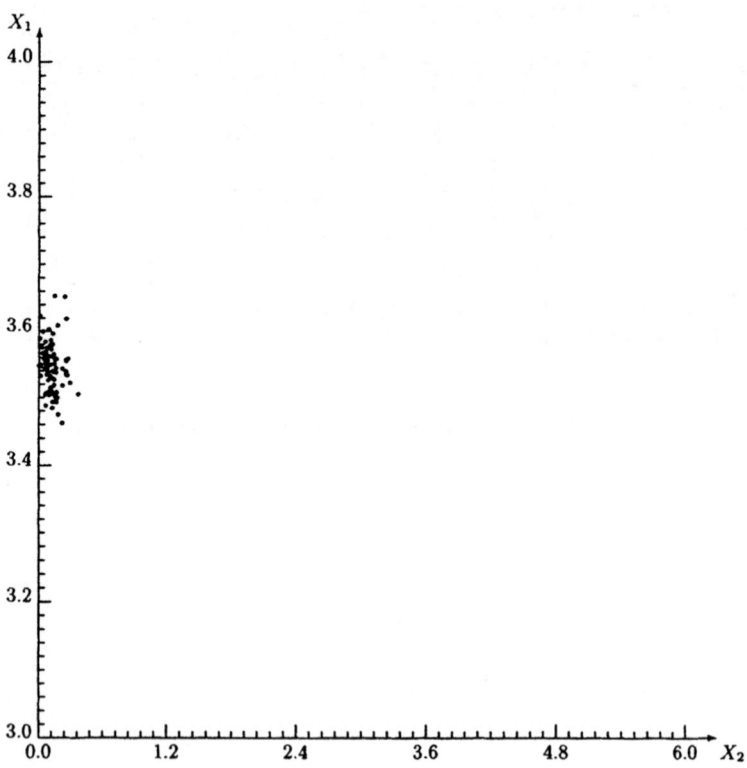

Fig. 4. Representations of the 90 Chinese characters in the central moment feature space.

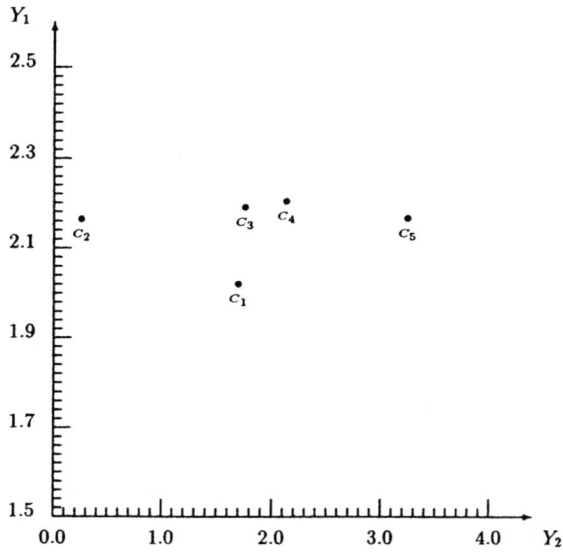

Fig. 5. Representations of the five Chinese characters in the Legendre moment feature space.

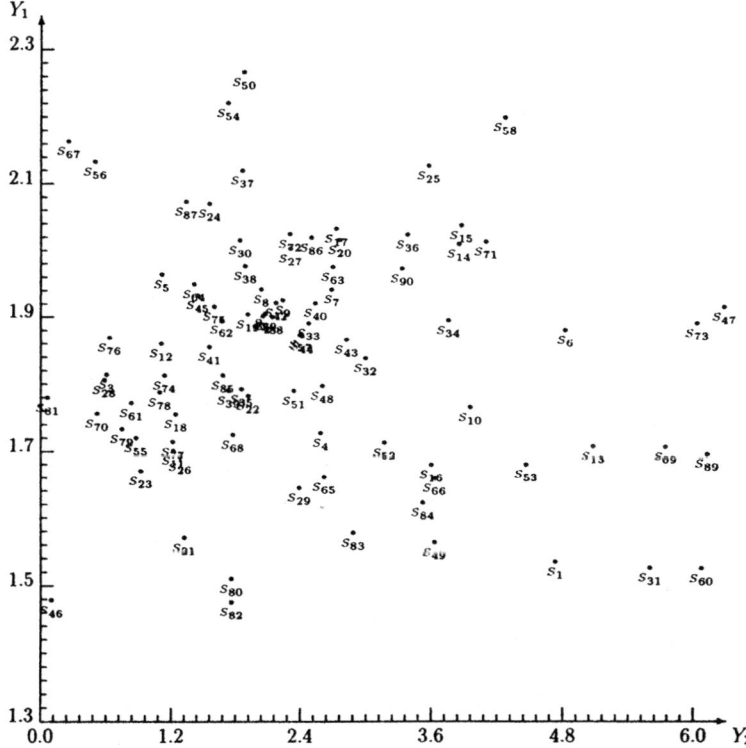

Fig. 6. Representations of the 90 Chinese characters in the Legendre moment feature space.

Sample	Y_1	Y_2	Y_3	Sample	Y_1	Y_2	Y_3
S_1	1.5346	4.7334	2.6002	S_{46}	1.4783	0.0995	1.0175
S_2	1.9007	2.0506	1.8604	S_{47}	1.9144	6.2822	1.9013
S_3	1.8135	0.6016	2.0815	S_{48}	1.7970	2.6013	2.5707
S_4	1.7271	2.5868	3.0566	S_{49}	1.5632	3.6298	1.3954
S_5	1.9632	1.1077	1.5408	S_{50}	2.2657	1.8740	2.8191
S_6	1.8792	4.8260	2.8729	S_{51}	1.7894	2.3395	2.3448
S_7	1.9411	2.6879	2.7054	S_{52}	1.7126	3.1672	0.5698
S_8	1.9408	2.0360	0.9139	S_{53}	1.6793	4.4645	2.8409
S_9	1.9248	2.2374	2.0778	S_{54}	2.2200	1.7281	2.2174
S_{10}	1.7645	3.9552	3.2093	S_{55}	1.7195	0.8730	1.5833
S_{11}	1.7005	1.2156	2.4936	S_{56}	2.1327	0.4964	2.7262
S_{12}	1.8596	1.1031	1.4362	S_{57}	1.8733	2.3991	1.6876
S_{13}	1.7067	5.0808	2.4617	S_{58}	2.1976	4.2732	2.5669
S_{14}	2.0086	3.8532	3.1573	S_{59}	1.9050	2.0736	1.9350
S_{15}	2.0362	3.8763	3.6982	S_{60}	1.5251	6.0759	2.4210
S_{16}	1.6787	3.5999	2.3467	S_{61}	1.7716	0.8286	1.7164
S_{17}	2.0319	2.7304	3.7947	S_{62}	1.8951	1.6680	1.1195
S_{18}	1.7548	1.2384	0.7203	S_{63}	1.9745	2.6982	1.4327
S_{19}	1.9031	1.9106	1.6195	S_{64}	1.9486	1.4085	2.4803
S_{20}	2.0153	2.7606	2.7951	S_{65}	1.6608	2.6182	1.8849
S_{21}	1.5705	1.3187	1.3711	S_{66}	1.6585	3.6280	1.2719
S_{22}	1.7817	1.9157	1.3440	S_{67}	2.1636	0.2535	1.3685
S_{23}	1.6696	0.9155	0.7421	S_{68}	1.7241	1.7722	2.7329
S_{24}	2.0691	1.5532	1.2411	S_{69}	1.7064	5.7469	3.1251
S_{25}	2.1259	3.5769	3.1406	S_{70}	1.7558	0.5146	1.4417
S_{26}	1.6907	1.2791	1.5703	S_{71}	2.0121	4.0992	3.4287
S_{27}	2.0025	2.3081	1.9807	S_{72}	2.0239	2.3041	2.6732
S_{28}	1.8056	0.5816	0.7381	S_{73}	1.8900	6.0365	3.4701
S_{29}	1.6444	2.3904	1.2175	S_{74}	1.8122	1.1317	1.6745
S_{30}	2.0144	1.8385	2.2021	S_{75}	1.9149	1.5990	1.5343
S_{31}	1.5260	5.6012	3.0475	S_{76}	1.8685	0.6289	1.5335
S_{32}	1.8379	2.9933	1.8440	S_{77}	1.7135	1.2120	1.6727
S_{33}	1.8906	2.4768	2.5081	S_{78}	1.7870	1.0894	1.4135
S_{34}	1.8940	3.7551	1.6790	S_{79}	1.7331	0.7432	0.8510
S_{35}	1.7915	1.8526	2.0289	S_{80}	1.5097	1.7610	2.0634
S_{36}	2.0227	3.3827	2.2726	S_{81}	1.7795	0.0637	1.2769
S_{37}	2.1192	1.8604	1.7906	S_{82}	1.4753	1.7641	1.2034
S_{38}	1.9756	1.8860	2.2459	S_{83}	1.5775	2.8834	1.7849
S_{39}	1.7896	1.7319	1.4725	S_{84}	1.6220	3.5224	2.3445
S_{40}	1.9206	2.5398	2.7447	S_{85}	1.8122	1.6799	1.8171
S_{41}	1.8540	1.5533	1.5474	S_{86}	2.0188	2.5025	2.2536
S_{42}	1.9207	2.1761	2.8264	S_{87}	2.0721	1.3334	1.8721
S_{43}	1.8654	2.8219	2.6819	S_{88}	1.8996	2.1385	1.6345
S_{44}	1.8708	2.4153	0.6933	S_{89}	1.6954	6.1273	3.7291
S_{45}	1.9319	1.4443	0.9003	S_{90}	1.9719	3.3317	2.5255

Table 5. Values of the 90 Chinese characters in the Legendre moment three-dimensional feature space.

Author Index

Lecture Notes in Computer Science

For information about Vols. 1–1080

please contact your bookseller or Springer-Verlag

Vol. 1115: P.W. Eklund, G. Ellis, G. Mann (Eds.), Conceptual Structures: Knowledge Representation as Interlingua. Proceedings, 1996. XIII, 321 pages. 1996. (Subseries LNAI).

Vol. 1116: J. Hall (Ed.), Management of Telecommunication Systems and Services. XXI, 229 pages. 1996.

Vol. 1117: A. Ferreira, J. Rolim, Y. Saad, T. Yang (Eds.), Parallel Algorithms for Irregularly Structured Problems. Proceedings, 1996. IX, 358 pages. 1996.

Vol. 1118: E.C. Freuder (Ed.), Principles and Practice of Constraint Programming — CP 96. Proceedings, 1996. XIX, 574 pages. 1996.

Vol. 1119: U. Montanari, V. Sassone (Eds.), CONCUR '96: Concurrency Theory. Proceedings, 1996. XII, 751 pages. 1996.

Vol. 1120: M. Deza. R. Euler, I. Manoussakis (Eds.), Combinatorics and Computer Science. Proceedings, 1995. IX, 415 pages. 1996.

Vol. 1121: P. Perner, P. Wang, A. Rosenfeld (Eds.), Advances in Structural and Syntactical Pattern Recognition. Proceedings, 1996. X, 393 pages. 1996.

Vol. 1122: H. Cohen (Ed.), Algorithmic Number Theory. Proceedings, 1996. IX, 405 pages. 1996.

Vol. 1123: L. Bougé, P. Fraigniaud, A. Mignotte, Y. Robert (Eds.), Euro-Par'96. Parallel Processing. Proceedings, 1996, Vol. I. XXXIII, 842 pages. 1996.

Vol. 1124: L. Bougé, P. Fraigniaud, A. Mignotte, Y. Robert (Eds.), Euro-Par'96. Parallel Processing. Proceedings, 1996, Vol. II. XXXIII, 926 pages. 1996.

Vol. 1125: J. von Wright, J. Grundy, J. Harrison (Eds.), Theorem Proving in Higher Order Logics. Proceedings, 1996. VIII, 447 pages. 1996.

Vol. 1126: J.J. Alferes, L. Moniz Pereira, E. Orlowska (Eds.), Logics in Artificial Intelligence. Proceedings, 1996. IX, 417 pages. 1996. (Subseries LNAI).

Vol. 1127: L. Böszörményi (Ed.), Parallel Computation. Proceedings, 1996. XI, 235 pages. 1996.

Vol. 1128: J. Calmet, C. Limongelli (Eds.), Design and Implementation of Symbolic Computation Systems. Proceedings, 1996. IX, 356 pages. 1996.

Vol. 1129: J. Launchbury, E. Meijer, T. Sheard (Eds.), Advanced Functional Programming. Proceedings, 1996. VII, 238 pages. 1996.

Vol. 1130: M. Haveraaen, O. Owe, O.-J. Dahl (Eds.), Recent Trends in Data Type Specification. Proceedings, 1995. VIII, 551 pages. 1996.

Vol. 1131: K.H. Höhne, R. Kikinis (Eds.), Visualization in Biomedical Computing. Proceedings, 1996. XII, 610 pages. 1996.

Vol. 1132: G.-R. Perrin, A. Darte (Eds.), The Data Parallel Programming Model. XV, 284 pages. 1996.

Vol. 1133: J.-Y. Chouinard, P. Fortier, T.A. Gulliver (Eds.), Information Theory and Applications II. Proceedings, 1995. XII, 309 pages. 1996.

Vol. 1134: R. Wagner, H. Thoma (Eds.), Database and Expert Systems Applications. Proceedings, 1996. XV, 921 pages. 1996.

Vol. 1135: B. Jonsson, J. Parrow (Eds.), Formal Techniques in Real-Time and Fault-Tolerant Systems. Proceedings, 1996. X, 479 pages. 1996.

Vol. 1136: J. Diaz, M. Serna (Eds.), Algorithms – ESA '96. Proceedings, 1996. XII, 566 pages. 1996.

Vol. 1137: G. Görz, S. Hölldobler (Eds.), KI-96: Advances in Artificial Intelligence. Proceedings, 1996. XI, 387 pages. 1996. (Subseries LNAI).

Vol. 1138: J. Calmet, J.A. Campbell, J. Pfalzgraf (Eds.), Artificial Intelligence and Symbolic Mathematical Computation. Proceedings, 1996. VIII, 381 pages. 1996.

Vol. 1139: M. Hanus, M. Rogriguez-Artalejo (Eds.), Algebraic and Logic Programming. Proceedings, 1996. VIII, 345 pages. 1996.

Vol. 1140: H. Kuchen, S. Doaitse Swierstra (Eds.), Programming Languages: Implementations, Logics, and Programs. Proceedings, 1996. XI, 479 pages. 1996.

Vol. 1141: H.-M. Voigt, W. Ebeling, I. Rechenberg, H.-P. Schwefel (Eds.), Parallel Problem Solving from Nature – PPSN IV. Proceedings, 1996. XVII, 1.050 pages. 1996.

Vol. 1142: R.W. Hartenstein, M. Glesner (Eds.), Field-Programmable Logic. Proceedings, 1996. X, 432 pages. 1996.

Vol. 1143: T.C. Fogarty (Ed.), Evolutionary Computing. Proceedings, 1996. VIII, 305 pages. 1996.

Vol. 1144: J. Ponce, A. Zisserman, M. Hebert (Eds.), Object Representation in Computer Vision. Proceedings, 1996. VIII, 403 pages. 1996.

Vol. 1145: R. Cousot, D.A. Schmidt (Eds.), Static Analysis. Proceedings, 1996. IX, 389 pages. 1996.

Vol. 1146: E. Bertino, H. Kurth, G. Martella, E. Montolivo (Eds.), Computer Security – ESORICS 96. Proceedings, 1996. X, 365 pages. 1996.

Vol. 1147: L. Miclet, C. de la Higuera (Eds.), Grammatical Inference: Learning Syntax from Sentences. Proceedings, 1996. VIII, 327 pages. 1996. (Subseries LNAI).

Vol. 1148: M.C. Lin, D. Manocha (Eds.), Applied Computational Geometry. Proceedings, 1996. VIII, 223 pages. 1996.

Vol. 1149: C. Montangero (Ed.), Software Process Technology. Proceedings, 1996. IX, 291 pages. 1996.

Vol. 1150: A. Hlawiczka, J.G. Silva, L. Simoncini (Eds.), Dependable Computing – EDCC-2. Proceedings, 1996. XVI, 440 pages. 1996.

Vol. 1151: Ö. Babaoğlu, K. Marzullo (Eds.), Distributed Algorithms. Proceedings, 1996. VIII, 381 pages. 1996.

Vol. 1153: E. Burke, P. Ross (Eds.), Practice and Theory of Automated Timetabling. Proceedings, 1995. XIII, 381 pages. 1996.

Vol. 1156: A. Bode, J. Dongarra, T. Ludwig, V. Sunderam (Eds.), Parallel Virtual Machine – EuroPVM '96. Proceedings, 1996. XIV, 362 pages. 1996.

Vol. 1157: B. Thalheim (Ed.), Entity-Relationship Approach – ER '96. Proceedings, 1996. XII, 489 pages. 1996.

Vol. 1158: S. Berardi, M. Coppo (Eds.), Types for Proofs and Programs. Proceedings, 1995. X, 296 pages. 1996.